D1348471

Finite Element Simulation
in Surface and Subsurface Hydrology

# FINITE ELEMENT SIMULATION IN SURFACE AND SUBSURFACE HYDROLOGY

*George F. Pinder*
*William G. Gray*
DEPARTMENT OF CIVIL ENGINEERING
PRINCETON UNIVERSITY
PRINCETON, NEW JERSEY

ACADEMIC PRESS, INC.
**Harcourt Brace Jovanovich, Publishers**
Orlando San Diego New York
Austin Boston London Sydney
Tokyo Toronto

COPYRIGHT © 1977, BY ACADEMIC PRESS, INC.
ALL RIGHTS RESERVED.
NO PART OF THIS PUBLICATION MAY BE REPRODUCED OR
TRANSMITTED IN ANY FORM OR BY ANY MEANS. ELECTRONIC
OR MECHANICAL, INCLUDING PHOTOCOPY, RECORDING, OR ANY
INFORMATION STORAGE AND RETRIEVAL SYSTEM, WITHOUT
PERMISSION IN WRITING FROM THE PUBLISHER.

ACADEMIC PRESS, INC.
Orlando, Florida 32887

*United Kingdom Edition published by*
ACADEMIC PRESS, INC. (LONDON) LTD.
**24/28 Oval Road, London NW1 7DX**

**Library of Congress Cataloging in Publication Data**

Pinder, George Francis, Date
Finite element simulation in surface and subsurface hydrology.

Includes bibliographical references.
1. Hydrology–Mathematical models. 2. Hydrology–Data processing. 3. Finite element method. I. Gray, William Guerin, Date joint author. II. Title.
GB656.2.M33P56 551.4′8′018 76-42977
ISBN 0–12–556950–5

PRINTED IN THE UNITED STATES OF AMERICA

86 87 88 89 9 8 7 6 5 4

87 08452

To Peter

# Contents

## Chapter 4 Finite Elements



## Chapter 5 Finite Element Method in Subsurface Hydrology



## Chapter 6 Lakes



## Chapter 7 Analysis of Model Behavior

## Chapter 8 Estuaries and Coastal Regions

# Preface

The finite element method has developed into an important tool for the simulation of water resource systems. It is the purpose of this text to introduce the reader to the fundamental concepts of finite element analysis and then demonstrate state of the art applications to groundwater and surface water modeling. The book is intended for senior undergraduate or graduate students in science and engineering and also for the practicing hydrologist.

The first chapter deals with computational considerations that are important to the numerical solution of differential equations. Classification of partial differential equations by type, errors encountered in approximating the equations discretely, and matrix manipulations useful in solving the discrete equations are all covered. Chapters 2 and 3 provide an introduction to finite difference and weighted residual integral techniques for obtaining approximate solutions to differential equations. The Galerkin finite element method as a special weighted residual procedure is illustrated in Chapter 4. To familiarize the reader with the method and as an indication of some of its flexibility, different types of basis functions and element shapes are used in solving some simple problems. In Chapter 5, the finite element method is applied to problems in subsurface hydrology. The governing partial differential equations are developed in sufficient detail to allow the reader familiar with only the most basic concepts (e.g., conservation of mass, Darcy's law) to understand their origin. Various forms of the equations which describe lake circulation are developed and solved using finite elements in Chapter 6. Chapter 7 addresses some of the considerations that go into making a numerical model a true representation of a physical system rather than a function made to fit through selected data points by parameter adjustment. In Chapter 8, applications of the finite element method to tidal regions are surveyed.

This book had its roots in notes we developed for one week short courses taught to professional scientists and engineers at Princeton over the past three years. We are grateful to those students who pointed out various errors in the notes. In the final stages of manuscript preparation, important suggestions were provided by M. Th. van Genuchten, E. O. Frind, and C. S. Desai. D. R. Lynch was an invaluable aid in screening out errors that we had doggedly carried through the various rewrites. Unfortunately, responsibility for errors which persist remains solely with the authors. We also wish to thank the many other people who made helpful suggestions and comments which we have incorporated into the book.

The patience and expert workmanship of K. Mueller, D. Hannigan, and P. Roman who typed the various versions of this text from our illegible scrawl and ambiguous directions is also gratefully acknowledged.

Finally we would like to thank our families for unbegrudgingly giving us the time, which should have been theirs, to write this book.

# Chapter 1 Introduction

## 1.1 Purpose and Scope of the Book

The objective of this book is not only to provide an introduction to the theoretical foundation on which the finite element method is built, but also to demonstrate how this approach can be applied to problems in surface and subsurface hydrology. While surface and subsurface hydrology have classically been treated as two separate entities, this subdivision is rather arbitrary. Because of the fundamental similarities in the physical laws governing these subsystems and in the numerical schemes used to simulate them, we have elected to consider them together.

The material is subdivided into three basic parts. In the first few chapters the basic concepts of the numerical methods and the finite element approach are introduced. Having developed the basic theory, we next demonstrate how finite elements can be applied to problems in subsurface hydrology such as groundwater flow and mass and energy transport. The third and final part is devoted to applying the finite element method to problems involving surface water dynamics. As the text proceeds from topic to topic, adequate background material concerning the fundamental concepts of subsurface flow and surface water dynamics is provided for the reader who is unfamiliar with either topic.

Simulation of hydrologic systems using numerical methods has evolved within a decade from an interesting curiosity to an accepted and established engineering and scientific approach to the study of water resource problems. Consequently, interesting new ideas are forthcoming with each new issue of pertinent professional journals and we can only provide herein a current appraisal of the state of the art.

In the next section several concepts are introduced which are fundamental both in a theoretical and practical sense: approximations, errors, and significant figures.

## 1.2 Approximations, Errors, and Significant Figures

In numerical analysis we are generally given information about a function, say $f(x)$, and wish to obtain additional information which will render the function more useful for a specified objective. Usually $f(x)$ is either assumed known or required to be continuous over a specified range [1]. To obtain the desired additional information about $f(x)$, a set of $(n + 1)$ coordinate functions $w_0(x), w_1(x), \ldots, w_n(x)$ are selected to have properties that allow us to extract additional information from them easily. If the function $f(x)$ is a member of the set $S_n$ generated by linear combinations of these coordinate functions, then the information obtained will be exactly representative of the behavior of $f(x)$. In the event that $f(x)$ is not represented in $S_n$ we must select from this set that one, say $y(x)$, which corresponds as closely as possible to $f(x)$. The function $f(x)$ is then approximated by the selected function $y(x)$. Additional information about $f(x)$, which was not employed in selecting $y(x)$, is now introduced to estimate the error associated with this approximation.

When the function to be approximated, $f(x)$, is continuous and the interval of approximation is finite, polynomial approximations are often used. Algebraic polynomials of degree $n$, described by the $(n + 1)$ functions $1, x, x^2, \ldots, x^n$, are not only easily integrated and differentiated but also generate a set $S_n$ of functions which will contain at least one member that approximates $f(x)$ within a specified tolerance. Consequently, it is not surprising that polynomial approximations are used almost exclusively in the finite element method.

In addition to errors introduced through the approximation of $f(x)$ by a suitable function $y(x)$, which are called truncation errors, round-off errors are also generated. These are generated by approximating with $n$ correct digits a number that requires more than $n$ digits to be correctly specified. To illustrate these two error types, we will consider an example presented by Hildebrand [1].

Consider a function $f(x) = e^{-x}$ which can be approximated in a region near point $a$ by an infinite Taylor series:

$$f(x) = f(a) + \frac{df(a)/dx}{1!}(x - a) + \frac{d^2f(a)/dx^2}{2!}(x - a)^2 + \cdots + \frac{d^nf(a)/dx^n}{n!}(x - a)^n + \cdots \tag{1.1}$$

When only $n$ terms of this series are considered, the error of the approximation is given by

$$E_T = \frac{d^n f(\xi)/dx^n}{n!}(x-a)^n$$

where $\xi$ is some number between $a$ and $x$. For example, if $a = 0$, the four-term approximation for $e^{-x}$ using the Taylor series becomes

$$e^{-x} = 1 - x + \tfrac{1}{2}x^2 - \tfrac{1}{6}x^3 + E_T(x) \tag{1.2}$$

where $E_T = \frac{1}{24}e^{-\xi}x^4$ ($\xi$ between 0 and $x$). If $x$ is positive, $\xi$ is also positive and $e^{-\xi} < 1$; consequently the truncation error must be smaller than $\frac{1}{24}x^4$. Specifically, if $x$ is chosen as $\frac{1}{3}$,

$$e^{-1/3} \simeq 1 - \tfrac{1}{3} + \tfrac{1}{18} - \tfrac{1}{162} = \tfrac{116}{162}$$

with a truncation error $0.00036 \leq E_T \leq 0.00052$, depending on the value of $\xi$. Rounding $\frac{116}{162}$ to four significant figures gives 0.7160 where the additional error introduced by the round-off is $4.9 \times 10^{-5}$. Consequently, the approximation of $e^{-1/3} = 0.7160$ has a total error of magnitude less than $5.7 \times 10^{-4}$ which is the sum of the truncation error $5.2 \times 10^{-4}$ and the round-off error $4.9 \times 10^{-5}$. The magnitude of the round-off error can be reduced arbitrarily by retaining additional digits and the truncation error can be reduced by taking additional terms of the convergent Taylor expansion of $e^{-x}$ around $x = 0$.

Because we are forced to represent numbers by a finite number of digits, some additional consideration should be given to the rounding process. Rounding a number consists of replacing this number by an $m$ digit approximation with minimum error. This is in contrast to truncating or chopping where the first $m$ digits are retained and the remainder discarded. Each correct digit of this approximation, except a zero that is used only to fix the decimal point, is termed a significant figure. Thus the numbers 1.131, 0.01114, and 10.00 all have four significant figures. Moreover, if a number $N$ and its approximation $\hat{N}$ both round to the same set of $m$ significant figures, then $\hat{N}$ is said to approximate $N$ to $m$ significant figures [1]. For example, if $N = 12.34500$ and $\hat{N} = 12.3466$, $\hat{N}$ approximates $N$ to three significant figures ($m = 3$). Here the convention has been adopted that when there are two equally admissable roundings, the one generating a number with an even $m$th digit is chosen.

One of the principal sources of error encountered in the application of numerical methods is the loss of significant figures in subtraction. If possible, it is advantageous to arrange calculations so as to avoid subtractions. When subtractions are necessary, some care should be taken in designing the sequence of operations. Consider for example [1] the problem of calculating $ab - ac \equiv a(b - c)$, where $b$ and $c$ are nearly equal. In this case, a decision to calculate the products first followed by the subtraction could lead to a loss of significant figures if many of the leading digits are common to each number. Alternatively, the difference $(b - c)$ could be computed first thus avoiding this difficulty.

There are occasions where the properties of a function can be utilized to avoid the loss of significant figures due to subtraction [1]. When the arguments $a$ and $b$ are nearly equal, it may be advantageous to replace $\log b - \log a$ by $\log(b/a)$, $\sin b - \sin a$ by $2 \sin \frac{1}{2}(b - a) \cos \frac{1}{2}(a + b)$, and $\sqrt{b} - \sqrt{a}$ by $(b - a)/(\sqrt{a} + \sqrt{b})$.

In the next few sections, we will review several topics in applied mathematics in order to set a foundation for the ensuing discussion of the finite element method and its applications.

## 1.3 Initial Value and Initial Boundary Value Problems

### *1.3.1 Ordinary Differential Equations*

Consider an $n$th order ordinary differential equation of the form

$$F\left[x, y(x), \frac{dy(x)}{dx}, \frac{d^2y(x)}{dx^2}, \ldots, \frac{d^ny(x)}{dx^n}\right] = 0 \tag{1.3}$$

where $y(x)$ is a real function. The general solution of this equation normally depends on $n$ parameters $c_1, c_2, \ldots, c_n$. In an initial value problem these parameters are determined through the specification of the values

$$y_0^{(p)} = y^{(p)}(x_0) \qquad (p = 0, 1, 2, \ldots, n - 1) \tag{1.4}$$

at the fixed point $x = x_0$. When the specification of these conditions involves more than one point, the problem is classified as a boundary value problem and the boundary conditions are of the form [2]

$$\begin{aligned} \forall_p\Bigg[ & y(x_1), \frac{dy(x_1)}{dx}, \ldots, \\ & \frac{d^{(n-1)}y(x_1)}{dx^{(n-1)}}, y(x_2), \frac{dy(x_2)}{dx}, \ldots, \frac{d^{(n-1)}y(x_2)}{dx^{(n-1)}}, \ldots, \\ & y(x_k), \frac{dy(x_k)}{dx}, \ldots, \frac{d^{(n-1)}y(x_k)}{dx^{(n-1)}}\Bigg] = 0 \qquad (p = 0, 1, \ldots, n - 1) \end{aligned} \tag{1.5}$$

where $x_s$ $(x_1 < x_2 < \cdots < x_k)$ are prescribed points which may include $\pm\infty$

and $d^{(\rho)}y(x_s)/dx^{(\rho)}$ denotes the value of the $\rho$th derivative of $y(x)$ at the point $x = x_s$. The solution to the boundary value problem is obtained when a function $y(x)$ is found that satisfies (1.3) and (1.5). It is assumed that $F$ and $\forall_p$ are given linear or nonlinear functions.

In the discussion of the variational approach to finite element formulations for linear boundary value problems we will find that when the differential equation is of even order, i.e., $n = 2m$, the boundary conditions can be divided into "essential" and "natural."† The natural boundary conditions are generated by forming as many as possible linearly independent linear combinations of the given $2m$ boundary conditions which have derivatives of the $m$th and higher orders. If $k$ boundary conditions are thus obtained, then there remain $(2m - k)$ linearly independent boundary conditions which contain only derivatives of up to the $(m - 1)$th order; these are termed essential boundary conditions.

Consider, for example, the problem outlined by Collatz [2] wherein the boundary conditions $y(0) + dy(1)/dx = 1$ and $y(1) + 2\,dy(1)/dx = 3$ are given with a second-order differential equation $(m = 1)$. These conditions can be combined linearly to give $2y(0) - y(1) = -1$ in which the first derivative no longer appears. Inasmuch as no additional linear combinations of these boundary conditions can be generated which contain no derivatives, we now have one essential boundary condition $2y(0) - y(1) = -1$ and one natural boundary condition which can be expressed either as $y(0) + dy(1)/dx = 1$ or as $y(1) + 2\,dy(1)/dx = 3$.

A second unrelated classification of boundary conditions commonly encountered in the literature can be summarized as

$$a_0\,y(x_s) + a_1\,dy(x_s)/dx = f(x_s)$$

where $a_1 = 0$, $a_0 \neq 0$ for Dirichlet or first-type boundary conditions, $a_0 = 0$, $a_1 \neq 0$ for Neumann or second-type boundary conditions, and $a_1 \neq 0$, $a_2 \neq 0$ for mixed or third-type boundary conditions. This classification is easily extended to partial differential equations.

### *1.3.2 Partial Differential Equations*

In partial differential equations, the concepts of ordinary differential equations are extended to consider functions of $n$ independent variables. Let us assume such a function has the form $u(x_1, x_2, \ldots, x_n)$ and satisfies the

† Collatz [2] refers to "natural boundary conditions" as "suppressible" boundary conditions, and other terminology is also found in the literature.

partial differential equation

$$F\left(x_1, x_2, \ldots, x_n, u, \frac{\partial u}{\partial x_1}, \ldots, \frac{\partial u}{\partial x_n}, \frac{\partial^2 u}{\partial x_1^2}, \frac{\partial^2 u}{\partial x_1 \, \partial x_2}, \ldots, \frac{\partial^2 u}{\partial x_n^2}, \ldots\right) = 0 \quad \text{in} \quad B \tag{1.6}$$

with boundary conditions

$$\forall_\mu\left(x_1, x_2, \ldots, x_n, u, \frac{\partial u}{\partial x_1}, \ldots, \frac{\partial u}{\partial x_n}, \frac{\partial^2 u}{\partial x_1^2}, \frac{\partial^2 u}{\partial x_1 \, \partial x_2}, \ldots, \frac{\partial^2 u}{\partial x_n^2}, \ldots\right) = 0 \quad \text{on} \quad \Gamma_\mu \tag{1.7}$$

where $B$ is a given region in $(x_1, x_2, \ldots, x_n)$ space, $\Gamma_\mu$ are $(n - 1)$-dimensional hypersurfaces, and $F$ and $\forall_\mu$ are given functions.

Linear and quasi-linear partial differential equations represent important subgroups of this general class of equations. A partial differential equation is defined as linear when it is linear in $u$ and the derivatives of $u$, and quasi-linear when it is linear in the highest-order derivatives. A boundary value problem is termed linear when the differential equation and the boundary conditions are linear.

For the case of constant coefficients, the concepts of initial and boundary value problems introduced earlier are applicable to partial differential equations. Variable coefficients, however, complicate the situation considerably because the equations may change their general properties within the region $B$. In the next section, these properties will be considered in the discussion of the classification of partial differential equations.

## 1.4 Classification of Partial Differential Equations

Partial differential equations are classified as parabolic, hyperbolic, or elliptic based on the properties of the equations. It is not surprising, therefore, that the selection of a numerical scheme for solving a given problem is generally based on the type of equation encountered. The method of characteristics, for example, is often applied to equations of hyperbolic type while parabolic equations can often be solved effectively with finite difference methods.

The classification of partial differential equations is formally developed utilizing the theory of characteristics such as was outlined by Crandall [3]. This development provides considerable insight into the theoretical foundation of the classification. The essential elements of the resulting cataloging procedure can be summarized by considering a second-order partial differential equation in the function $u$ for the two independent variables $x$ and $y$ such as

$$a(x, y)\frac{\partial^2 u}{\partial x^2} + b(x, y)\frac{\partial^2 u}{\partial x\, \partial y} + c(x, y)\frac{\partial^2 u}{\partial y^2} + f\left(x, y, u, \frac{\partial u}{\partial x}, \frac{\partial u}{\partial y}\right) = 0 \quad (1.8)$$

which may be either linear or quasi-linear. Equation (1.8) is classified as hyperbolic, parabolic, or elliptic in a region $B$ when the discriminant $b^2 - 4ac$ is positive, zero, or negative, respectively. Because the coefficients $a$, $b$, and $c$ are, in general, functions of the independent variables, the classification of an equation may change at different locations in the region $B$ in which the equation is defined.

### 1.4.1 *Hyperbolic Equations*

A hyperbolic partial differential equation defined in the region $B$ is characterized by having a positive discriminant everywhere within $B$. Equations of this type require both initial and boundary conditions (see Fig. 1.1). The

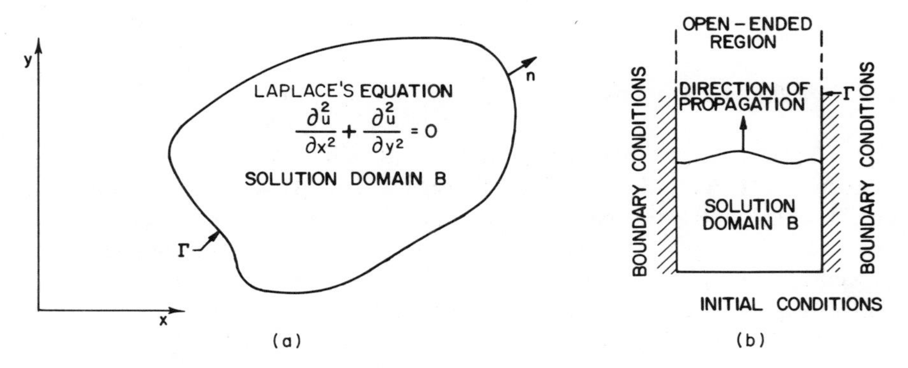


**Fig. 1.1.** Schematic representation of solution domain. (a) Elliptic problem; (b) parabolic or hyperbolic problem.

initial conditions are the values of the function $u$ and its first time derivative defined at some time $t_0$. The boundary conditions may consist of the value of the function (Dirichlet type), its normal derivative (Neumann type), or a combination of the function and its normal derivative (mixed type) on the region of definition. An important equation of this type is the one-dimensional wave equation which describes the longitudinal velocity $u$ of water in a channel with a wave being generated at one end

$$a^2\, \partial^2 u/\partial x^2 = \partial^2 u/\partial t^2 \tag{1.9}$$

where $a$ is the celerity of the wave.

### 1.4.2 *Parabolic Equations*

A parabolic partial differential equation is characterized by a zero discriminant at all points within the region $B$ over which the equation is defined. Initial and boundary values are required for a properly posed problem (see Fig. 1.1). The initial value consists of the function $u$ defined at some time $t_0$ and the boundary conditions are either the value of the function, its normal derivative, or a linear combination of the function and its normal derivative on the boundary.

An important equation of parabolic type is the one-dimensional heat flow equation

$$K\, \partial^2 u/\partial x^2 = \partial u/\partial t \tag{1.10}$$

where $K = k/\rho C$, $C$ is the specific heat, $\rho$ the density of the material, and $k$ the thermal conductivity. A similar equation describes the potential flow problem in saturated porous media.

### 1.4.3 *Elliptic Equations*

Elliptic partial differential equations are characterized by a negative discriminant everywhere within the region over which the equations are defined. In contrast to parabolic and hyperbolic equations, which are propagated in an open domain, elliptic equations require boundary conditions specified over a closed boundary of the region $B$. Either the function $u$, its normal derivative, or a combination of the function and its normal derivative must be specified to assure a unique solution.

An important elliptic equation is Laplace's equation which describes the steady state temperature distribution in two space dimensions as

$$(\partial^2 u/\partial x^2) + (\partial^2 u/\partial y^2) = 0 \tag{1.11}$$

The equivalent expression with a source term is called the Poisson equation and has the form

$$(\partial^2 u/\partial x^2) + (\partial^2 u/\partial y^2) = f(x, y) \tag{1.12}$$

Before turning our attention to numerical methods, we consider in the next few sections some of the fundamental concepts of matrix algebra. This brief summary not only provides us with the vocabulary of the finite element method, but also gives an insight into techniques available for the solution of the large systems of algebraic equations common to all numerical schemes.

## 1.5 Matrix Operations Related to the Finite Element Method

### *1.5.1 Matrix Notation*

The application of numerical techniques to the solution of partial differential equations generates large systems of approximating algebraic equations. A typical set of $n$ equations in $m$ unknowns would have the form

$$\begin{aligned} a_{1,1}x_1 + a_{1,2}x_2 + \cdots + a_{1,m}x_m &= b_1 \\ a_{2,1}x_1 + a_{2,2}x_2 + \cdots + a_{2,m}x_m &= b_2 \\ &\vdots \\ a_{n,1}x_1 + a_{n,2}x_2 + \cdots + a_{n,m}x_m &= b_n \end{aligned} \tag{1.13}$$

where $x_i$, $i = 1, 2, \ldots, m$, are $m$ unknown parameters, $b_i$, $i = 1, 2, \ldots, n$, are known values, and $a_{i,j}$ are known coefficients.

This set of equations can be written in matrix form as

$$\begin{bmatrix} a_{1,1} & a_{1,2} & \cdots & a_{1,m} \\ a_{2,1} & a_{2,2} & \cdots & a_{2,m} \\ & & \vdots & \\ a_{n,1} & a_{n,2} & \cdots & a_{n,m} \end{bmatrix} \begin{Bmatrix} x_1 \\ x_2 \\ \vdots \\ x_m \end{Bmatrix} = \begin{Bmatrix} b_1 \\ b_2 \\ \vdots \\ b_m \end{Bmatrix} \tag{1.14}$$

where the elements of a row of rectangular array $a_{i,j}$, $j = 1, 2, \ldots, m$, denote the coefficients of equation $i$. A column of the array $a_{i,j}$, $i = 1, 2, \ldots, n$, contains the coefficients associated with the $j$th unknown for each of the $i$ equations. This array is defined as an $n \times m$ rectangular matrix and, in general, does not have the same number of rows as columns. The array of unknown parameters in the set of equations is arranged as a column matrix or vector, as are the known values on the right-hand side of the equation.

Equation (1.14) can be rewritten using capital letters to denote each array, brackets to denote a rectangular matrix, and braces to denote a column matrix:

$$[A]\{X\} = \{B\}$$

When elements of the array are identified individually such as the element $a_{i,j}$ in $[A]$, the order of the subscripts is important. The first subscript (in this example $i$) denotes the row containing the element and the second (i.e., $j$) denotes the column.

### 1.5.2 *Definitions*

**Matrix** An $(n \times m)$ matrix is a rectangular array of numbers of the form

$$[A] = \begin{bmatrix} a_{1,1} & a_{1,2} & a_{1,3} & \cdots & a_{1,m} \\ a_{2,1} & a_{2,2} & a_{2,3} & \cdots & a_{2,m} \\ & & & \vdots & \\ a_{n,1} & a_{n,2} & a_{n,3} & \cdots & a_{n,m} \end{bmatrix}$$

having $n$ rows and $m$ columns.

**Square matrix** A matrix is called square when it has the same number of rows as columns (that is, $m = n$). The order of a square matrix is $m$ where $m$ is the number of rows or columns.

**Symmetric matrix** This is a square matrix whose elements are symmetric with respect to the diagonal. In a symmetric matrix, the element of the $i$th row and $j$th column equals the element of the $j$th row and $i$th column, or in indicial notation $a_{i,j} = a_{j,i}$.

**Skew symmetric matrix** This is a square matrix that has a negative symmetry with respect to the diagonal. In indicial notation (such that $a_{i,j}$ is the element of $[A]$ in row $i$ and column $j$), a matrix is skew symmetric if $a_{i,j} = -a_{j,i}$ for $i \neq j$.

**Diagonal matrix** This is a square matrix with nonzero elements only along the diagonal. A matrix is a diagonal matrix if $a_{i,j} = 0$, $i \neq j$.

**Identity matrix** (*unit matrix*) This is a diagonal matrix with its diagonal elements equal to unity. A matrix is a unit matrix when $a_{i,j} = 0, i \neq j$; $a_{i,j} = 1$, $i = j$.

**Triangular matrix** There are upper and lower triangular matrices and they have all zero elements below and above the diagonal, respectively.

**Conjugate matrix** This matrix is obtained by replacing each element of a matrix by its complex conjugate. Therefore if $[A]$ has elements $a_{i,j}$ that are all real, the conjugate matrix, denoted $[A]^*$, is equal to $[A]$.

**Transpose of a matrix** This matrix, denoted $[A]^T$, is obtained from $[A]$ by interchanging its rows and columns such that $a^T_{i,j} = a_{j,i}$.

**Hermitian matrix** A matrix whose transpose is its conjugate is Hermitian. Thus if all the elements of [A] are real, [A] is Hermitian if and only if it is symmetric.

### *1.5.3 Matrix Operations*

**Matrix addition** Two matrices can be added (or subtracted) if they each have the same dimensions. Addition of suitable matrices requires the addition of corresponding elements to form a new matrix. If $[C] = [A] + [B]$, then $c_{i,j} = a_{i,j} + b_{i,j}$ for all values of $i$ and $j$. For example,

$$\begin{bmatrix} c_{1,1} & c_{1,2} & c_{1,3} \\ c_{2,1} & c_{2,2} & c_{2,3} \end{bmatrix} = \begin{bmatrix} a_{1,1} & a_{1,2} & a_{1,3} \\ a_{2,1} & a_{2,2} & a_{2,3} \end{bmatrix} + \begin{bmatrix} b_{1,1} & b_{1,2} & b_{1,3} \\ b_{2,1} & b_{2,2} & b_{2,3} \end{bmatrix}$$
$$= \begin{bmatrix} (a_{1,1} + b_{1,1}) & (a_{1,2} + b_{1,2}) & (a_{1,3} + b_{1,3}) \\ (a_{2,1} + b_{2,1}) & (a_{2,2} + b_{2,2}) & (a_{2,3} + b_{2,3}) \end{bmatrix} \tag{1.15}$$

Equation (1.15) may also be written more simply as

$$\underset{2\times 3}{[C]} = \underset{2\times 3}{[A]} + \underset{2\times 3}{[B]} = \underset{2\times 3}{[A + B]}$$

**Matrix multiplication** Matrix [A] may be multiplied by matrix [B] to form the product [A][B] only if the number of columns of [A] is identical to the number of rows of [B]. If [A] has $m$ columns and [B] has $m$ rows, post multiplication of [A] by [B] can be concisely stated as

$$\underset{n\times m}{[A]}\ \underset{m\times r}{[B]} = \underset{n\times r}{[C]}$$

where

$$c_{i,j} = \sum_{k=1}^{m} a_{i,k} b_{k,j} \qquad (i = 1, \ldots, n, \quad j = 1, \ldots, r)$$

Consider the example

$$\underset{2\times 2}{\begin{bmatrix} 1 & 2 \\ 3 & 4 \end{bmatrix}} \underset{2\times 3}{\begin{bmatrix} 5 & 6 & 7 \\ 8 & 9 & 0 \end{bmatrix}}$$
$$= \underset{2\times 3}{\begin{bmatrix} (1\times 5) + (2\times 8) & (1\times 6) + (2\times 9) & (1\times 7) + (2\times 0) \\ (3\times 5) + (4\times 8) & (3\times 6) + (4\times 9) & (3\times 7) + (4\times 0) \end{bmatrix}}$$
$$= \underset{2\times 3}{\begin{bmatrix} 21 & 24 & 7 \\ 47 & 54 & 21 \end{bmatrix}}$$

It is important to note that the commutative law does not hold for matrix multiplication: $[A][B] \neq [B][A]$.

**Determinant of a matrix** Although there are several schemes for calculating the determinant (which is defined only for a square matrix), we will consider the method known as expansion by cofactors [6]. In this approach, the determinant is the scalar obtained as

$$|A| = \sum_{i=1}^{n} a_{i,j} \cdot [\text{cof}(a_{i,j})] = \sum_{j=1}^{n} a_{i,j} \cdot [\text{cof}(a_{i,j})] \tag{1.16}$$

for any relevant value of $j$ or $i$, respectively, and where the cofactor (cof) is defined as

$$\text{cof}(a_{i,j}) = (-1)^{i+j}(\text{minor } a_{i,j})$$

and the minor is the determinant of the submatrix obtained by deleting the $i$th row and $j$th column of the original matrix. The process is hierarchical in the sense that it may be reapplied until the required determinant involves only a $2 \times 2$ matrix which can easily be obtained from the definition

$$|A| = \begin{vmatrix} a_{1,1} & a_{1,2} \\ a_{2,1} & a_{2,2} \end{vmatrix} = a_{1,1}a_{2,2} - a_{2,1}a_{1,2}$$

To demonstrate the method of expansion by cofactors let us calculate the determinant of a $3 \times 3$ matrix:

$$\det[A] = |A| = \begin{vmatrix} a_{1,1} & a_{1,2} & a_{1,3} \\ a_{2,1} & a_{2,2} & a_{2,3} \\ a_{3,1} & a_{3,2} & a_{3,3} \end{vmatrix}$$

$$= a_{1,1}(-1)^{(1+1)} \begin{vmatrix} a_{2,2} & a_{2,3} \\ a_{3,2} & a_{3,3} \end{vmatrix}$$

$$+ a_{1,2}(-1)^{(1+2)} \begin{vmatrix} a_{2,1} & a_{2,3} \\ a_{3,1} & a_{3,3} \end{vmatrix} + a_{1,3}(-1)^{(1+3)} \begin{vmatrix} a_{2,1} & a_{2,2} \\ a_{3,1} & a_{3,2} \end{vmatrix}$$

$$= a_{1,1}(1)(a_{2,2}a_{3,3} - a_{3,2}a_{2,3}) + a_{1,2}(-1)(a_{2,1}a_{3,3} - a_{3,1}a_{2,3})$$

$$+ a_{1,3}(1)(a_{2,1}a_{3,2} - a_{3,1}a_{2,2})$$

**Adjoint of a matrix** The adjoint of a square matrix $[A]$ is the transpose of the cofactor matrix of $[A]$ which can be written

$$\text{adj}[A] = [\text{cof } A]^{\text{T}} \tag{1.17}$$

To illustrate the calculation of the adjoint of a matrix, we will calculate the adjoint of the $3 \times 3$ matrix $[A]$ of the previous example. The cofactor matrix of $[A]$ is defined as

$$\operatorname{cof}[A] = \operatorname{cof}[a_{i,j}] = \begin{bmatrix} \operatorname{cof}(a_{1,1}) & \operatorname{cof}(a_{1,2}) & \operatorname{cof}(a_{1,3}) \\ \operatorname{cof}(a_{2,1}) & \operatorname{cof}(a_{2,2}) & \operatorname{cof}(a_{2,3}) \\ \operatorname{cof}(a_{3,1}) & \operatorname{cof}(a_{3,2}) & \operatorname{cof}(a_{3,3}) \end{bmatrix}$$

which, from the definition of the cofactor, becomes

$$\operatorname{cof}[A] = \begin{bmatrix} (a_{2,2}a_{3,3} - a_{3,2}a_{2,3}) & -(a_{2,1}a_{3,3} - a_{3,1}a_{2,3}) & (a_{2,1}a_{3,2} - a_{3,1}a_{2,2}) \\ -(a_{1,2}a_{3,3} - a_{3,2}a_{1,3}) & (a_{1,1}a_{3,3} - a_{3,1}a_{1,3}) & -(a_{1,1}a_{3,2} - a_{3,1}a_{1,2}) \\ (a_{1,2}a_{2,3} - a_{2,2}a_{1,3}) & -(a_{1,1}a_{2,3} - a_{2,1}a_{1,3}) & (a_{1,1}a_{2,2} - a_{2,1}a_{1,2}) \end{bmatrix}$$

To obtain the adjoint we take the transpose of $\operatorname{cof}[A]$ and obtain

$$\operatorname{adj}[A] = [\operatorname{cof} A]^{\mathrm{T}} = \begin{bmatrix} (a_{2,2}a_{3,3} - a_{3,2}a_{2,3}) & -(a_{1,2}a_{3,3} - a_{3,2}a_{1,3}) & (a_{1,2}a_{2,3} - a_{2,2}a_{1,3}) \\ -(a_{2,1}a_{3,3} - a_{3,1}a_{2,3}) & (a_{1,1}a_{3,3} - a_{3,1}a_{1,3}) & -(a_{1,1}a_{2,3} - a_{2,1}a_{1,3}) \\ (a_{2,1}a_{3,2} - a_{3,1}a_{2,2}) & -(a_{1,1}a_{3,2} - a_{3,1}a_{1,2}) & (a_{1,1}a_{2,2} - a_{2,1}a_{1,2}) \end{bmatrix}$$

**Matrix inversion** As in the calculation of determinants, there are many schemes for calculating the inverse of a square matrix. The matrix inverse $[A]^{-1}$ is defined by the relationship

$$[A][A]^{-1} = [A]^{-1}[A] = [I]$$

where $[I]$ is the identity matrix. Referring back to (1.14), which defined the original set of equations, we observe that if

$$[A]\{X\} = \{B\}$$

the column matrix of unknowns may be obtained directly using the inverse matrix (when $n = m$)

$$\{X\} = [A]^{-1}\{B\}$$

Consequently, a set of simultaneous linear algebraic equations is solved through the use of the inverse matrix.

The scheme for inverting a matrix which we have chosen for illustration yields the inverse matrix through division of each member of the adjoint of $[A]$ by the determinant of $[A]$:

$$[A]^{-1} = (\operatorname{adj}[A])/|A| \tag{1.18}$$

To demonstrate not only the calculation of the inverse but also the solution of simultaneous algebraic equations, we will consider the example

$$\begin{aligned} 2x_1 + x_3 &= 7 \\ x_2 + 5x_3 &= 4 \\ 4x_1 + 6x_2 - 5x_3 &= 2 \end{aligned} \tag{1.19}$$

This set of equations can be written in matrix form as

$$\begin{bmatrix} 2 & 0 & 1 \\ 0 & 1 & 5 \\ 4 & 6 & -5 \end{bmatrix} \begin{Bmatrix} x_1 \\ x_2 \\ x_3 \end{Bmatrix} = \begin{Bmatrix} 7 \\ 4 \\ 2 \end{Bmatrix} \tag{1.20}$$

*Step 1* Compute the determinant of the coefficient matrix:

$$\begin{vmatrix} 2 & 0 & 1 \\ 0 & 1 & 5 \\ 4 & 6 & -5 \end{vmatrix} = 2(-5-30) - 0(0-20) + 1(0-4) = -74$$

*Step 2* Calculate the cofactor matrix of $[A]$:

$$\text{cof}[A] = \begin{bmatrix} (-5-30) & -(0-20) & (0-4) \\ -(0-6) & (-10-4) & -(12-0) \\ (0-1) & -(10-0) & (2-0) \end{bmatrix} = \begin{bmatrix} -35 & 20 & -4 \\ 6 & -14 & -12 \\ -1 & -10 & 2 \end{bmatrix}$$

*Step 3* Compute the adjoint of $[A]$:

$$\text{adj}[A] = [\text{cof } A]^{\mathrm{T}} = \begin{bmatrix} -35 & 6 & -1 \\ 20 & -14 & -10 \\ -4 & -12 & 2 \end{bmatrix}$$

*Step 4* Compute the inverse matrix $[A]^{-1}$:

$$[A]^{-1} = \frac{\text{adj}[A]}{|A|} = \frac{1}{-74} \begin{bmatrix} -35 & 6 & -1 \\ 20 & -14 & -10 \\ -4 & -12 & 2 \end{bmatrix}$$

*Step 5* Compute the vector of unknowns using the inverse matrix:

$$\{X\} = [A]^{-1}\{B\} = \frac{1}{-74} \begin{bmatrix} -35 & 6 & -1 \\ 20 & -14 & -10 \\ -4 & -12 & 2 \end{bmatrix} \begin{Bmatrix} 7 \\ 4 \\ 2 \end{Bmatrix} = \begin{Bmatrix} 3.01 \\ -0.865 \\ 0.973 \end{Bmatrix} = \begin{Bmatrix} x_1 \\ x_2 \\ x_3 \end{Bmatrix}$$

Although the solution of linear algebraic equations with the preceding technique is of fundamental importance from a theoretical point of view, in most practical problems involving a large systems of equations, the method is highly inefficient. Consequently, many other methods have been devised, and several of these will be considered in the following sections.

## 1.6 Direct Methods for the Solution of Linear Algebraic Equations

While many methods are available for the direct solution of systems of algebraic equations, we will focus our attention here on the class of solution algorithms based on the efficient elimination methods of C. F. Gauss. Inasmuch as this topic has received considerable attention in the past and pertinent theory as well as documented computer codes are readily available elsewhere [7–10], we will not consider these techniques in great detail.

### *1.6.1 Gaussian Elimination*

Many direct methods for solving linear algebraic equations are variants of Gaussian elimination. Although they are algebraically identical, they differ in how matrices are stored, the elimination sequence, or the approach taken to minimize large round-off errors. The algebraic basis for Gaussian elimination is the following theorem [7].

**Theorem 1** Given a square matrix $[A]$ of order $n$, let $[A_k]$ denote the principal minor matrix made from the first $k$ rows and columns. Assume that $\det[A_k] \neq 0$ for $k = 1, 2, \ldots, n-1$. Then there exists a unique lower triangular matrix $[L]$ containing elements $l_{i,j}$ with $l_{1,1} = l_{2,2} = \cdots = l_{n,n} = 1$, and a unique upper triangular matrix $[U]$ containing elements $u_{i,j}$ such that $[L][U] = [A]$. Moreover, $\det[A] = u_{1,1}u_{2,2}\cdots u_{n,n}$.

An important corollary to this theorem, which will be utilized in the discussion of the Crout algorithm, is the following [7].

**Corollary** If $[A]$ is a symmetric, positive definite matrix, then it can be decomposed uniquely into $[G][G]^{\mathrm{T}}$, where $[G]$ is a lower triangular matrix with positive diagonal elements.†

The factorization of the matrix $[A]$ into an upper and lower triangular matrix is the essential idea in all Gaussian elimination schemes. Consider the system of linear equations

$$[A]\{X\} = \{B\} \tag{1.21}$$

From the above theorem we know that $[A]$ can be written as the product $[L][U]$, which, when substituted into Eq. (1.21), gives

$$[L][U]\{X\} = \{B\} \tag{1.22}$$

† See Hildebrand [11, p. 47] for a discussion of positive definite matrices.

An intermediate solution $\{Y\}$ is defined as

$$\{Y\} = [U]\{X\}$$

Equation (1.22) can now be written as

$$[L]\{Y\} = \{B\}$$

Given the lower triangular matrix $[L]$, the $\{Y\}$ matrix is easily calculated because the first algebraic equation involves only one unknown $y_1$. By solving the equations sequentially, the unknown value in each equation can be readily obtained using a forward substitution scheme. Similarly, given $[U]$ and having obtained $\{Y\}$, then $\{X\}$ can be easily calculated by first solving the $n$th equation and back substituting in the $(n - 1)$ remaining equations.

The question which thus far remains unanswered is how to obtain the triangular matrices $[L]$ and $[U]$. This process, known as triangular decomposition, may be combined with the calculation of $\{Y\}$ and the overall procedure termed forward elimination. The major effort in applying the Gaussian elimination procedure is concentrated in this forward elimination step. The order in which the sequence of operations is performed in executing this step defines other related schemes.

In the ordinary Gaussian elimination approach one renders the matrix $[A]$ upper triangular through a sequence of elementary row operations. To demonstrate that this approach is, indeed, a form of $LU$ decomposition, we will follow a development similar to that of Forsythe and Moler [7, p. 31]. Consider once again the set of equations (1.19) as an illustrative example:

$$\begin{bmatrix} 2 & 0 & 1 \\ 0 & 1 & 5 \\ 4 & 6 & -5 \end{bmatrix} \begin{Bmatrix} x_1 \\ x_2 \\ x_3 \end{Bmatrix} = \begin{Bmatrix} 7 \\ 4 \\ 2 \end{Bmatrix} \quad \text{or} \quad [A]\{X\} = \{B\}$$

Define $l_{i,1} = a_{i,1}/a_{1,1}$ for $i$ greater than 1. Next form the lower triangular matrix $[L_1]$ of the form

$$[L_1] = \begin{bmatrix} 1 & 0 & 0 \\ -l_{2,1} & 1 & 0 \\ -l_{3,1} & 0 & 1 \end{bmatrix} = \begin{bmatrix} 1 & 0 & 0 \\ 0 & 1 & 0 \\ -2 & 0 & 1 \end{bmatrix}$$

Now premultiplication of the matrix equation by $[L_1]$ yields

$$[A]^{(2)}\{X\} = \{B\}^{(2)}$$

where $[A]^{(2)} = [L_1][A]$ and $\{B\}^{(2)} = [L_1]\{B\}$. Carrying through the multiplication, we obtain

$$\begin{bmatrix} 2 & 0 & 1 \\ 0 & 1 & 5 \\ 0 & 6 & -7 \end{bmatrix} \begin{Bmatrix} x_1 \\ x_2 \\ x_3 \end{Bmatrix} = \begin{Bmatrix} 7 \\ 4 \\ -12 \end{Bmatrix}$$

Now consider the second diagonal element of $[A]^{(2)}$ and let $l_{i,2} = a_{i,2}^{(2)}/a_{2,2}^{(2)}$ for $i$ greater than 2. The new matrices $A^{(3)}$ and $B^{(3)}$ are obtained by premultiplying by the new lower triangular matrix $[L_2]$ defined by

$$[L_2] = \begin{bmatrix} 1 & 0 & 0 \\ 0 & 1 & 0 \\ 0 & -l_{3,2} & 1 \end{bmatrix} = \begin{bmatrix} 1 & 0 & 0 \\ 0 & 1 & 0 \\ 0 & -6 & 1 \end{bmatrix}$$

Then $[A]^{(3)} = [L_2][A]^{(2)}$, $[B]^{(3)} = [L_2][B]^{(2)}$, and the equation now has the form

$$\begin{bmatrix} 2 & 0 & 1 \\ 0 & 1 & 5 \\ 0 & 0 & -37 \end{bmatrix} \begin{Bmatrix} x_1 \\ x_2 \\ x_3 \end{Bmatrix} = \begin{Bmatrix} 7 \\ 4 \\ -36 \end{Bmatrix} \tag{1.23}$$

The matrix $[A]^{(3)} = [L_2][A]^{(2)} = [L_2][L_1][A]$ is in upper triangular form, and will be denoted by $[U]$. The right-hand side of (1.23) is the intermediate solution $\{Y\}$. By back substitution, we obtain $x_3 = 36/37 = 0.973$, $x_2 = -5x_3 + 4 = -0.865$, and $x_1 = \frac{1}{2}(-x_3 + 7) = 3.01$ which is, of course, identical to the solution obtained using the inverse matrix.

Although the primary objective of solving a set of algebraic equations has been achieved, it is useful to demonstrate that we have, in fact, applied $LU$ decomposition. Let $[L^*] = [L_2][L_1]$. Then since $[L^*][A] = [U]$, we have $[A] = [L^*]^{-1}[U]$. But, $[L^*]^{-1}$ can also be written $[L_1]^{-1}[L_2]^{-1}$, and it is easily demonstrated that $[L_k]^{-1}$ is simply $[L_k]$ with the sign of the off-diagonal elements reversed. Finally, if the product $[L_1]^{-1}[L_2]^{-1}$ is computed, we obtain

$$[L] = [L^*]^{-1} = \begin{bmatrix} 1 & 0 & 0 \\ l_{2,1} & 1 & 0 \\ l_{3,1} & l_{3,2} & 1 \end{bmatrix} = \begin{bmatrix} 1 & 0 & 0 \\ 0 & 1 & 0 \\ 2 & 6 & 1 \end{bmatrix}$$

which shows that $[L^*]^{-1}$ is the lower triangular matrix of the decomposition of $[A]$ into a product of lower and upper triangular matrices. This can be further demonstrated by taking the product $[L][U]$ as

$$\begin{bmatrix} 1 & 0 & 0 \\ 0 & 1 & 0 \\ 2 & 6 & 1 \end{bmatrix} \begin{bmatrix} 2 & 0 & 1 \\ 0 & 1 & 5 \\ 0 & 0 & -37 \end{bmatrix} = \begin{bmatrix} 2 & 0 & 1 \\ 0 & 1 & 5 \\ 4 & 6 & -5 \end{bmatrix} = [A]$$

which is, as required, the original coefficient matrix.

From a computational point of view, it is important to point out that, as the elimination process proceeds, the below diagonal elements $l_{i,j}$ are stored in place of the original below diagonal elements of $[A]$ and the elements $u_{i,j}$ of $[U]$ are stored in place of the diagonal and above diagonal elements of

$[A]$. Consequently, at the end of the decomposition process, the original matrix $A$ contains the elements

$$\begin{bmatrix} u_{1,1} & u_{1,2} & u_{1,3} \\ l_{2,1} & u_{2,2} & u_{2,3} \\ l_{3,1} & l_{3,2} & u_{3,3} \end{bmatrix}$$

Inasmuch as the $l_{i,j}$ elements have been retained, solutions for other known matrices $\{B\}$ are readily computed by premultiplication of the new $\{B\}$ by $[L]$ and back substitution. Consequently, other problems with the same coefficient matrix $[A]$ are solved without repetition of the triangular decomposition process.

### *1.6.2 Pivoting to Minimize Round-Off Errors*

In each stage of Gaussian elimination one element of the coefficient matrix $[A]^{(k)}$, say $a_{k,k}^{(k)}$, is selected as the divisor for all other elements in the $k$th column. This element is called the "pivot" for the $k$th stage of elimination. It was assumed in the earlier example that the pivot element was nonzero, but this is not always the case. Division by zero would cause, of course, the elimination to fail. However, if $\det[A] \neq 0$, we know that there is at least one nonzero element in the column, i.e., $a_{i,k} \neq 0$ for some value of $(k \leq i \leq n)$. Consequently, by simply interchanging the $i$th and $k$th rows of $[A]^{(k)}$ and $\{B\}^{(k)}$, a nonzero pivot is thereby obtained in an equivalent equation system.

Naturally, one might ask the significance of using a pivot that, although nonzero, has an absolute value very small relative to other elements of the matrix. Selecting a pivot that has a small absolute value generally introduces round-off errors that may provide totally erroneous solutions. It is advisable to select pivot elements that will generate values of $l_{i,j}$ less than or equal to one in absolute value [7]. This may be accomplished by choosing as a suitable pivot the largest, in absolute value, of the elements in the $k$th column $(|a_{p,k}^{(k)}|) = \max_i |a_{i,k}^{(k)}|$ $(k \leq i \leq n)$. As in the case of a zero pivot element this improved pivot may be obtained by interchanging the $p$th and $k$th rows of $[A]^{(k)}$ and $\{B\}^{(k)}$. This procedure, which involves taking the largest absolute value in a column as the pivot element, is called partial pivoting.

A second technique for minimizing round-off errors involves selecting, as pivot element for the $k$th elimination, the element of largest absolute value in the remaining $(n - k + 1)$ equations. Consequently, this scheme, known as complete pivoting, requires the pivot element to be selected from the last $(n - k + 1)$ rows and columns of $[A]$. In general, the row and column permutations inherent in pivoting are recorded for use later in interpreting the solution.

When solving large systems of equations, it may suffice to replace partial or complete pivoting with a requirement that any pivot element exceed, in absolute value, a predetermined criterion $\varepsilon$. Tewarson [12] suggests that for most practical problems involving large *sparse* matrices $\varepsilon = 10^{-3}$ has been found satisfactory when the nonzero elements are recorded to nine or ten decimal digits.

### 1.6.3 The Crout Method

Gaussian elimination has been shown above to be one form of *LU* decomposition. An important variant of this fundamental approach is the method introduced in the United States by P. D. Crout in 1941.

As in Gaussian elimination, the coefficient matrix $[A]$ is decomposed into upper and lower triangular matrices $[U]$ and $[L]$, respectively. The principal advantage to the Crout method (and several closely related schemes which will be mentioned briefly later) is that the intermediate reduced matrices $[A]^{(k)}$ need not be recorded. As a result the elements $l_{i,j}$ and $u_{i,j}$ can be calculated in a single machine operation by a continuous accumulation of products.

Consider a generalized three-equation system. We first formulate the augmented $[A]$ matrix by including the known vector $\{B\}$ as a fourth column, i.e., $a_{i,4} = b_i$ $(i = 1, 2, 3)$. Because $[A]$ can be decomposed into an $[L][U]$ product, we can write

$$\underset{[A]\qquad\qquad\{B\}}{\left[\begin{array}{ccc|c} a_{1,1} & a_{1,2} & a_{1,3} & b_1 \\ a_{2,1} & a_{2,2} & a_{2,3} & b_2 \\ a_{3,1} & a_{3,2} & a_{3,3} & b_3 \end{array}\right]}$$

$$= \underset{[L]}{\begin{bmatrix} l_{1,1} & 0 & 0 \\ l_{2,1} & l_{2,2} & 0 \\ l_{3,1} & l_{3,2} & l_{3,3} \end{bmatrix}} \underset{[U]\qquad\qquad\{Y\}}{\left[\begin{array}{ccc|c} 1 & u_{1,2} & u_{1,3} & y_1 \\ 0 & 1 & u_{2,3} & y_2 \\ 0 & 0 & 1 & y_3 \end{array}\right]} \tag{1.24}$$

where, in contrast to the earlier example, the unit diagonal is associated with the upper triangular matrix and $\{Y\}$ is the matrix of intermediate values. Expanding (1.24), we obtain the relationship [4]

$$a_{i,1} = l_{i,1}$$

that is, the first column of $[L]$ is identical to the first column of $[A]$;

$$a_{1,j} = l_{1,1} u_{1,j} = a_{1,1} u_{1,j}$$

that is, the first row of [U] equals the first row of [A] divided by $a_{1,1}$;

$$a_{2,2} = l_{2,1}u_{1,2} + l_{2,2} \quad \text{or} \quad l_{2,2} = a_{2,2} - l_{2,1}u_{1,2}$$

$$a_{2,3} = l_{2,1}u_{1,3} + l_{2,2}u_{2,3} \quad \text{or} \quad u_{2,3} = (a_{2,3} - l_{2,1}u_{1,3})/l_{2,2}$$

$$a_{3,2} = l_{3,1}u_{1,2} + l_{3,2} \quad \text{or} \quad l_{3,2} = a_{3,2} - l_{3,1}u_{1,2}$$

$$b_2 = l_{2,1}y_1 + l_{2,2}y_2 \quad \text{or} \quad y_2 = (b_2 - l_{2,1}y_1)/l_{2,2}$$

and so forth.

The forward elimination step thus proceeds stepwise moving horizontally from element $l_{22}$. The general equations for the Crout algorithm are [4]

$$l_{i,j} = a_{i,j} - \sum_{r=1}^{j-1} l_{i,r}u_{r,j}, \qquad l_{i,1} = a_{i,1} \tag{1.25}$$

$$u_{i,j} = (1/l_{i,i})\left(a_{i,j} - \sum_{r=1}^{i-1} l_{i,r}u_{r,j}\right), \qquad u_{1,j} = a_{1,j}/a_{1,1} \tag{1.26}$$

where $a_{i,j}$ and $u_{i,j}$ are elements of the augmented matrices.

As an illustrative example let us consider once again the system of equations represented by (1.20):

$$\underset{[A]}{\begin{bmatrix} 2 & 0 & 1 \\ 0 & 1 & 5 \\ 4 & 6 & -5 \end{bmatrix}} \underset{\{X\}}{\begin{Bmatrix} x_1 \\ x_2 \\ x_3 \end{Bmatrix}} = \underset{\{B\}}{\begin{Bmatrix} 7 \\ 4 \\ 2 \end{Bmatrix}}$$

*Step 1* Generate the augmented $A$ matrix:

$$\begin{bmatrix} 2 & 0 & 1 & 7 \\ 0 & 1 & 5 & 4 \\ 4 & 6 & -5 & 2 \end{bmatrix}$$

*Step 2* Write the first column of [L] using (1.25) and the first row of the augmented [U] matrix using (1.26):

$$\begin{bmatrix} 2 & 0 & 0 \\ 0 & l_{2,2} & 0 \\ 4 & l_{3,2} & l_{3,3} \end{bmatrix} \begin{bmatrix} 1 & 0 & \frac{1}{2} & \frac{7}{2} \\ 0 & 1 & u_{2,3} & y_2 \\ 0 & 0 & 1 & y_3 \end{bmatrix}$$

*Step 3* Calculate sequentially, using (1.25) and (1.26), $l_{2,2}$, $u_{2,3}$, and $y_2$:

$$\begin{bmatrix} 2 & 0 & 0 \\ 0 & 1 & 0 \\ 4 & l_{3,2} & l_{3,3} \end{bmatrix} \begin{bmatrix} 1 & 0 & \frac{1}{2} & \frac{7}{2} \\ 0 & 1 & 5 & 4 \\ 0 & 0 & 1 & y_3 \end{bmatrix}$$

*Step 4* Calculate sequentially $l_{3,2}$, $l_{3,3}$, and $y_3$:

$$\underset{[L]}{\begin{bmatrix} 2 & 0 & 0 \\ 0 & 1 & 0 \\ 4 & 6 & -37 \end{bmatrix}} \underset{[U]}{\begin{bmatrix} 1 & 0 & \frac{1}{2} \\ 0 & 1 & 5 \\ 0 & 0 & 1 \end{bmatrix}} \quad \underset{\{Y\}}{\begin{bmatrix} \frac{7}{2} \\ 4 \\ \frac{36}{37} \end{bmatrix}}$$

From $[U]\{X\} = \{Y\}$ we obtain, by back substitution $x_3 = \frac{36}{37} = 0.973$, $x_2 = -5x_3 + 4 = -0.865$, and $x_1 = -\frac{1}{2}x_3 + \frac{7}{2} = 3.01$, which is, once again, identical to the earlier solution. To verify that an accurate decomposition on $[A]$ has been obtained, the $[L][U]$ product is calculated as

$$\begin{bmatrix} 2 & 0 & 0 \\ 0 & 1 & 0 \\ 4 & 6 & -37 \end{bmatrix} \begin{bmatrix} 1 & 0 & \frac{1}{2} \\ 0 & 1 & 5 \\ 0 & 0 & 1 \end{bmatrix} = \begin{bmatrix} 2 & 0 & 1 \\ 0 & 1 & 5 \\ 4 & 6 & -5 \end{bmatrix}$$

and found to be identical to the original coefficient matrix $[A]$.

In a comparison between the Crout method and Gaussian elimination, Forsythe and Moler [7] state that the amount of arithmetic is the same for both methods since the same calculations are carried out in each, although in a different order. However, if one considers only multiplications, divisions,† and recording of numbers (the time-consuming operations on a digital computer), Gaussian elimination requires $\frac{1}{3}n^3 + O(n^2)$ operations while the Crout algorithm requires $n^2 + O(n)$ operations to solve the same system of $n$ equations [4]. Generally the Crout method has the smaller round-off error although this depends on the storage characteristics of the digital computer.

### 1.6.4 *The Doolittle Method*

The Doolittle method is a minor modification of the Crout algorithm designed to facilitate the forward elimination step when the coefficient matrix $[A]$ is stored by rows. In this scheme only the $k$th rows of $[L]$ and $[U]$ are generated at the $k$th stage. The computations proceed from left to right along each row; for the $k$th row the formula [12] is

$$l_{k,j} = a_{k,j} - \sum_{r=1}^{j-1} l_{k,r} u_{r,j} \qquad (j = 1, 2, \ldots, k)$$

and the elements of $[U]$ are calculated using the Crout formula (1.26). A matrix stored by columns can also be decomposed directly from the

† It is assumed in this analysis that the computer can perform two multiplications in the time it takes to do one division.

relationship

$$u_{i,k} = \left(a_{i,k} - \sum_{r=1}^{i-1} l_{i,r} u_{r,k}\right) \Big/ l_{i,i} \qquad (i = 1, 2, \ldots, k-1)$$

and, for $i \geq k$, Eq. (1.25) can be used.

### 1.6.5 *The Cholesky Method* (*Square Root Method*)

This method was used by A. L. Cholesky in France prior to 1916 to solve problems involving symmetric matrices [4]. It is sometimes called the "Banachiewicz method" after Th. Banachiewicz who derived the method in matrix form in Poland in 1938. The method is applicable to symmetric, positive definite matrices, and for such matrices is the best method for triangular decomposition [13]. Positive definiteness implies the nonsingularity of the minors necessary for the *LU* decomposition theorem presented earlier [7].

The corollary of this theorem states that symmetric positive definite matrices can be decomposed in the form

$$[A] = [L][L]^{\mathrm{T}} \tag{1.27}$$

where $[L]$ is a lower triangular matrix with positive diagonal elements. Expansion of (1.27) in terms of its individual elements yields for the main diagonal

$$a_{j,j} = l_{j,1}^2 + l_{j,2}^2 + \cdots + l_{j,j}^2 \tag{1.28}$$

and for $a_{i,j}$ below the diagonal

$$a_{i,j} = l_{i,1} l_{j,1} + l_{i,2} l_{j,2} + \cdots + l_{i,j} l_{j,j} \qquad (j < i) \tag{1.29}$$

One can solve (1.28) and (1.29) for the elements of $[L]$ provided all elements for the $k$th column are obtained, beginning with the diagonal, before proceeding to the $(k+1)$th. The appropriate formulas are

$$l_{j,j} = \left(a_{j,j} - \sum_{k=1}^{j-1} l_{j,k}^2\right)^{1/2}$$

for the diagonal elements and

$$l_{i,j} = \left(a_{i,j} - \sum_{k=1}^{j-1} l_{i,k} l_{j,k}\right) \Big/ l_{j,j}$$

for those elements below the diagonal. The required solution can now be obtained by forward substitution.

An important advantage of Cholesky's scheme is the small round-off error generated without row or column interchanges. Moreover, because of the symmetry of $[A]$, it is necessary to store only $n(n + 1)/2$ of its elements, an important reduction in computer storage.

In the discussions thus far, specific reference has not been made as to whether the matrix is full, sparse, or banded. Considerable saving in core storage and computational effort can be achieved by coding a program to utilize the structure of sparse matrices. Currently, considerable effort is being expended in developing such techniques and the interested reader is referred to the recent monograph by Tewarson [12] for a review of sparse matrix methods.

The matrices generated using the finite element approach are generally sparse and banded. The band structure should be utilized to optimize the efficiency of equation-solving codes. Fortunately, the techniques mentioned above and other similar direct methods can be efficiently applied to banded matrices by simply specifying the range of the subscripts so as to eliminate unnecessary calculations outside the band. Nevertheless, care should be exercised to utilize the band structure to minimize the computational effort in decomposition. In fact, decomposition can be reduced from $n^3/3$ operations for the full matrix to $nm^2$ operations for the banded matrix (where $2m + 1$ is the band width as defined by the relationship $a_{i,j} = 0$ ($|i - j| > m$). For example, a $100 \times 100$ matrix with a band width of 10 requires less than 1% of the computational effort required to decompose the full matrix. Further savings may be gained by using envelope methods that require that only the band width of each row be stored in the computer.

### 1.6.6 *The Thomas Algorithm*

Possibly the most frequently encountered matrix in the application of numerical methods is the tridiagonal band matrix which may be written as

$$\begin{bmatrix} e_1 & f_1 & & & & 0 \\ d_2 & e_2 & f_2 & & & \\ & d_3 & e_3 & f_3 & & \\ & & \cdot & \cdot & \cdot & \\ & & & \cdot & \cdot & \cdot \\ 0 & & & & d_n & e_n \end{bmatrix}$$

where the elements of $A$ have been arranged as three vectors $\mathbf{d}$, $\mathbf{e}$, and $\mathbf{f}$ because they can be retained in this compact form during the equation-solving procedure. This follows from the following theorem [7].

**Theorem 2** If a band matrix with band width $2m + 1$ has an $LU$ decomposition, then $L = (l_{ij})$ and $U = (u_{ij})$ are triangular band matrices. That is

$$u_{ij} \neq 0 \quad \text{only for} \quad i = j - m, \ldots, j$$

$$l_{ij} \neq 0 \quad \text{only for} \quad i = j, \ldots, j + m$$

In the tridiagonal case $m = 1$ and the $LU$ matrices have the form

$$[L] = \begin{bmatrix} 1 & & & & 0 \\ l_2 & 1 & & & \\ & l_3 & 1 & & \\ & & \cdot & \cdot & \\ & & & \cdot & \cdot \\ 0 & & & l_n & 1 \end{bmatrix}, \qquad [U] = \begin{bmatrix} u_1 & f_1 & & & 0 \\ & u_2 & f_2 & & \\ & & u_3 & f_3 & \\ & & & \cdot & \cdot \\ & & & & \cdot \\ 0 & & & & u_n \end{bmatrix}$$

An efficient three-step algorithm for solving a tridiagonal set of equations has been developed using a decomposition step which proceeds according to

$$u_1 = e_1$$

$$l_i = d_i/u_{i-1} \qquad (i = 2, 3, \ldots, n)$$

$$u_i = e_i - l_i f_{i-1} \qquad (i = 2, 3, \ldots, n)$$

The intermediate solution $[L]\{Y\} = \{B\}$ is obtained from

$$y_1 = b_1$$

$$y_i = b_i - l_i y_{i-i} \qquad (i = 2, \ldots, n)$$

Finally, the back substitution procedure generates the solution to the system. We have from $[U]\{X\} = \{Y\}$ that

$$x_n = y_n/u_n$$

$$x_i = (y_i - f_i x_{i+1})/u_i \qquad (i = n - 1, n - 2, \ldots, 1)$$

If $[A]$ is factored such that the diagonal of the upper, rather than lower, triangular matrix is unity, the resulting algorithm for decomposition is

$$l_1 = e_1$$

$$u_i = f_i/l_i \qquad (i = 1, 2, \ldots, n - 1)$$

$$l_i = e_i - d_i u_{i-1} \qquad (i = 2, 3, \ldots, n)$$

The intermediate solution is obtained from

$$y_1 = b_1/l_1$$

$$y_i = (b_i - d_i y_{i-1})/l_i \qquad (i = 2, 3, \ldots, n)$$

and the back substitution procedure is carried out according to the relationship

$$x_n = y_n$$

$$x_i = y_i - u_i x_{i+1} \qquad (i = n-1, n-2, \ldots, 1)$$

This form of the *LU* decomposition algorithm is generally attributed to Thomas [14] and is termed the Thomas algorithm.

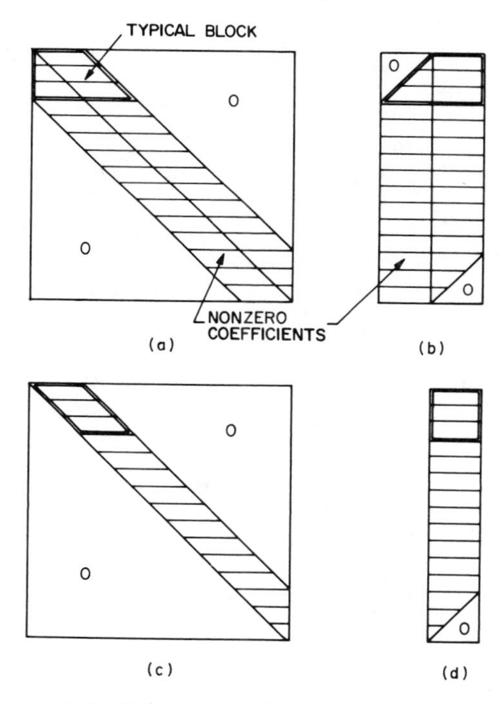


**Fig. 1.2.** Nonsymmetric banded matrix partition for out-of-core equation-solving scheme. (a) Original matrix; (b) matrix modified to minimize storage; (c) matrix in upper triangular form; (d) triangularized matrix modified to minimize storage.

Before leaving the discussion of direct solution methods, mention should be made of "out-of-core" equation solvers. In some practical problems it is necessary to solve exceedingly large systems of equations. When the storage requirements of the resulting matrices exceed the capacity of the computer memory, it becomes necessary to resort to the utilization of peripheral storage devices. In this approach, the matrix is partitioned as shown in Fig. 1.2 and a minimum number of submatrices are retained in the computer memory at any given instant. The remaining elements of the coefficient matrix are retained in peripheral storage such as a disk. The most effective programs of this kind combine the coefficient generation process with the solution scheme and stand-alone equation-solving codes are not generally available. Moreover, those that are available (see Segui [10], for appropriate listings) are generally written for symmetric matrices. Unfortunately, symmetric matrices are the exception rather than the rule in the application of finite element methods in hydrologic analysis.

## 1.7 Iterative Methods for the Solution of Linear Algebraic Equations

Whereas iterative methods for solving linear algebraic equations have found widespread application in conjunction with finite difference methods, this approach has not been widely used in finite element analysis. While the approximating finite difference equations generate coefficient matrices which are amenable to efficient solution by specific iterative techniques, the finite element method is generally associated with an irregular net, or grid, which characteristically leads to a sparse, banded, but less rigidly structured matrix. Nevertheless, iterative methods have been used to advantage in finite element analyses and considerable effort is currently being expended to find new schemes that are specifically suited to finite element coefficient matrices. Consequently, we will consider in the following sections a few of the most commonly encountered matrix schemes: point iterative and block iterative methods.

### *1.7.1 Point Iterative Methods*

Consider once again a set of linear algebraic equations which can be expressed in matrix notation as

$$[A]\{X\} = \{B\}$$

We will assume that $[A]$ is nonsingular and that no zero elements appear on the diagonal, i.e., $a_{i,i} \neq 0$, $i = 1, 2, \ldots, n$. Two new matrices $[D]$ and $[C]$ can be defined such that

$$[A] = [D] - [C] \tag{1.30}$$

where $[D]$ is a diagonal matrix and the elements of $[C]$ are the off-diagonal elements of $[A]$ but of opposite sign.

*Point Jacobi Method*

This early method was apparently introduced by C. G. J. Jacobi in 1845 and is variously known as the method of simultaneous displacements, the point total-step iterative method, and the Richardson iterative method [15].

Define a scheme whereby the off-diagonal elements of $[A]$ operate at the $k$th iteration level and the diagonal elements at the $(k+1)$th level such as

$$[D]\{X\}^{(k+1)} = [C]\{X\}^{(k)} + \{B\} \qquad (k \geq 0) \tag{1.31}$$

and $\{X\}^{(0)}$ are the initial estimates of the unknown matrix $\{X\}$. When $[D]$ has nonzero diagonal elements, the $k$th iterative approximation to $\{X\}$ is obtained from

$$\{X\}^{(k+1)} = [D]^{-1}[C]\{X\}^{(k)} + [D]^{-1}\{B\} \qquad (k \geq 0) \tag{1.32}$$

This formulation defines the point Jacobi method and matrix $[D]^{-1}[C]$ is the point Jacobi matrix. An algorithm for the evaluation of (1.32) can be obtained by expanding (1.31) and dividing through by the diagonal element:

$$x_i^{(k+1)} = -\sum_{j=1,\, j\neq i}^{n} \frac{(a_{i,j})}{a_{i,i}} x_j^{(k)} + \frac{b_i}{a_{i,i}} \qquad (1 \leq i \leq n, \quad k \geq 0)$$

Having obtained $\{X\}^{(k+1)}$, one substitutes this value into the right-hand side of (1.32) and calculates $\{X\}^{(k+2)}$.

*Point Gauss–Seidel Method*

Examination of (1.32) reveals that none of the elements of $\{X\}^{(k+1)}$ are used in calculating this matrix. However, it would intuitively seem advantageous to use updated information whenever possible and the point Gauss–Seidel method incorporates this information into the iteration procedure. Let us define two new matrices $[L]$ and $[U]$ that are lower and upper triangular, respectively, and contain the elements of $[A]$ but with opposite signs such that

$$[A] = [D] - [L] - [U]$$

where

$$[L] = \begin{bmatrix} 0 & & & \\ -a_{2,1} & 0 & & \\ \vdots & & \ddots & \\ -a_{n,1} & -a_{n,2} & -a_{n,n-1} & 0 \end{bmatrix}$$

and

$$[U] = \begin{bmatrix} 0 & -a_{1,2} & \cdots & -a_{1,n} \\ & 0 & & -a_{2,n} \\ & & \ddots & \vdots \\ & & & -a_{n-1,n} \\ & & & 0 \end{bmatrix}$$

We can now define the following iterative scheme alternatively known as the point Gauss–Seidel iteration method, the Liebermann method, the point single-step iterative method, or the method of successive displacements which employs updated information:

$$[D]\{X\}^{(k+1)} = [L]\{X\}^{(k+1)} + [U]\{X\}^{(k)} + \{B\} \qquad (k \geq 0) \qquad (1.33)$$

Because $[D - L]$ is a nonsingular lower triangular matrix, (1.33) can be rewritten as

$$\{X\}^{(k+1)} = [D - L]^{-1}[U]\{X\}^{(k)} + [D - L]^{-1}\{B\} \qquad (k \geq 0) \qquad (1.34)$$

where $[D - L]^{-1}[U]$ is the point Gauss–Seidel matrix. A scheme for evaluating (1.34) is obtained by expanding (1.33) and once again dividing by the diagonal element:

$$x_i^{(k+1)} = -\sum_{j=1}^{i-1}\left(\frac{a_{i,j}}{a_{i,i}}\right)x_j^{(k+1)} - \sum_{j=i+1}^{n}\left(\frac{a_{i,j}}{a_{i,i}}\right)x_j^{(k)} + \frac{b_i}{a_{i,i}} \qquad (1 \leq i \leq n, \quad k \geq 0)$$

*Point Successive Overrelaxation Iterative Method*

In this approach [15a], it is assumed that the calculated values obtained at the $(k + 1)$th iteration using the Gauss–Seidel method can be improved by considering a weighted mean of $x_i^{(k+1)}$ and $x_i^{(k)}$.

Initially a new vector $\{\tilde{X}\}^{(k+1)}$ is defined by

$$[D]\{\tilde{X}\}^{(k+1)} = [L]\{X\}^{(k+1)} + [U]\{X\}^{(k)} + \{B\} \qquad (k \geq 0) \qquad (1.35)$$

The vector $\{X\}^{(k+1)}$ is then obtained from the relationship

$$\{X\}^{(k+1)} = \{X\}^{(k)} + \omega(\{\tilde{X}\}^{(k+1)} - \{X\}^{(k)})$$

which can be rearranged as

$$\{X\}^{(k+1)} = (1 - \omega)\{X\}^{(k)} + \omega\{\tilde{X}\}^{(k+1)} \qquad (1.36)$$

This form clearly shows $\{X\}^{(k+1)}$ to be a weighted mean of $\{X\}^{(k)}$ and $\{\tilde{X}\}^{(k+1)}$. The parameter $\omega$ is the relaxation factor. The procedure is called underrelaxation when $0 \leq \omega \leq 1$ and overrelaxation when $\omega > 1$. Substitution of (1.36) into (1.35) yields

$$([D] - \omega[L])\{X\}^{(k+1)} = [(1 - \omega)[D] + \omega[U]]\{X\}^{(k)} + \omega\{B\} \qquad (k \geq 0) \qquad (1.37)$$

Inasmuch as $[D] - [L]$ is nonsingular for any choice of $\omega$, this equation can be solved uniquely for $\{X\}^{(k+1)}$. Once again, an efficient algorithm can be used to solve (1.37) through an expansion using indicial notation:

$$x_i^{(k+1)} = (1 - \omega)x_i^{(k)} - (\omega/a_{i,i})\left\{\sum_{j=1}^{i-1} a_{i,j}x_j^{(k+1)} + \sum_{j=i+1}^{n} a_{i,j}x_j^{(k)} - b_i\right\}$$

### 1.7.2 *Convergence of Point Iterative Methods*

Consider, for example, the system of equations (1.20) that was solved earlier using direct methods:

$$\begin{bmatrix} 2 & 0 & 1 \\ 0 & 1 & 5 \\ 4 & 6 & -5 \end{bmatrix} \begin{Bmatrix} x_1 \\ x_2 \\ x_3 \end{Bmatrix} = \begin{Bmatrix} 7 \\ 4 \\ 2 \end{Bmatrix}$$

From the definitions of $[C]$ and $[D]$ given in Eq. (1.32), we have

$$[D] = \begin{bmatrix} 2 & 0 & 0 \\ 0 & 1 & 0 \\ 0 & 0 & -5 \end{bmatrix} \quad \text{and} \quad [C] = \begin{bmatrix} 0 & 0 & -1 \\ 0 & 0 & -5 \\ -4 & -6 & 0 \end{bmatrix}$$

Let us assume an initial guess for $\{X\}^{(0)}$ of unity and evaluate (1.32) for $k = 0$:

$$\begin{Bmatrix} x_1 \\ x_2 \\ x_3 \end{Bmatrix}^{(1)} = \begin{bmatrix} \frac{1}{2} & 0 & 0 \\ 0 & 1 & 0 \\ 0 & 0 & -\frac{1}{5} \end{bmatrix} \begin{bmatrix} 0 & 0 & -1 \\ 0 & 0 & -5 \\ -4 & -6 & 0 \end{bmatrix} \begin{Bmatrix} 1 \\ 1 \\ 1 \end{Bmatrix}$$

$$+ \begin{bmatrix} \frac{1}{2} & 0 & 0 \\ 0 & 1 & 0 \\ 0 & 0 & -\frac{1}{5} \end{bmatrix} \begin{Bmatrix} 7 \\ 4 \\ 2 \end{Bmatrix} = \begin{Bmatrix} 3 \\ -1 \\ \frac{8}{5} \end{Bmatrix}$$

The point Jacobi matrix is

$$\begin{bmatrix} 0 & 0 & -\frac{1}{2} \\ 0 & 0 & -5 \\ \frac{4}{5} & \frac{6}{5} & 0 \end{bmatrix}$$

Subtraction of the first iteration from the correct solution yields a residual matrix

$$\{R\}^{(1)} = \begin{Bmatrix} 3 \\ -1 \\ \frac{8}{5} \end{Bmatrix} - \begin{Bmatrix} 3.01 \\ -0.865 \\ 0.973 \end{Bmatrix} = \begin{Bmatrix} -0.01 \\ -0.135 \\ 0.627 \end{Bmatrix}$$

In order to improve this solution, and thereby minimize the residual matrix, a second iteration is performed using (1.32):

$$\begin{Bmatrix} x_1 \\ x_2 \\ x_3 \end{Bmatrix} = \begin{bmatrix} 0 & 0 & -\frac{1}{2} \\ 0 & 0 & -5 \\ \frac{4}{5} & \frac{6}{5} & 0 \end{bmatrix} \begin{Bmatrix} 3 \\ -1 \\ \frac{8}{5} \end{Bmatrix} + \begin{Bmatrix} \frac{7}{2} \\ 4 \\ -\frac{2}{5} \end{Bmatrix} = \begin{Bmatrix} 2.7 \\ -4 \\ \frac{4}{5} \end{Bmatrix}$$

The residual matrix is now $\{R\}^{(2)} = \{-0.31, \ -3.14, \ -0.173\}^{\mathrm{T}}$. Clearly the second iteration provided answers that were less accurate than the first and the solution is not converging. What then is the criterion which guarantees convergence of the Jacobi method? To answer this question, the general convergence requirements of point iterative methods must be examined. To do this, it is necessary to introduce two new definitions.

**Reducible matrix** [15] For $n \geq 2$, an $n \times n$ complex matrix $[A]$ is reducible if there exists an $n \times n$ permutation matrix $[P]$ such that

$$[P][A][P]^{\mathrm{T}} = \begin{bmatrix} A_{1,1} & A_{1,2} \\ 0 & A_{2,2} \end{bmatrix}$$

where $A_{1,1}$ is an $r \times r$ submatrix, $A_{2,2}$ is an $(n - r) \times (n - r)$ submatrix, and $1 \leq r \leq n$. A permutation matrix $[P]$ contains exactly one value of unity and $(n - 1)$ zeros in each row and column. It is used to rearrange rows and columns: $[P][A]$ is $[A]$ with its rows permuted and $[A][P]$ is $[A]$ with its columns permuted [7]. If no such permutation matrix exists, then $[A]$ is irreducible. If $[A]$ is a $1 \times 1$ complex matrix,† then $[A]$ is irreducible if its single entry is nonzero, and reducible otherwise.

The question of whether or not a matrix is irreducible can easily be answered using a geometrical interpretation derived from the theory of graphs [15]. In this approach $n$ points $P_1, P_2, \ldots, P_n$ in a plane are associated with an $n \times n$ complex matrix $A$. For every nonzero entry $a_{ij}$ of the matrix, we connect the node $P_i$ to the node $P_j$ by means of a path $P_i P_j$ directed from $P_i$ to $P_j$. A diagonal entry $a_{i,i}$ requires a path which joins $P_i$ to itself: such a path is called a *loop*. If a graph constructed according to this procedure generates a path from every point $P_i$ to every other point $P_j$, then the matrix is irreducible.

As an example, consider the following matrices and associated directed graphs:

$$\begin{bmatrix} 1 & 2 \\ 0 & 3 \end{bmatrix}$$

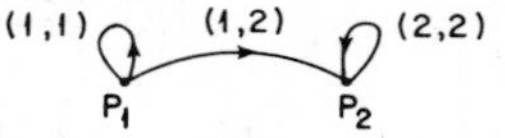

$$\begin{bmatrix} 1 & 2 \\ 4 & 3 \end{bmatrix}$$

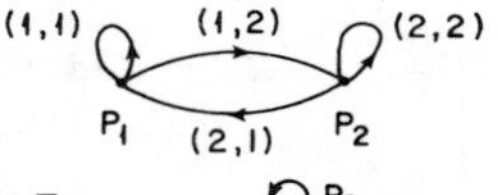

$$\begin{bmatrix} 12 & 1 & 1 \\ 1 & 4 & 1 \\ 1 & 1 & 4 \end{bmatrix}$$

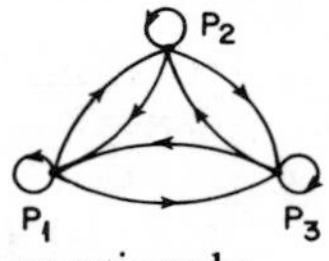

and, finally, the example matrix used previously

$$\begin{bmatrix} 2 & 0 & 1 \\ 0 & 1 & 5 \\ 4 & 6 & -5 \end{bmatrix}$$

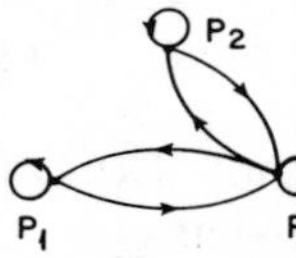

† A complex matrix is one in which the elements are complex numbers.

Because the points $P_i$ and $P_j$ can be connected by a directed path in the second, third, and fourth instances, these matrices are irreducible. In the first instance, however, there is no directed path from $P_2$ to $P_1$ and this matrix is reducible.

**Diagonally dominant matrix** [15] An $n \times n$ complex matrix is diagonally dominant if

$$|a_{i,i}| \geq \sum_{j=1,\, j \neq i}^{n} |a_{i,j}| \tag{1.38}$$

for all $1 \leq i \leq n$. An $n \times n$ matrix is strictly diagonally dominant if the strict inequality in (1.38) is valid for all $1 \leq i \leq n$. Similarly, $[A]$ is irreducibly diagonally dominant if $[A]$ is irreducible and diagonally dominant with strict inequality in (1.38) holding for at least one $i$.

Convergence of the point iterative schemes is closely tied to the irreducibility and diagonal dominance of the coefficient matrix as indicated in the following theorem.

**Theorem 3** [15] Let $[A] = (a_{i,j})$ be a strictly or irreducibly diagonally dominant $n \times n$ complex matrix. Then both the associated point Jacobi and the point Gauss–Seidel matrices are convergent, and the iterative methods of (1.34) and (1.37) for the matrix problem $[A]\{X\} = \{B\}$ are convergent for any initial vector approximation $\{X\}^{(0)}$.

Examination of the coefficient matrix of the example problem which failed to converge reveals that the matrix was neither strictly diagonally dominant nor irreducibly diagonally dominant. Consequently, it did not fulfill the necessary convergence requirements.

Let us now consider another example wherein the coefficient matrix satisfies the convergence criteria:

$$\begin{bmatrix} 12 & 1 & 1 \\ 1 & 4 & 1 \\ 1 & 1 & 4 \end{bmatrix} \begin{Bmatrix} x_1 \\ x_2 \\ x_3 \end{Bmatrix} = \begin{Bmatrix} 1 \\ 0 \\ 0 \end{Bmatrix} \tag{1.39}$$

Because the coefficient matrix in this equation is both strictly diagonally dominant and irreducibly diagonally dominant the system of equations should be amenable to solution by the point Jacobi method. Calculation of the values of $x_i^{(k+1)}$ using an expansion of (1.33) directly yields the values appearing in Table 1.1. It is apparent that in this example the solution procedure converges. However, at this point, nothing can be said about the rate of convergence.

**Table 1.1**

Solution to Eq. (1.39) Obtained through Seven Iterations Using Point Jacobi Iterative Scheme

| variable | Iteration ($k$) 0 | 1 | 2 | 3 | 4 | 5 | 6 | 7 | Exact solution |
|---|---|---|---|---|---|---|---|---|---|
| $x_1$ | 1.00 | −0.0833 | 0.1666 | 0.0590 | 0.0963 | 0.0825 | 0.0875 | 0.0857 | 0.08638 |
| $x_2$ | 1.00 | −0.5000 | 0.1458 | −0.0781 | −0.0047 | −0.0252 | −0.0143 | −0.0183 | −0.01685 |
| $x_3$ | 1.00 | −0.5000 | 0.1458 | −0.0781 | −0.0047 | −0.0252 | −0.0143 | −0.0183 | −0.01685 |

### *1.7.3 Block Iterative Methods*

Whereas only one unknown value at a time has been considered in the point iterative schemes, in the block iterative methods the coefficient matrix is partitioned into blocks and all elements of a block are operated on during one computational step. Because these methods are designed primarily for rectangular nets such as encountered in finite difference approximations we will discuss only two of the many block iterative schemes: the successive line overrelaxation method (SLOR) and the alternating direction process (ADI).

*Successive Line Overrelaxation Method*

In the block iterative schemes one utilizes the structure of the coefficient matrix to partition it into smaller block matrices. Consider, for example, the following matrix equation derived as the finite difference approximation to the equation $(\partial^2 u/\partial x^2) + (\partial^2 u/\partial y^2) = 2$ with boundary conditions $u(0, y) = u(5, y) = u(x, 0) = u(x, 4) = 0$.

$$\left[\begin{array}{cccc|cccc|cccc}
4 & -1 & 0 & 0 & -1 & 0 & 0 & 0 & & & & \\
-1 & 4 & -1 & 0 & 0 & -1 & 0 & 0 & & 0 & & \\
0 & -1 & 4 & -1 & 0 & 0 & -1 & 0 & & & & \\
0 & 0 & -1 & 4 & 0 & 0 & 0 & -1 & & & & \\
\hline
-1 & 0 & 0 & 0 & 4 & -1 & 0 & 0 & -1 & 0 & 0 & 0 \\
0 & -1 & 0 & 0 & -1 & 4 & -1 & 0 & 0 & -1 & 0 & 0 \\
0 & 0 & -1 & 0 & 0 & -1 & 4 & -1 & 0 & 0 & -1 & 0 \\
0 & 0 & 0 & -1 & 0 & 0 & -1 & 4 & 0 & 0 & 0 & -1 \\
\hline
& & & & -1 & 0 & 0 & 0 & 4 & -1 & 0 & 0 \\
& & & & 0 & -1 & 0 & 0 & -1 & 4 & -1 & 0 \\
& 0 & & & 0 & 0 & -1 & 0 & 0 & -1 & 4 & -1 \\
& & & & 0 & 0 & 0 & -1 & 0 & 0 & -1 & 4
\end{array}\right]
\left\{\begin{array}{c} u_1 \\ u_2 \\ u_3 \\ u_4 \\ \hline u_5 \\ u_6 \\ u_7 \\ u_8 \\ \hline u_9 \\ u_{10} \\ u_{11} \\ u_{12} \end{array}\right\}
=
\left\{\begin{array}{c} 2 \\ 2 \\ 2 \\ 2 \\ \hline 2 \\ 2 \\ 2 \\ 2 \\ \hline 2 \\ 2 \\ 2 \\ 2 \end{array}\right\}
\tag{1.40}$$

The system of equations (1.40) may be written in compact notation as

$$\begin{bmatrix} A_{1,1} & A_{1,2} & 0 \\ A_{2,1} & A_{2,2} & A_{2,3} \\ 0 & A_{3,2} & A_{3,3} \end{bmatrix} \begin{Bmatrix} \{X_1\} \\ \{X_2\} \\ \{X_3\} \end{Bmatrix} = \begin{Bmatrix} \{B_1\} \\ \{B_2\} \\ \{B_3\} \end{Bmatrix} \tag{1.41}$$

where the blocks $A_{1,1}$, $A_{1,2}$, $A_{2,1}$, etc. are the square, nonempty matrices outlined in Eq. (1.40) and the $X_i$ and $B_i$ blocks are vectors also outlined in Eq. (1.40). By a procedure analogous to the one used in the point iterative schemes, three new matrices are defined such that

$$[D] = \begin{bmatrix} A_{1,1} & 0 & 0 \\ 0 & A_{2,2} & 0 \\ 0 & 0 & A_{3,3} \end{bmatrix}, \qquad [L] = -\begin{bmatrix} 0 & 0 & 0 \\ A_{2,1} & 0 & 0 \\ 0 & A_{3,2} & 0 \end{bmatrix}$$

$$[U] = -\begin{bmatrix} 0 & A_{1,2} & 0 \\ 0 & 0 & A_{2,3} \\ 0 & 0 & 0 \end{bmatrix}$$

where $[D]$ is a diagonal matrix, $[L]$ and $[U]$ are lower and upper triangular matrices, respectively, and

$$[A] = [D] - [L] - [U]$$

As can be seen from (1.40), $[A]$ and $[D]$ are symmetric and $[L]^{\mathrm{T}} = [U]$. If we can further assume that $[D]$ is positive definite, then for all values of $\omega$, $[D - \omega L]$ is nonsingular. Block successive overrelaxation is defined by

$$\{X\}^{(k+1)} = ([D] - \omega[L])^{-1}[(\omega[U] + (1 - \omega)[D])\{X\}^{(k)} + \omega\{B\}] \tag{1.42}$$

where $\{B\}$ is the matrix of known information (in this example, $b_i = 2$). To carry out the necessary calculations (1.42) is written as

$$X_i^{(k+1)} = (1 - \omega)X_i^{(k)} - \omega A_{i,i}^{-1}\left\{\sum_{j=1}^{i-1} A_{i,j}X_j^{(k+1)} + \sum_{j=i+1}^{N} A_{i,j}X_j^{(k)} - B_i\right\} \tag{1.43}$$

where $A_{i,i}^{-1}$ is the inverse of a diagonal block matrix and $N$ the number of blocks. The block iterative scheme thus exhibits characteristics of both direct and iterative methods. Requirements for convergence of the method can be specified through the following two theorems [16].

**Theorem 4** If $[A]$ is symmetric and $A_{i,i}$ is positive definite for each $i = 1, 2, \ldots, N$, then the method of block successive overrelaxation converges for $\omega = 1$ if and only if $[A]$ is positive definite.

**Theorem 5** If $[A]$ is Hermitian and $[D]$ is Hermitian positive definite, then the method of block successive overrelaxation converges for all $\{X\}^{(0)}$ if and only if $0 < \omega < 2$ and $[A]$ is positive definite.

Before illustrating the application of this technique to Eq. (1.41) a value for the relaxation factor $\omega$ must be selected. It can be shown that the optimal value for $\omega$ is given by

$$\omega_{\text{opt}} = 2/[1 + (1 - \lambda^2)^{1/2}]$$

where $\lambda$ is the largest absolute value of the eigenvalues of the matrix $([D]^{-1}[A] + [I])$, which is the block Jacobi iteration matrix. The eigenvalue is often called the spectral radius of the matrix and denoted by $\rho(A)$. The task remains to determine efficiently the spectral radius of the matrix and suitable techniques can be found in Varga [15, p. 284] and Wachspress [17, p. 105].

For the present problem, the first block iteration is obtained by solving (1.43) for $k = 0$, $N = 3$. This substitution generates the three equations

$$X_1^{(1)} = (1 - \omega)X_1^{(0)} - \omega A_{1,1}^{-1}(A_{1,2}X_2^{(0)} + A_{1,3}X_3^{(0)} - B_1)$$
$$X_2^{(1)} = (1 - \omega)X_2^{(0)} - \omega A_{2,2}^{-1}(A_{2,1}X_1^{(1)} + A_{2,3}X_3^{(0)} - B_2)$$
$$X_3^{(1)} = (1 - \omega)X_3^{(0)} - \omega A_{3,3}^{-1}(A_{3,1}X_1^{(1)} + A_{3,2}X_2^{(1)} - B_3)$$

If $A_{i,i}^{-1}$ are calculated and $X_1^{(0)}$, $X_2^{(0)}$, $X_3^{(0)}$ are all assumed equal to $\{2.0, 2.0, 2.0, 2.0\}^{\text{T}}$, the first iterate solution for $\{X_1\}$ obtained using $\omega = 1.1$ is

$$X_1^{(1)} = -0.10 \begin{Bmatrix} 2 \\ 2 \\ 2 \\ 2 \end{Bmatrix} - \frac{1.10}{209} \begin{bmatrix} 56 & 15 & 4 & 1 \\ 15 & 60 & 16 & 4 \\ 4 & 16 & 60 & 15 \\ 1 & 4 & 15 & 56 \end{bmatrix}$$
$$\cdot \left( \begin{bmatrix} -1 & 0 & 0 & 0 \\ 0 & -1 & 0 & 0 \\ 0 & 0 & -1 & 0 \\ 0 & 0 & 0 & -1 \end{bmatrix} \begin{Bmatrix} 2 \\ 2 \\ 2 \\ 2 \end{Bmatrix} - \begin{Bmatrix} 2 \\ 2 \\ 2 \\ 2 \end{Bmatrix} \right) = \begin{Bmatrix} 1.40 \\ 1.80 \\ 1.80 \\ 1.40 \end{Bmatrix}$$

In a similar way one obtains $X_2^{(1)}$ and $X_3^{(1)}$ equal to $\{1.96, 2.70, 2.70, 1.96\}^{\text{T}}$ and $\{1.38, 2.15, 2.15, 1.38\}^{\text{T}}$, respectively. Subsequent iterates may be obtained from these solutions, and the relaxation procedure may be continued until the converged solution is obtained.

For computational purposes, Eq. (1.43) is often written in the form

$$X_i^{(k+1)} = (1 - \omega)X_i^{(k)} + \omega\hat{X}_i^{(k+1)}$$

where

$$A_{i,i}\hat{X}_i^{(k+1)} = B_i - \sum_{j=1}^{i-1} A_{i,j}X_j^{(k+1)} - \sum_{j=i+1}^{N} A_{i,j}X_j^{(k)} \tag{1.44}$$

The matrices $A_{i,i}$ obtained from common finite difference approximations to differential equations are tridiagonal (or block tridiagonal if the differential equation is higher than second order). Thus the matrix problems of (1.44) can be readily solved for $X_i^{(k+1)}$ using the efficient Thomas algorithm developed earlier and the need to invert all the $A_{i,i}$ matrices is eliminated.

*Alternating Direction Procedure*

The particular version of the alternating direction procedure to be considered in this section is known as the Peaceman–Rachford method and was introduced by Peaceman and Rachford in 1955 [18]. As discussed here, it is particularly applicable in solving the matrix equations which arise in the discrete representation of differential equations in two space variables. In this approach, the coefficient matrix $[A]$ is expressed as the sum of a nonnegative diagonal matrix $[D]$, and two symmetric, positive definite matrices $[A_1]$ and $[A_2]$ which have positive diagonal elements and nonpositive off-diagonal elements, i.e.,

$$[A] = [A_1] + [A_2] + [D]$$

The matrix problem $[A]\{X\} = \{B\}$ may now be restated as [16]

$$([A_1] + [D] + [E_1])\{X\} = \{B\} - ([A_2] - [E_1])\{X\}$$

$$([A_2] + [D] + [E_2])\{X\} = \{B\} - ([A_1] - [E_2])\{X\}$$

and solved provided $([A_1] + [D] + [E_1])$ and $([A_2] + [D] + [E_2])$ are nonsingular. To obtain the Peaceman–Rachford scheme we first let $[E_1]^{(k)}$ be $\omega_k[I]$ and $[E_2]^k$ be $\hat{\omega}_k[I]$. Then assume that one complete iteration requires the sequential solution of two matrix equations:

$$([A_1] + [D] + \omega_k[I])\{X\}^{(k+\frac{1}{2})} = \{B\} - ([A_2] - \omega_k[I])\{X\}^{(k)}$$

and

$$([A_2] + [D] + \hat{\omega}_k[I])\{X\}^{(k+1)} = \{B\} - ([A_1] - \hat{\omega}_k[I])\{X\}^{(k+\frac{1}{2})}$$

Each of these equations can be rapidly solved using the Thomas algorithm provided the governing coefficient matrices are tridiagonal and positive definite. The matrices $[A_1] + [D]$ and $[A_2] + [D]$ contain coefficients derived from numerical approximations to spatial derivatives. Moreover, they are defined such that only the derivatives in one space variable are represented in each matrix. Consequently, one solves the equations implicitly in one space dimension while using known values in the other. This sequential sweeping of the matrix horizontally, then vertically, gives rise to the term "alternating direction." This procedure may, with slight modification, be extended to more than two space dimensions.

It is important to note that the matrix $\{X\}^{(k+\frac{1}{2})}$ is an auxiliary matrix which is not necessarily an accurate solution and which is not retained from one iteration to the next. In contrast to the block successive overrelaxation procedure the most recent iteration values do not generally appear in the equations.

If an iteration matrix is defined for the above scheme (assuming $\omega_k = \omega = \hat{\omega}$) as [16]

$$[T_\omega] = ([A_2] + [D] + \omega[I])^{-1}([A_1] - \omega[I]) \cdot ([A_1] + [D] + \omega[I])^{-1}([A_2] - \omega[I])$$

the rate of convergence of the method is determined by the largest absolute eigenvalue of this matrix and this value must be less than unity for convergence. Moreover, the following convergence theorem is available for this particular (ADI) procedure.

**Theorem 6** [16] The Peaceman–Rachford method is always convergent for $E_1 = E_2$ $(\omega = \hat{\omega})$, when $(\omega[I] + \frac{1}{2}[D])$ is positive definite and symmetric and $([A_1] + [A_2] + [D])$ is positive definite.

As in the block successive overrelaxation procedure, it is once again necessary to calculate an optimal iteration parameter. Whereas this formulation has assumed one $\omega$, in general a sequence of parameters are used and techniques for evaluating them can be found in Westlake [16, p. 78].

### *1.7.4 Comparison of Iterative Methods*

Extensive comparisons of selected iterative schemes have been made by various researchers over the years. In summary, these tests have revealed the following:

(i) The point Gauss–Seidel method converges twice as fast as the point Jacobi method.

(ii) For Laplace's equation solved over several different regions the Peaceman–Rachford ADI method with cyclic iteration parameters is superior to point successive overrelaxation. However, when only one parameter was used SOR proved superior to ADI.

(iii) Block successive overrelaxation is superior to point successive overrelaxation whenever the following conditions are satisfied:

(a) $|A| \neq 0$,

(b) $\sum_{j=1,\, j\neq i}^{n} |a_{i,j}| \leq a_{i,i}$,

(c) $a_{i,j} \leq 0$ for $i \neq j$.

An overview of the various convergence characteristics can be obtained through examination of Table 1.2 which summarizes the results of experiments by Forsythe and Wasow [19] using Poisson's equation with Dirichlet boundary conditions defined on a square region.

**Table 1.2**

Convergence Rates of Selected Iterative Schemes[a]

| Method | Approximate rate of convergence |
|---|---|
| Point Jacobi | $(\Delta x)^2/2$ |
| Point Gauss–Seidel | $(\Delta x)^2$ |
| Optimum point SOR | $2.0\ \Delta x$ |
| Block SOR (two line) | $2\sqrt{2}\ \Delta x$ |
| Peaceman–Rachford ADI | $> -0.777/\ln(\Delta x)$ |

[a] After Westlake [16].

## References


1. F. B. Hildebrand, "Introduction to Numerical Analysis." McGraw-Hill, New York, 1956.
2. L. Collatz, "The Numerical Treatment of Differential Equations," 3rd ed. Springer-Verlag, Berlin and New York, 1967.
3. S. H. Crandall, "Engineering Analysis," McGraw-Hill, New York, 1956.
4. M. G. Salvadori and M. L. Baron, "Numerical Methods in Engineering." Prentice-Hall, Englewood Cliffs, New Jersey, 1961.
5. K. S. Miller, "Partial Differential Equations in Engineering Problems." Prentice-Hall, Englewood Cliffs, New Jersey, 1953.
6. O. Ural, "Finite Element Method." Intext Education Publ., New York, 1973.
7. G. Forsythe and C. B. Moler, "Computer Solution of Linear Algebraic Systems." Prentice-Hall, Englewood Cliffs, New Jersey, 1967.
8. B. Carnahan, H. A. Luther, and J. O. Wilkes, "Applied Numerical Methods." Wiley, New York, 1969.
9. F. R. Gantmacher, "The Theory of Matrices," Vol. 1. Chelsea, New York, 1960.
10. W. T. Segui, Computer Programs for the Solution of Systems of Linear Algebraic Equations. NASA Contractor Rep. CR-2173 (1973).
11. F. B. Hildebrand, "Methods of Applied Mathematics," 2nd ed. Prentice-Hall, Englewood Cliffs, New Jersey, 1965.
12. R. P. Tewarson, "Sparse Matrices." Academic Press, New York, 1973.
13. J. H. Wilkinson, "The Algebraic Eigenvalue Problem," Oxford Univ. Press, London, 1965.
14. L. H. Thomas, Elliptic Problems in Linear Difference Equations over a Network. Watson Sci. Comput. Lab. Rep., Columbia Univ., New York (1949).
15. R. S. Varga, "Matrix Iterative Analysis." Prentice-Hall, Englewood Cliffs, New Jersey, 1962.

15a. G. D. Smith, "Numerical Solution of Partial Differential Equations." Oxford Univ. Press, London and New York, 1965.
16. J. R. Westlake, "A Handbook of Numerical Matrix Inversion and Solution of Linear Equations." Wiley, New York, 1968.
17. E. L. Wachspress, "Iterative Solution of Elliptic Systems." Prentice-Hall, Englewood Cliffs, New Jersey, 1966.
18. D. W. Peaceman and H. H. Rachford, Jr., The numerical solution of parabolic and elliptic differential equations, *J. Soc. Ind. Appl. Math.* **3**, 28–41 (1955).
19. G. E. Forsythe and W. R. Wasow, "Finite-Difference Methods for Partial Differential Equations." Wiley, New York, 1960.

# Chapter 2 Introduction to Finite Difference Theory

## 2.1 Why Consider Finite Difference Theory?

While the finite element approach to solving partial differential equations is a relatively recent development, finite difference methods were known and studied by such early scientists as Newton, Gauss, Bessel, and Laplace. Naturally, a vast body of knowledge has been built up over the years concerning both the theoretical and applied aspects of these methods. We will utilize finite difference techniques not only in approximation schemes but also in the development of the finite element theory. Inasmuch as only selected elements of finite difference theory will be required, the introduction provided in the following sections is brief and limited in scope. More complete developments may be found in several excellent monographs [1–6].

## 2.2 Finite Difference Approximations

### 2.2.1 *Discretization*

Fundamental to both the finite element and finite difference approaches to solving partial differential equations is the concept of discretization, wherein a continuous domain $D$ is represented as a number of adjacent subareas, as indicated in Fig. 2.1. While approximations to a continuous solution are defined at isolated points by finite differences, with finite elements, the approximate solution is defined over the entire domain $D$. Consequently, it is unnecessary to apply additional interpolation schemes to obtain a solution at an arbitrary point in $D$.

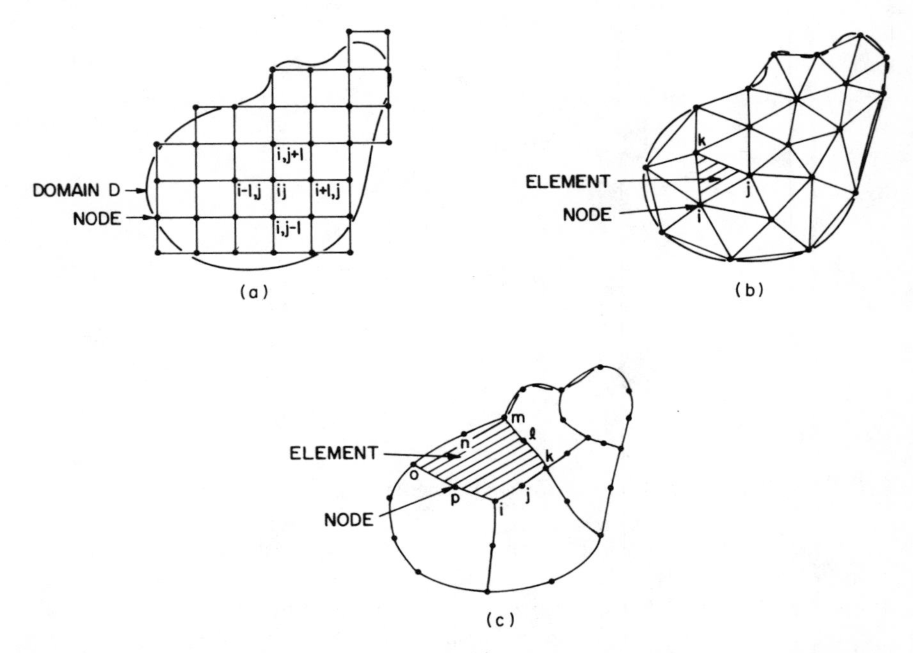


**Fig. 2.1.** Finite difference and finite element discretization schemes. (a) Discretization using regular finite difference net; (b) discretization using triangular finite elements; (c) discretization using quadratic isoparametric elements.

### 2.2.2 *Differentiation Formulas*

Considerable insight into the principles underlying the finite element method can be obtained through examination of the approximating equations in light of established finite difference theory. Fundamental to such an approach are differentiation formulas and their errors of approximation. We develop these formulas first using interpolating polynomials and second using Taylor series expansions. While the latter approach is less intuitive, it provides error estimates that are useful when making comparisons among different numerical schemes.

#### *Formulas by Interpolating Polynomials*

In this approach, to approximate the $n$th derivative of a function, a polynomial of degree $m$ where $m \geq n$ must be selected to model the function. The polynomial is then forced to be exactly equal to the function at $m + 1$ pivotal points or nodes. Then the $n$th derivative of the polynomial is an approximation to the $n$th derivative of the function. For example, one can develop an approximation to $d^2u/dx^2$ using a second degree polynomial and three nodes. The polynomial is restricted by requiring that it exactly equal $u$ at the arbitrary pivotal points indicated in Fig. 2.2. Thus $u$ is approximated by $\hat{u}$ such that

$$\hat{u} = Ax^2 + Bx + C \tag{2.1}$$

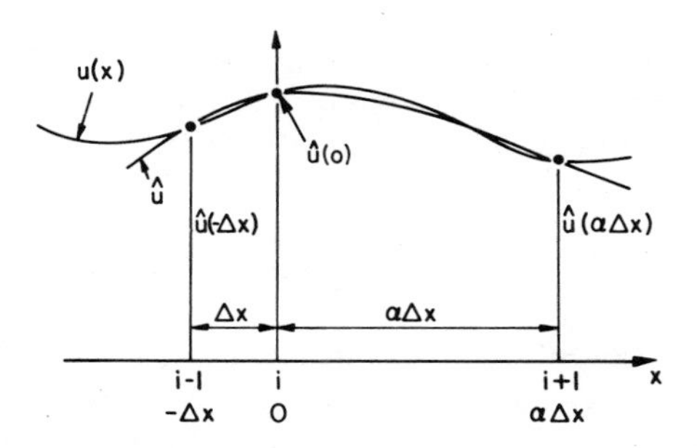


**Fig. 2.2.** Graphical representation of unequally spaced pivotal points [7].

If, for simplicity, the origin is taken at point $i$, Eq. (2.1) can be evaluated at the three pivotal points to obtain

$$\begin{aligned} \hat{u}(-\Delta x) &= A(\Delta x)^2 - B\,\Delta x + C \\ \hat{u}(0) &= C \\ \hat{u}(\alpha\,\Delta x) &= A(\alpha\,\Delta x)^2 + \alpha B\,\Delta x + C \end{aligned} \tag{2.2}$$

These equations may be solved for the unknown coefficients

$$A = \frac{1}{\alpha(\alpha+1)(\Delta x)^2}[\alpha\hat{u}(-\Delta x) - (1+\alpha)\hat{u}(0) + \hat{u}(\alpha\,\Delta x)]$$

$$B = \frac{1}{\alpha(\alpha+1)\,\Delta x}[\hat{u}(\alpha\,\Delta x) - (1-\alpha^2)\hat{u}(0) - \alpha^2\hat{u}(-\Delta x)]$$

$$C = \hat{u}(0)$$

An approximation for $du/dx$ and $d^2u/dx^2$ at any $x$ can now be generated by differentiating (2.1) to give

$$d\hat{u}/dx = 2Ax + B \qquad \text{and} \qquad d^2\hat{u}/dx^2 = 2A$$

Thus at point $i$ where $x = 0$, one obtains

$$\frac{d\hat{u}_i}{dx} = 2A(0) + B = \frac{1}{\alpha(\alpha + 1)\,\Delta x}[\hat{u}_{i+1} - (1 - \alpha^2)\hat{u}_i - \alpha^2\hat{u}_{i-1}] \quad (2.3)$$

$$\frac{d^2\hat{u}_i}{dx^2} = 2A = \frac{2}{\alpha(\alpha + 1)(\Delta x)^2}[\hat{u}_{i+1} - (1 + \alpha)\hat{u}_i + \alpha\hat{u}_{i-1}] \quad (2.4)$$

Similar expressions for higher-order derivatives can be developed if a higher-degree polynomial is used in conjunction with an appropriate number of pivotal points.

If the pivotal points in the preceding example are equally spaced (corresponding to $\alpha = 1.0$), the derivative approximations obtained reduce to the well-known central difference approximations

$$\frac{d\hat{u}_i}{dx} = \frac{1}{2\,\Delta x}(\hat{u}_{i+1} - \hat{u}_{i-1}) \quad \text{and} \quad \frac{d^2\hat{u}_i}{dx^2} = \frac{1}{(\Delta x)^2}(\hat{u}_{i+1} - 2\hat{u}_i + \hat{u}_{i-1})$$

Although this straightforward approach using interpolating polynomials provides approximations to the derivatives, it does not give any insight into the accuracy of the approximation. However, the accuracy may be obtained by deriving the approximating formulas using Taylor series expansions.

*Formulas by Taylor Series Expansions*

A Taylor's series expansion around $u = u(x, y)$ can be written as

$$\begin{aligned} u(x + \Delta x, y + \Delta y) &= u(x, y) + \left(\Delta x \frac{\partial}{\partial x} + \Delta y \frac{\partial}{\partial y}\right) u(x, y) \\ &\quad + \frac{1}{2!}\left(\Delta x \frac{\partial}{\partial x} + \Delta y \frac{\partial}{\partial y}\right)^2 u(x, y) + \cdots \\ &\quad + \frac{1}{(n-1)!}\left(\Delta x \frac{\partial}{\partial x} + \Delta y \frac{\partial}{\partial y}\right)^{n-1} u(x, y) + R_n \end{aligned} \quad (2.5)$$

where $u$ is assumed to be sufficiently differentiable in the range $(x, y)$ to $(x + \Delta x, y + \Delta y)$ and $R_n$ is a remainder term defined as

$$R_n = \frac{1}{n!}\left(\Delta x \frac{\partial}{\partial x} + \Delta y \frac{\partial}{\partial y}\right)^n u(x + \xi\,\Delta x, y + \eta\,\Delta y) \qquad (0 < \xi < 1, 0 < \eta < 1)$$

An approximating finite difference expression similar to that obtained in the previous section is generated by writing the Taylor series (2.5) about the point $i$ of Fig. 2.2 as

$$u_{i-1} = u_i - \frac{\partial u_i}{\partial x} \Delta x + \frac{\partial^2 u_i}{\partial x^2} \frac{(\Delta x)^2}{2!} - \frac{\partial^3 u_i}{\partial x^3} \frac{(\Delta x)^3}{3!} + \frac{\partial^4 u_i}{\partial x^4} \frac{(\Delta x)^4}{4!} - \cdots \tag{2.6}$$

$$u_{i+1} = u_i + \frac{\partial u_i}{\partial x} \alpha \, \Delta x + \frac{\partial^2 u_i}{\partial x^2} \frac{(\alpha \, \Delta x)^2}{2!} + \frac{\partial^3 u_i}{\partial x^3} \frac{(\alpha \, \Delta x)^3}{3!} + \frac{\partial^4 u_i}{\partial x^4} \frac{(\alpha \, \Delta x)^4}{4!} + \cdots \tag{2.7}$$

where the notation $\partial u_i / \partial x$ indicates the partial of $u$ with respect to $x$ evaluated at node $i$. Subtraction of (2.7) from (2.6) yields, as an approximation for $\partial u_i / \partial x$,

$$\frac{\partial u_i}{\partial x} = \frac{u_{i+1} - u_{i-1}}{(1+\alpha)\,\Delta x} + \frac{\partial^2 u_i}{\partial x^2} \frac{(1-\alpha)\,\Delta x}{2!} - \frac{\partial^3 u_i}{\partial x^3} \frac{(1 - \alpha + \alpha^2)(\Delta x)^2}{3!} + \cdots \tag{2.8}$$

which can be interpreted as

$$\frac{\partial u_i}{\partial x} = \frac{u_{i+1} - u_{i-1}}{(1+\alpha)\,\Delta x}$$

with a remainder of

$$\frac{\partial^2 u_i}{\partial x^2} \frac{(1-\alpha)\,\Delta x}{2!} - \frac{\partial^3 u_i}{\partial x^3} \frac{(1 + \alpha^3)(\Delta x)^2}{(1+\alpha)3!} + \cdots$$

Because the remainder approaches zero proportionally as $\Delta x$ approaches zero, the error of the approximation is of order $\Delta x$ or $O(\Delta x)$. It is interesting to note that for evenly spaced nodes, $\alpha$ is equal to 1 and the approximation is of order $(\Delta x)^2$.

The approximation to $\partial u_i / \partial x$, which is analogous to (2.3), can be obtained by eliminating $\partial^2 u_i / \partial x^2$ between (2.6) and (2.7) to give

$$\frac{\partial u_i}{\partial x} = \frac{u_{i+1} - (1 - \alpha^2) u_i - \alpha^2 u_{i-1}}{\alpha(\alpha + 1)\,\Delta x} - \frac{\partial^3 u_i}{\partial x^3} \frac{\alpha(\Delta x)^2}{3!} + \cdots \tag{2.9}$$

This development of the approximation by Taylor series shows that the expression is correct up to second order for any specified $\alpha$. Comparison of (2.8) and (2.9) reveals that, for an irregular grid, increased accuracy is gained using (2.9), but at the cost of an expression more tedious to evaluate.

An approximation to $\partial^2 u_i/\partial x^2$ is obtained by elimination of $\partial u_i/\partial x$ between (2.6) and (2.7) to give

$$\frac{\partial^2 u_i}{\partial x^2} = \frac{2[u_{i+1} - (1+\alpha)u_i + \alpha u_{i-1}]}{\alpha(\alpha+1)(\Delta x)^2} - \frac{\partial^3 u_i}{\partial x^3}\frac{(\alpha-1)\,\Delta x}{3} + \frac{\partial^4 u_i}{\partial x^4}\frac{(1-\alpha+\alpha^2)(\Delta x)^2}{12} + \cdots \tag{2.10}$$

This finite difference expression is analogous to (2.4) but, in addition, it can be seen that the approximation is first-order correct for unequally spaced pivotal points. If the points are equally spaced, it is apparant that the approximation given in (2.10) is second-order correct. Additional finite difference formulas can be generated by combining (2.6) and (2.7) in a variety of ways.

### 2.2.3 *Integration Formulas*

In the analysis of the finite element method, we must evaluate not only errors due to approximating derivatives by finite difference formulas but also errors inherent in approximate integration formulas. In the next two sections approximation schemes for integration are developed using techniques analogous to those used to obtain differentiation formulas.

#### *Formulas by Interpolating Polynomials*

In this development, an interpolating polynomial is required to pass through a given number of pivotal points of the integrand and the area under the polynomial is taken to be an approximation to the area under the integrand [7]. Consider a set of evenly spaced pivotal values of $f(x)$ as shown in Fig. 2.3. As a first approximation, the function can be represented by a series of straight lines of the form

$$\hat{f}(x) = Ax + B \tag{2.11}$$

For instance, the two points $(\Delta x, f_0)$ and $(2\,\Delta x, f_1)$ will lie on this line provided $A = (f_1 - f_0)/\Delta x$ and $B = 2f_0 - f_1$. Substituting into (2.11), we obtain the first-degree polynomial

$$\hat{f}(x) = \frac{f_1 - f_0}{\Delta x}x + 2f_0 - f_1 \tag{2.12}$$

and the area under $f(x)$ between $\Delta x$ and $2\,\Delta x$ can be approximated by the integration

$$A_1 = \int_{\Delta x}^{2\,\Delta x} f(x)\,dx \simeq \int_{\Delta x}^{2\,\Delta x} \hat{f}(x)\,dx = \left[\frac{f_1 - f_0}{2\,\Delta x}x^2 + 2f_0 x - f_1 x\right]\Bigg|_{\Delta x}^{2\,\Delta x}$$

$$= \frac{(f_0 + f_1)\,\Delta x}{2} \tag{2.13}$$

This formula is known as the "trapezoidal rule" because the area of integration is represented by a trapezoid.

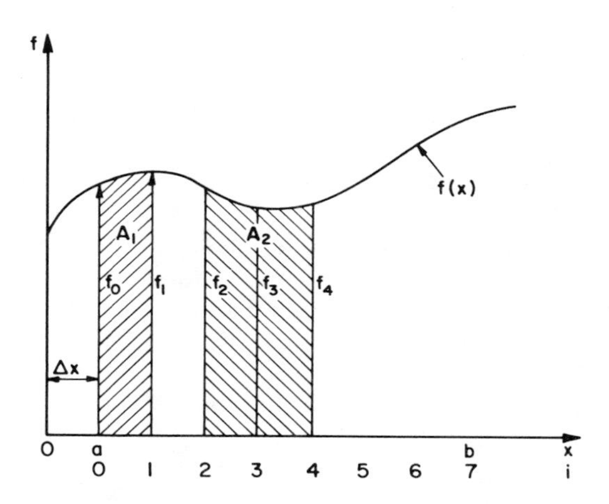


**Fig. 2.3.** Pivotal points for integration formulas [7].

As with differentiation formulas, it is possible to generate additional integration schemes using higher-degree polynomials and additional pivotal points. Simpson's $\frac{1}{3}$ rule can be derived by assuming a series of quadratic interpolations

$$f(x) \simeq \hat{f}(x) = Ax^2 + Bx + C \tag{2.14}$$

Three undetermined coefficients applying to the curve from $x_2$ to $x_4$ can be obtained by requiring (2.14) to pass through the points $(3\,\Delta x, f_2)$, $(4\,\Delta x, f_3)$, and $(5\,\Delta x, f_4)$. Solving for $A$, $B$, and $C$ and substituting into (2.14), we obtain

$$\hat{f}(x) = \frac{f_4 - 2f_3 + f_2}{2(\Delta x)^2}x^2 - \frac{7f_4 - 16f_3 + 9f_2}{2\,\Delta x}x + (6f_4 - 15f_3 + 10f_2)$$

Integration of this quadratic approximation between 3 $\Delta x$ and 5 $\Delta x$ gives

$$A_2 = \int_{3\,\Delta x}^{5\,\Delta x} f(x)\,dx \simeq \int_{3\,\Delta x}^{5\,\Delta x} \hat{f}(x)\,dx$$

$$= \left[\frac{f_4 - 2f_3 + f_2}{6(\Delta x)^2}x^3 - \frac{7f_4 - 16f_3 + 9f_2}{4\,\Delta x}x^2 + (6f_4 - 15f_3 + 10f_2)x\right]\Bigg|_{3\,\Delta x}^{5\,\Delta x}$$

$$= \frac{\Delta x}{3}[f_4 + 4f_3 + f_2]$$

which is Simpson's rule. Note that application of the trapezoidal rule to the region between 3 $\Delta x$ and 4 $\Delta x$ followed by another application to the region 4 $\Delta x$ and 5 $\Delta x$ would yield

$$A_2 \simeq \frac{\Delta x}{2}[f_4 + 2f_3 + f_2]$$

which is a somewhat different approximation from that obtained with the quadratic interpolation functions.

While it has been shown that suitable integration formulas can be developed using the interpolating polynomial approach, once again there is no measure of the accuracy of the approximating scheme. Appropriate error estimates may be obtained by reexamining the integration approximations using Taylor's series expansions.

*Formulas by Taylor's Series*

To develop integration formulas from Taylor's series, we note that the successive derivatives of $u$ can be obtained from the integral [7]

$$u(x) = \int_a^x f(z)\,dz$$

$$\frac{du}{dx} = f(x), \quad \frac{d^2u}{dx^2} = \frac{df(x)}{dx}, \quad \ldots, \quad \frac{d^nu}{dx^n} = \frac{d^{n-1}f(x)}{dx^{n-1}}$$

A Taylor series expansion for $u(x \pm \Delta x)$ about the point $x$ yields

$$u(x \pm \Delta x) = u(x) \pm f(x)\,\Delta x + \frac{df(x)}{dx}\frac{(\Delta x)^2}{2!} \pm \frac{d^2f(x)}{dx^2}\frac{(\Delta x)^3}{3!} + \cdots \tag{2.15}$$

Now to examine the error inherent in the trapezoidal rule, consider the area of one of the strips in Fig. 2.3,

$$I_1 = \int_x^{x+\Delta x} f(z)\,dz = u(x + \Delta x) - u(x) \tag{2.16}$$

Substitution of (2.15) into (2.16) yields

$$I_1 = \Delta x\left[f(x) + \frac{df(x)}{dx}\frac{\Delta x}{2!} + \frac{d^2f(x)}{dx^2}\frac{(\Delta x)^2}{3!} + \cdots\right] \tag{2.17}$$

To evaluate (2.17), we may introduce a first-order correct finite difference expression which can be obtained directly from Eq. (2.7):

$$\frac{df_i}{dx} = \frac{f_{i+1} - f_i}{\Delta x} - \frac{d^2f_i}{dx^2}\frac{(\Delta x)}{2!} - \frac{d^3f_i}{dx^3}\frac{(\Delta x)^2}{3!} - \cdots$$

and substitute this into (2.17) to give the approximate integration formula

$$I_1 = \Delta x\left[f_i + \frac{1}{2}(f_{i+1} - f_i) - \frac{d^2f_i}{dx^2}\frac{(\Delta x)^2}{2 \cdot 2!} - \frac{d^3f_i}{dx^3}\frac{(\Delta x)^3}{2 \cdot 3!} - \cdots \right.$$
$$\left. + \frac{d^2f_i}{dx^2}\frac{(\Delta x)^2}{3!} + \cdots\right]$$

or

$$I_1 = \frac{\Delta x}{2}(f_{i+1} + f_i) - \frac{d^2f_i}{dx^2}\frac{(\Delta x)^3}{12} - \frac{d^3f_i}{dx^3}\frac{(\Delta x)^4}{24} + \cdots \tag{2.18}$$

Comparison of this expression with (2.13) reveals that (2.18) is, indeed, the trapezoidal formula. Because $\Delta x$ appears to the third degree in the error term but to the first degree in the approximation, we say that the trapezoidal rule is only second-order correct. A similar analysis applied to Simpson's $\frac{1}{3}$ rule indicates that it is fourth-order accurate. By combining various polynomial approximations with other finite difference schemes, it is possible to generate formulas with errors of different order. For a similar analysis using unevenly spaced pivoted points, see the work of Salvadori and Baron [7, p. 95].

Heretofore, those aspects of finite difference theory which will assist in understanding the fundamental principles of the finite element theory have been considered. However, finite differences also may play a more direct role in finite element analysis.

## 2.3 Temporal Finite Difference Approximations

The majority of practical problems in surface and subsurface hydrology require time-dependent simulation for analysis. While finite elements are very effective in the approximation of spatial derivatives, finite difference schemes are generally called on to handle the time derivative. Typically, the analysis of the transient system results in a system of equations of the form

$$[A]\{\Phi\} + [B]\{d\Phi/dt\} = \{F\} \tag{2.19}$$

where the spatial dependence of the system is carried by the finite element approximation in the coefficient matrices $[A]$ and $[B]$, and $\{F\}$ contains known information such as boundary conditions or forcing functions. To solve this system of ordinary differential equations, the time derivative is generally replaced by an appropriate difference formula. As was observed in the previous section, the first derivative in time may be approximated using a variety of formulas of various orders of accuracy. Having selected a suitable formula one must still consider the temporal characteristics of the matrix $\{\Phi\}$. This important aspect of finite difference theory will be examined in the following section.

### 2.3.1 *Weighted Average Approximations*

When the time derivative is replaced by a weighted finite difference approximation, the general matrix equation (2.19) becomes

$$\begin{aligned}[A](\varepsilon\{\Phi\}_{t+\Delta t} + (1-\varepsilon)\{\Phi\}_t) + (1/\Delta t)[B](\{\Phi\}_{t+\Delta t} - \{\Phi\}_t) \\ = \varepsilon\{F\}_{t+\Delta t} + (1-\varepsilon)\{F\}_t \qquad (0 \le \varepsilon \le 1)\end{aligned} \tag{2.20}$$

Whether this approximation is first- or second-order accurate depends on $\varepsilon$ which dictates the reference point in time of the difference formula. Furthermore, the stability of the numerical solution will also depend on $\varepsilon$. We will consider three specific formulations which are in general use: $\varepsilon = 0$, $\varepsilon = \frac{1}{2}$, and $\varepsilon = 1$.

*Explicit Method*

The explicit method generally requires a minimum of computational effort but, as will be shown later, is usually only conditionally stable. This first-order scheme in time is generated by writing (2.20) for $\varepsilon = 0$ as†

$$([A] - (1/\Delta t)[B])\{\Phi\}_t + (1/\Delta t)[B]\{\Phi\}_{t+\Delta t} = \{F\}_t \tag{2.21}$$

Because the first term in (2.21) is dependent only on the old time level $t$, it is always known. Hence, it is convenient to transfer this term to the right-hand side of the equation and combine it with the known vector $\{F\}_t$. Once the $[B]$ coefficient matrix of the resulting set of equations has been reduced to upper triangular form by methods such as those discussed in Chapter 1, the solu-

† Note that in the finite element scheme $[B]$ is a symmetric banded matrix and the use of the term "explicit" may be misleading.

tion for each new time step can be obtained by a simple back substitution procedure. It is important to note that if $[A]$ and $[B]$ are unchanged, upper triangularization (or matrix inversion) is required only once even if the time step $\Delta t$ is modified.

*Implicit Method*

The implicit method is usually unconditionally stable, first-order correct in time, but requires additional computational effort when a variable time step is used. The scheme is obtained by substituting $\varepsilon = 1$ in Eq. (2.20) to give

$$([A] + (1/\Delta t)[B])\{\Phi\}_{t+\Delta t} - (1/\Delta t)[B]\{\Phi\}_t = \{F\}_{t+\Delta t} \tag{2.22}$$

Because the solution of (2.22) requires the upper triangularization of the matrix $([A] + [B]/\Delta t)$, which contains $\Delta t$, this procedure must be repeated whenever the time step is modified. Moreover, the coefficient matrix is, in general, not symmetric and thus requires more calculation time as well as additional computer storage when compared with the explicit formulation. For simplicity, the analysis has been restricted to two time levels. Higher-order schemes can be generated using three or more time levels.

*Crank–Nicolson Implicit Method*

The Crank–Nicolson method is a second-order accurate scheme which is centered in time. This scheme is usually neutrally stable and may be obtained from (2.20) by setting $\varepsilon = \frac{1}{2}$:

$$\tfrac{1}{2}[A](\{\Phi\}_{t+\Delta t} + \{\Phi\}_t) + (1/\Delta t)[B](\{\Phi\}_{t+\Delta t} - \{\Phi\}_t) = \tfrac{1}{2}(\{F\}_{t+\Delta t} + \{F\}_t) \tag{2.23}$$

This method requires slightly more computational effort than the implicit method, but this disadvantage is often offset by the higher-order accuracy of the calculated solution.

### 2.3.2 *Stability and Convergence*

Let us assume that the accurate solution to the approximating finite difference equations can be written as $\phi_{i,k}$, where $i$ is the pivotal point or node and $k$ the time level. Due principally to round-off errors encountered during algebraic manipulations, a solution $\hat{\phi}_{i,k}$ is obtained which will differ from $\phi_{i,k}$ by a numerical error $\rho_{i,k}$ such that

$$\rho_{i,k} \equiv \phi_{i,k} - \hat{\phi}_{i,k}$$

A numerical scheme will be considered stable if with increasing $k$, $\rho_{i,k}$ tends uniformly to zero for all values of $i$.

Substitution of a finite difference approximation into the homogeneous set of equations obtained from (2.19) by setting $\{F\}$ equal to zero gives an expression of the form

$$\{\Phi\}_{k+1} = [A^*]\{\Phi\}_k \qquad \text{and} \qquad \{\Phi\}_0 = \{\mathscr{F}\} \tag{2.24}$$

where $\{\Phi\}_0$ is the matrix of initial conditions and $[A^*]$ an $n \times n$ square matrix called the amplification matrix. Two stability criteria may now be considered.

For the first criterion, developed by Saul'yev [5], the following norms† are defined which are consistent inasmuch as they satisfy the Schwarz inequality‡:

$$\|\{\Phi\}_k\| = \max_{1 \le i \le n} |\phi_{i,k}|, \qquad \|[A^*]\| = \max_{1 \le i \le n} \sum_{j=1}^{n} |a^*_{i,j}|$$

Let us consider a matrix $\{\bar{\mathscr{F}}\}$ composed of the initial conditions slightly modified due, for example, to rounding off of its components. Using the overbar notation, an equation analogous to (2.24) can be written as

$$\{\bar{\Phi}\}_{k+1} = [A^*]\{\bar{\Phi}\}_k \qquad (k = 1, 2, \ldots, p)$$

where $\{\bar{\Phi}\}_0 = \{\bar{\mathscr{F}}\}$ and $p$ is the total number of time steps considered. If the error vector is defined by $\{\bar{\rho}\}_k = \{\Phi\}_k - \{\bar{\Phi}\}_k$, then

$$\{\bar{\rho}\}_{k+1} = [A^*]\{\bar{\rho}\}_k$$

where

$$\{\bar{\rho}\}_0 = \{\mathscr{F}\} - \{\bar{\mathscr{F}}\} \qquad \text{or} \qquad \{\bar{\rho}\}_k = [A^*]^k\{\bar{\rho}\}_0 \equiv [A^*]^k(\{\mathscr{F}\} - \{\bar{\mathscr{F}}\}) \tag{2.25}$$

From the Schwarz inequality and (2.25) we obtain

$$\|\{\bar{\rho}\}_k\| \le \|[A^*]\|^k \|\{\bar{\rho}\}_0\| \tag{2.26}$$

It is apparent from (2.26) that the error $\{\bar{\rho}_0\}$ in the initial data will not grow with an increase in $k$ (i.e., the solution will remain stable) only if

$$\|[A^*]\| \le 1 \tag{2.27}$$

This same relationship holds when an error is assumed at any arbitrary time

† See Westlake [8, p. 132] for a concise discussion of norms and their properties.

‡ The Cauchy–Schwarz–Bunyakovskij inequality states that [9], given two vectors $\{X\}$ and $\{Y\}$, $|\{X\}^T\{Y\}| \le \|\{X\}\| \cdot \|\{Y\}\|$, from which it can be shown that for an $n \times n$ matrix $A$, $\|[A]\{X\}\| \le \|[A]\| \cdot \|\{X\}\|$.

level, because this level can always be considered as a new set of initial conditions.

Possibly the best known stability criterion is the von Neumann necessary condition for stability. This condition is generally stated as

$$|\lambda_i| \leq 1 \tag{2.28}$$

where $\lambda_i$ is the $i$th eigenvalue of the amplification matrix. Richtmeyer and Morton [3], present a modification of this criterion

$$|\lambda_i| \leq 1 + O(\Delta t)$$

which allows for the possibility that the true solution may be exponential [a solution of this form would be incompatible with the condition expressed by Eq. (2.28)]. For many difference schemes, the maximum absolute eigenvalue, or spectral radius, of $[A^*]$ can be easily calculated. When the coefficients of the differential equation are spatially dependent and $\lambda_i$ can no longer be computed exactly, the inequality

$$\max_i |\lambda_i| \leq \|[A^*]\|$$

can be helpful in practical problems since $\|[A^*]\|$ can be readily calculated [5]. The stability of specific approximations will be considered after the structure of the coefficient matrix $[A]$ has been defined using finite element techniques in subsequent chapters.

Although two stability conditions have been considered, the question of convergence must still be addressed. If $u_{i,k}$ is the exact solution to the partial differential equation and $\Phi_{i,k}$ the numerical solution, convergence of the numerical scheme is achieved when the error in the solution, defined as

$$E_{i,k} = u_{i,k} - \phi_{i,k}$$

tends to zero uniformly with mesh refinement, i.e., as $\Delta x \to 0$ and $\Delta t \to 0$. In this discussion, stability was considered prior to convergence because stability, in a certain sense, is a stronger condition than convergence. Several authors have shown that, if certain supplementary conditions are fulfilled, stability is both a necessary and sufficient condition for convergence. Moreover, inequality (2.27) by guaranteeing stability also implies convergence of the method [5]. In other words, in some instances if a computational scheme is stable, it is convergent.

### 2.3.3 *Propogation Characteristics of the Numerical Solution*

In the simulation problems to be considered herein, important information regarding the accuracy of the numerical solution may be obtained through an analysis based on the behavior of the components of a Fourier series representation of the computed and real solutions. Specifically, the magnitude and phase of the components of the Fourier series calculated by numerical schemes are compared with the actual magnitude and phase computed analytically. While it is apparent that an analysis of this kind is essential for evaluating the ability of a numerical technique to simulate a surface water system properly, such an analysis is also very useful for examining the relative merits of numerical methods used to solve porous media problems, particularly those involving transport phenomena. Because this type of analysis will be applied later for specific problems, the procedure will only be introduced at this point.

Given one or more linear partial differential equations describing a transient one-dimensional problem, a general solution may be assumed of the form

$$u \simeq \bar{u} \exp(\hat{i}\beta t + \hat{i}\sigma x) \tag{2.29}$$

where $\hat{i} = (-1)^{1/2}$ and $\bar{u}$ is an undetermined coefficient. Substitution of (2.29) into the governing equations yields a general expression which describes the behavior of $u$ as a function of space and time. Similarly, one can approximate the governing equations numerically, and then substitute (2.29) into the resulting difference formula. The expression obtained will denote the behavior of $u$ as described by the numerical scheme. By comparing analytically and numerically, one can evaluate the relative accuracy of numerical techniques.

As indicated at the outset of this section, finite difference methods have been exhaustively examined in the literature; consequently, only those topics that are to be used directly in the ensuing discussion of finite element techniques have been considered.

## References


1. L. V. Kantorovich and V. I. Krylov, "Approximate Methods of Higher Analysis." Noordhoff, Groningen, The Netherlands, 1964.
2. L. Collatz, "The Numerical Treatment of Differential Equations," 3rd ed. Springer-Verlag, Berlin and New York, 1967.
3. R. D. Richtmeyer and K. W. Morton, "Difference Methods for Initial-Value Problems," 2nd ed. Wiley, New York, 1967.
4. G. D. Smith, "Numerical Solution of Partial Differential Equations." Oxford Univ. Press, London and New York, 1965.
5. V. K. Saul'yev, "Integration of Equations of Parabolic Type by the Method of Nets." Pergamon, Oxford, 1964.
6. D. U. Von Rosenberg, "Methods for the Numerical Solution of Partial Differential Equations." American Elsevier, New York, 1969.

7. M. G. Salvadori and M. L. Baron, "Numerical Methods in Engineering." Prentice-Hall, Englewood Cliffs, New Jersey, 1961.
8. J. R. Westlake, "A Handbook of Numerical Matrix Inversion and Solution of Linear Equations." Wiley, New York, 1968.
9. G. Forsythe and C. B. Moler, "Computer Solution of Linear Algebraic Systems." Prentice-Hall, Englewood Cliffs, New Jersey, 1967.

# Chapter 3 | The Method of Weighted Residuals

## 3.1 Finite Element Applications

There are several avenues of approach to formulating the approximating integral equations which are the foundation of the finite element method. Of the various alternatives, the Rayleigh–Ritz procedure based on the calculus of variations has been used most extensively. Recently, Galerkin's method, which is more general in application, has received considerable attention as an alternative to Rayleigh–Ritz. In fact, those problems that can be solved using the Rayleigh–Ritz procedure represent only a subclass of those amenable to solution by the Galerkin method. Because of its generality, only Galerkin's procedure will be considered in the theoretical development of this chapter and in the applications in later chapters.

## 3.2 The Fundamentals of Weighted Residual Procedures

B. G. Galerkin, a Russian engineer, first introduced his method, in its current form, in 1915.† It is actually a special case of a more general approach called the method of weighted residuals (MWR) [2]. To introduce MWR, let us consider the time-independent equation

$$Lu = f \qquad \text{in } B \tag{3.1}$$

where $B$ is a bounded domain and the operator $L$ acts on the unknown function $u$ to generate the known function $f$. Let $u$ be replaced by a function

† Mikhlin [1] notes that essentially the same concept was introduced by Bubnov in 1913 and therefore refers to the scheme as the Bubnov–Galerkin method.

$\hat{u}(x)$ which is made up of a linear combination of suitable functions and satisfies the principal or essential boundary conditions of the boundary value problem. A suitable trial function might be

$$\hat{u}(x) = \phi_1(x) + \sum_{j=2}^{M} a_j\phi_j(x) \tag{3.2}$$

where $\phi_1(x)$ is chosen to satisfy the essential boundary condition and the coordinate or basis functions $\phi_j(x)$ satisfy the corresponding homogeneous boundary condition. This ensures that for any value of the constant $a_j$, $\hat{u}(x)$ satisfies the essential boundary condition. Generally in the finite element method, the term $\phi_1(x)$ is not written explicitly but is incorporated in the series through a specification on the $a_j$; Eq. (3.2) becomes

$$\hat{u}(x) = \sum_{j=1}^{M} a_j\phi_j(x) \tag{3.3}$$

Let a residual $\mathscr{R}$ be defined by

$$\mathscr{R}(x) = L\hat{u} - f(x) \qquad \text{or} \qquad \mathscr{R}(x) = L\left[\sum_{j=1}^{M} a_j\phi_j(x)\right] - f(x)$$

If the trial solution were the exact solution, the residual would vanish. In the method of weighted residuals, an attempt is made to force this residual to zero, in an average sense, through selection of the constants $a_j$ ($j = 1, 2, \ldots, M$). The $a_j$ are calculated by satisfying the constraints which arise when setting the weighted integrals of the residual to zero, i.e.,

$$\int_B \mathscr{R}(x)w_i(x)\,dx = 0 \qquad (i = 1, 2, \ldots, M) \tag{3.4}$$

To simplify notation, use will be made of the following definition:

**Inner product** The inner product (or scalar product) of two functions $v$ and $w$ is written as $\langle v, w\rangle \equiv \int_B v \cdot w\,dB$.

Equation (3.4) can now be rewritten as

$$\langle \mathscr{R}(x), w_i\rangle = 0 \qquad (i = 1, 2, \ldots, M) \tag{3.5}$$

From (3.5), $M$ equations are obtained which, in conjunction with the boundary conditions, can be solved for $M$ values of $a_j$.

There are several weighted residual methods and each is distinguished by the choice of weighting function $w_i$. Several commonly used schemes and their weighting functions follow.

*Subdomain Method*

The domain $B$ is divided into $M$ smaller domains $B_i$, and the weighting functions become [2]

$$w_i = \begin{cases} 1, & x \text{ in } B_i \\ 0, & x \text{ not in } B_i \end{cases}$$

Thus the differential equation $L\hat{u}$ is integrated over each subdomain and set to zero.

*Collocation Method*

$M$ points $x_i$ (known as collocation points) are specified in $B$ and the weighting functions are Dirac delta functions:

$$w_i = \delta(x - x_i)$$

which have the property that

$$\int_B \mathscr{R}(x) w_i \, dx = \mathscr{R}(x_i) = 0$$

Thus the procedure consists of simply evaluating the residual at the collocation points and therefore involves a minimum of computational effort. Unfortunately, the method is very sensitive to the manner in which the collocation points are selected and consequently has not been widely used. For nonlinear problems, a special case of this method known as orthogonal collocation has been found particularly effective [2].

*Least Squares Method*

In this approach, the weighting function is $p(x)\, \partial\mathscr{R}/\partial a_i$, where $p(x)$ is an arbitrary positive function. This is equivalent to minimizing the integrated square residual [3]

$$I = \int p(x) \mathscr{R}^2(x) \, dx$$

with respect to the trial parameters $a_i$, i.e.,

$$\partial I / \partial a_i = 0 \qquad (i = 1, 2, \ldots, M)$$

This method generally leads to complicated expressions but is often used in error analysis since error bounds can be derived in terms of it [2].

## 3.3 Galerkin's Method

Galerkin's method is formulated by selecting the basis functions (also known as coordinate functions and bases) $\phi_i(x)$ as the weighting functions. Thus the weighted residual equations become

$$\langle \mathscr{R}(x), \phi_i(x) \rangle = 0 \qquad (i = 1, 2, \ldots, M) \tag{3.6}$$

There is a second interpretation of the Galerkin method that may provide additional insight into the important role played by the coordinate functions. As in the most general case of Ritz' procedure, the $\phi_j(x)$ are formally required to satisfy the boundary conditions imposed on the governing equation (3.1). Moreover, these functions must be linearly independent and represent the first $M$ functions of some system of functions $\phi_j$ $(j = 1, 2, \ldots, M, \ldots)$ which is complete in the given region. To obtain an exact solution, the residual $\mathscr{R}(x)$ must vanish and, assuming $L\hat{u}$ to be continuous, this is equivalent to requiring the orthogonality of the expression $\mathscr{R}(x)$ to all the functions $\phi_j(x)$ $(j = 1, 2, \ldots, M, \ldots)$. Since only $M$ coordinate functions are used, only $M$ conditions of orthogonality can be satisfied. This is, of course, precisely the message contained in Eq. (3.6). The question of completeness is considered subsequently in the discussion on convergence.

## 3.4 Convergence of the Finite Element Method

The purpose of this section is to establish sufficient conditions for convergence of the finite element method based on the Galerkin method of approximation. The development borrows heavily from the classical work of Mikhlin [1] and a paper of Hutton and Anderson [4]. The following definitions are important in the ensuing discussion.

**Symmetric operator** An operator $L$ is termed symmetric if, for any two functions $v$ and $w$ in the field of definition of the operator $L$, $\langle Lv, w \rangle = \langle Lw, v \rangle$.

**Positive definite operator** An operator $L$ is termed positive definite if $\langle Lv, v \rangle \geq 0$ in the field of definition of $L$, where the equality holds only if $v$ is identically zero.

**Bounded operator** An operator $L$ is termed bounded if $\Gamma\langle v, v \rangle \geq \langle Lv, v \rangle \geq \gamma\langle v, v \rangle$ in the field of definition of $L$, where $\Gamma$ and $\gamma$ are real finite constants.

**Positive, bounded below operator** The operator $L$ is termed positive, bounded below if $\langle Lv, v \rangle \geq \gamma^2 \langle v, v \rangle$ in the field of definition of $L$, where $\gamma$ is some real constant.

**Completely continuous operator** The operator $L$ is termed completely continuous if in the field of definition of $L$, there exists a sequence of operators $L_n$ such that

$$\lim_{n \to \infty} \langle (L - L_n)v, (L - L_n)v \rangle / \langle v, v \rangle = 0$$

Consider the problem of solving Eq. (3.1) together with an appropriate set of homogeneous boundary conditions.† Let $L$ be a linear bounded differential operator defined for some dense linear subset $M$ of the space of functions $H$. The set of functions $M$ will be considered dense in $H$ in the sense that each element $v$ in $H$ can be obtained as a limit of a sequence $v_n$ where the elements $v_n$ lie in $M$. The term $H$ refers to a real Hilbert space which requires, among other things, that the member functions be square summable over $B$ and that the inner product $\langle v, w \rangle$ be defined (where $v$ and $w$ are in $H$). For subsequent discussion, let the subset $M$ of $H$ be defined such that it contains functions with continuous $2m$th order derivatives where $2m$ is the order of the operator $L$.

Consider now the question of whether or not Eq. (3.1) has a solution in the subset $M$ of $H$. It is possible that there does not exist in $M$ an element that will provide a solution for an arbitrary element $f$ of $H$. However, when the operator $L$ is required to be not only symmetric and positive definite but also positive bounded below, it is possible to define a new Hilbert space $H_L$ which guarantees the existence of a solution. Let the field of definition of $L$ be extended from $M$, as originally required, to form this new Hilbert space $H_L$. This new space is defined in terms of a new scalar product (in contrast to the inner product requirement of $H$) which is called the energy product. For any two elements, $v$ and $w$ in $H_L$, the energy product is of the form

$$[v, w] = \lim_{n \to \infty} \langle Lv_n, w_n \rangle \tag{3.7}$$

The functions $v_n$ and $w_n$ are typical members of a sequence in which each member is in the field of definition of $L$. Because $L$ is, by definition, symmetric and $v_n$ and $w_n$ lie within its field of definition, then Eq. (3.7) can be restated in the form

$$[v, w] = \lim_{n \to \infty} \langle Rv_n, Rw_n \rangle = [w, v] \tag{3.8}$$

† The discussion is readily extended to the case of nonhomogeneous boundary conditions as demonstrated by Hutton and Anderson [4].

where $R$ is a differential operator obtained through the application of integration by parts to (3.7).

As in any Hilbert space, it is possible to define a norm. The particular norm in the space $H_L$ is called the energy norm and is denoted by the relationship

$$|v| \equiv [v, v]^{1/2} \tag{3.9}$$

The energy norm can now be used to establish convergence in the following sense: if the function $v$ and the sequence $v_n$ lie within the field of definition of $L$, then the sequence $v_n$ converges in energy to $v$ if

$$\lim_{n \to \infty} |v_n - v| \to 0 \tag{3.10}$$

If the operator $L$ is positive bounded below, it can be shown [1] that convergence in energy also implies convergence in the mean.

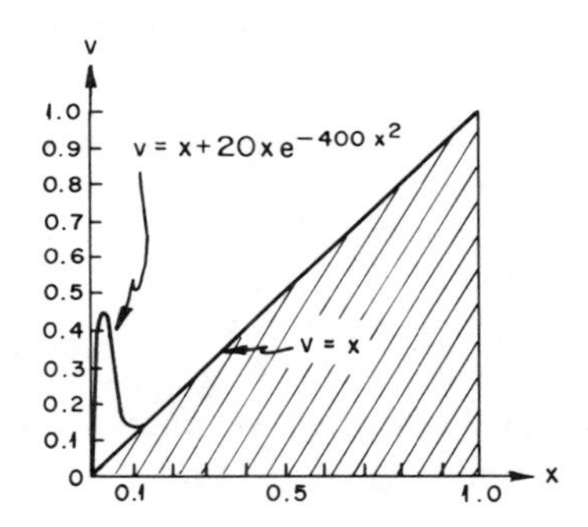


**Fig. 3.1.** Plot the function $v = x + 20x \exp(-400x^2)$ and $v = x$ for the interval $0 \leq x \leq 1$ [1].

While the concept of convergence in energy is relatively difficult to visualize except, perhaps, in the specific case of displacement due to an external stress, convergence in the mean can be demonstrated rather clearly. Because of the importance of this concept, consider the following example borrowed from Mikhlin [1]. It is apparent from Fig. 3.1 that the functions $v = x + 20x \exp(-400x^2)$ and $v = x$ are significantly different only in the interval $0 \leq x \leq 0.13$. If the criterion chosen as a measure of closeness for these functions is the maximum separation of the functions anywhere along the interval $0 \leq x \leq 1$, the functions are essentially different. The maximum deviation actually occurs at $x = \sqrt{2}/40$ and is approximately 0.43. On the other hand, if the difference between the areas under the two curves over the specified interval is the measure of comparison, the functions differ by very little. In fact, the difference in area under the two curves is 0.025 which represents only 5% of the hatched area indicated in Fig. 3.1. Functions that are close in this sense are designated as close in the mean. If notation is introduced such that

$$\|v\| \equiv \langle v, v \rangle^{1/2}$$

convergence in the mean is indicated by

$$\lim_{n \to \infty} \|v_n - v\| \to 0$$

where $v_n$ and $v$ are defined in $B$. In the present example with domain $B$ defined by $0 \leq x \leq 1$, let $v = x$ and the sequence $v_n$ be defined by

$$v_n = x + nx \exp(-n^2x^2) \qquad (n = 1, 2, \ldots, \infty)$$

To test for convergence in the mean, we must consider

$$\lim_{n \to \infty} \|v_n - v\| = \lim_{n \to \infty} \left[ \int_0^1 n^2x^2 \exp(-2n^2x^2)\, dx \right]^{1/2}$$

Because this integral goes to zero as $n$ approaches the limit of the sequence, $v_n$ and $v$ are convergent in the mean.

In the extension of the field of definition of the operator $L$ to create the new Hilbert space $H_L$, it may happen that some of the functions in $H_L$ do not satisfy all of the boundary conditions of the problem. Boundary conditions that are necessarily satisfied by functions from the original field of definition, but not necessarily satisfied by functions from $H_L$, are known as natural boundary conditions. Those boundary conditions that must be satisfied by $H_L$ in order to ensure that the energy norm is positive definite are known as principal, forced, or essential boundary conditions. It is important to recognize that for convergence in energy there is no necessity to subject the coordinate or basis functions to the natural boundary conditions but rather it suffices that they satisfy the principal boundary conditions. For a positive bounded below operator of order $2m$, the natural boundary conditions will be those into which there enter derivatives of order $m$ and higher. The principal boundary conditions will contain derivatives up to order $(m - 1)$.

It is now possible to consider sufficient conditions for the convergence of the finite element method when Galerkin's procedure is used to generate the approximating integral equations. Let the operator $L$, as defined by Eq. (3.1),

$$Lu - f = 0 \tag{3.1a}$$

have the form

$$L = L_0 + K$$

or, rewriting (3.1a),

$$L_0 u + Ku - f = 0 \tag{3.11}$$

where $L_0$ is a symmetric, positive, bounded below operator in the Hilbert space $H$. Let the field of definition of $K$ be wider than that of $L_0$ such that the expression $Ku$ is meaningful whenever $L_0 u$ is meaningful. The coordinate or basis functions indicated in Eq. (3.6) will be required to form a sequence within the field of definition of $L_0$ and be complete in $H_0$ (the question of completeness will be considered shortly). Given these constraints, Mikhlin [1] proves the following theorem.

**Theorem 6a** The approximate solution of Eq. (3.11) constructed by the Galerkin method converges in the energy of operator $L_0$ (in other words, in the sense of convergence in the space $H_L$) to the exact solution of this equation if the following conditions are satisfied:

A. Eq. (3.11) has not more than one solution in $H_L$;
B. the operator $\mathscr{L} = L_0^{-1}K$ is completely continuous in $H_L$.

A considerably more restrictive but simpler convergence criterion is also presented by Mikhlin.

**Theorem 6b** The Galerkin method converges if:
A. the equation $Lu - f = 0$ has no more than one solution;
B. the operator $L_0^{-1}$ is completely continuous and the operator $K$ is bounded in $H$.

For the case under consideration where $L_0$ is a symmetric, positive, bounded below operator, $L_0^{-1}$ can generally be shown to be completely continuous. Whether $K$ is a bounded operator, as required by requirement B of Theorem 6b, can be established from the definition of a bounded operator.

It has now been established that convergence in the sense of the energy norm is ensured provided the basis or coordinate functions $\phi_j$ are complete in $H_L$. Completeness in $H_L$, in the sense of complete in energy, is established for the set of functions $\phi_j$, $j = 1, 2, \ldots, M$, which are within the field of definition of $L_0$, if for any $v$ within $H_L$, it is possible to choose any $\varepsilon > 0$ and find an integer $N$ and constants $a_1, \ldots, a_M$ such that for $M > N$,

$$\left| v - \sum_{j=1}^{M} a_j \phi_j^M \right| < \varepsilon$$

Oliveira [5] proves that completeness is obtained with respect to $H_L$ if continuity of the $(m-1)$th derivatives is ensured throughout the domain $B$

and if the assumed approximation within each element is based on a complete polynomial of order not less than $m$. By their definition, it is apparent that natural boundary conditions are not necessarily satisfied by elements of $H_L$.

It should be noted that convergence in the energy norm rather than the norm was chosen because it requires lower-order continuity of the derivatives of the coordinate functions. One can show that the approximating functions $\phi_j$ would require $(2m - 1)$th order continuous derivatives and satisfy all boundary conditions to fulfill the more general completeness requirement. While such elements are certainly possible, they are more difficult to generate. This higher-order continuous set should be thought of as a smaller class of functions than $H_L$ within which the exact solution of the problem is known to exist.

## 3.5 Approximations in the Time Domain

Heretofore we have considered the solution of equations that are time independent. While variational principles can be obtained for transient problems using adjoint variational principles, in application the choice of trial functions for the Ritz method will generally reduce the variational problem to an equivalent Galerkin expression. The time-dependent approximation equations can also be obtained directly using the Galerkin approach. We will demonstrate in a later section that a judicious choice of coordinate functions will enhance the computational aspects of this technique considerably.

A second approach to solving the transient equations using Galerkin's method, discretizes the time derivative using a finite difference scheme. The fundamental idea is to discretize only the space variable first, by means of a Galerkin scheme, leaving the time derivative continuous. The resulting system of ordinary differential equations are then approximated using an appropriate finite difference technique. Although in the case of the relatively simple situations generally encountered in porous media flow, the choice of a time approximation scheme is more or less arbitrary, we will see that in the highly nonlinear systems encountered in estuary dynamics the success or failure of a simulation scheme often depends on how the equations are handled in the time domain.

In yet another approach, the variational principle for a linear problem is formulated using the concept of convolution. Gurtin [6] used convolution integrals directly to reduce an initial value problem to a form for which he could obtain a variational principle. The same equations can be generated by taking the Laplace transform, solving the resulting equations approximately, and then taking the inverse Laplace transform. When coordinate functions of the form

$$\hat{u} = \sum_{j=1}^{M} a_j(t)\phi(x)$$

are used in the approximation, the convolution schemes are equivalent to the Galerkin method [2].

## 3.6 Summary of Approximation Methods

It has been shown in this chapter that one can generate approximate integral equations using Galerkin's method. For symmetric (self-adjoint) operators, it can be shown that the Ritz procedure and Galerkin's method generate identical equations. For nonsymmetric and nonlinear operators, one can apply Galerkin's procedure which is universal in its applicability and can be shown to converge for a broader spectrum of operators than the variational procedure.

Transient problems can be treated using the variational approach in conjunction with either convolution integrals or the adjoint method. The Galerkin procedure may also be applied directly to time-dependent problems, or can be used to obtain ordinary differential equations which can be discretized using a variety of finite difference schemes.

## References

1. S. G. Mikhlin, "Variational Methods in Mathematical Physics." Macmillan, New York, 1964.
2. B. A. Finlayson, "The Method of Weighted Residuals and Variational Principles." Academic Press, New York, 1972.
3. J. T. Oden, "Finite Elements of Nonlinear Continua." McGraw-Hill, New York, 1972.
4. S. G. Hutton and D. L. Anderson, Finite element methods: A Galerkin approach, *J. Eng. Mech. Div. ASCE* **7(5)**, 1503–1520 (1971).
5. E. R. A. Oliveira, Theoretical foundations of the finite element method, *Int. J. Solids Struct.* **4**, 929–952 (1968).
6. M. E. Gurtin, Variational principles for linear initial-value problems, *Quart. Appl. Math.* **22**, 252–256 (1964).

# Chapter 4 | Finite Elements

In the previous chapter an approach to solving differential equations using approximate integral methods was described. This scheme involved the selection of trial functions that must satisfy certain requirements for completeness. We will now demonstrate how the finite element method of specifying these trial functions leads to a flexible and computationally efficient numerical scheme.

## 4.1 The Finite Element Concept

In the finite element method the basis functions adopted are polynomials which are piecewise continuous over subdomains called finite elements. Nodes are located along the boundaries of each subdomain and each basis function is identified with a specific node. Depending on the forms of the bases, certain of their derivatives may also be continuous across element boundaries.

## 4.2 Linear Basis Functions

The basic features of the finite element concept can be easily visualized using the information illustrated in Fig. 4.1. Let the closed interval $x_1 < x < x_5$ represent the domain of interest and let this interval be subdivided by nodes into four subintervals or elements. The function $u(x)$ can be represented approximately by the piecewise linear function $\hat{u}(x)$ which is defined by a series of the form

$$u \simeq \hat{u} = \sum_{j=1}^{N} u_j \phi_j(x) \tag{4.1}$$

where, for the linear scheme of Figs. 4.1 and 4.2 with five nodes, $N = 5$. The form of the functions $\phi_j(x)$, which are in fact the basis functions for this one-dimensional example, can be established by imposing appropriate conditions on $\hat{u}(x)$.

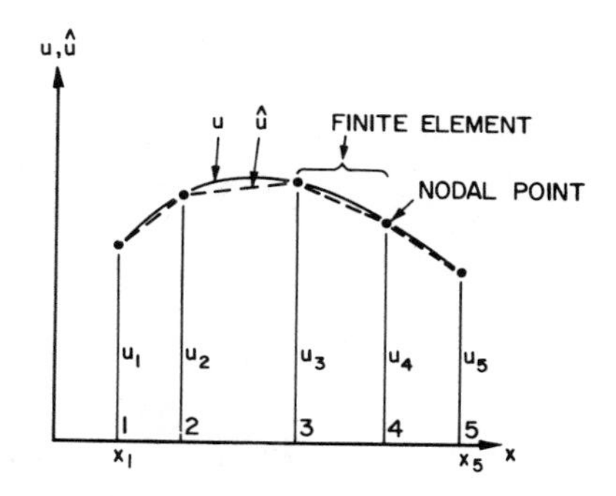


**Fig. 4.1.** One-dimensional finite element in global coordinates.

### *4.2.1 Linear Basis Functions in Global Coordinates*

Let the interpolating polynomial $\hat{u}(x)$ be represented within each element by the general linear polynomial

$$\hat{u}(x) = a + bx \tag{4.2}$$

To solve for the two constants $a$ and $b$, it is only necessary to require that $\hat{u}(x_i) = u_i$ and $\hat{u}(x_{i+1}) = u_{i+1}$, where $x_i$ and $x_{i+1}$ define the end points of the element. These constraints, in combination with (4.2), yield the matrix equation

$$\begin{Bmatrix} u_i \\ u_{i+1} \end{Bmatrix} = \begin{bmatrix} 1 & x_i \\ 1 & x_{i+1} \end{bmatrix} \begin{Bmatrix} a \\ b \end{Bmatrix}$$

which can be solved for the coefficients

$$\begin{Bmatrix} a \\ b \end{Bmatrix} = \frac{1}{x_{i+1} - x_i} \begin{bmatrix} x_{i+1} & -x_i \\ -1 & 1 \end{bmatrix} \begin{Bmatrix} u_i \\ u_{i+1} \end{Bmatrix}$$

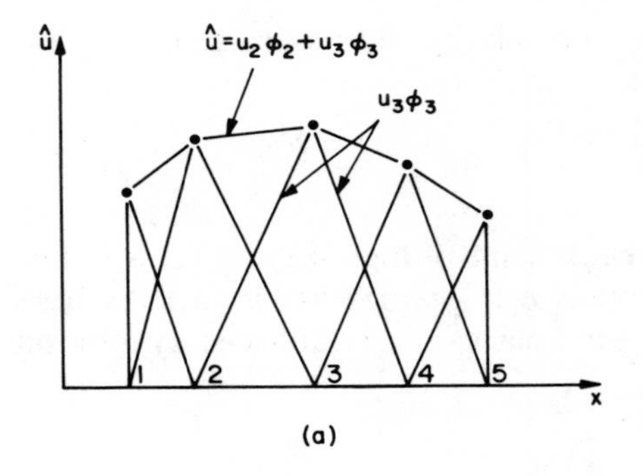


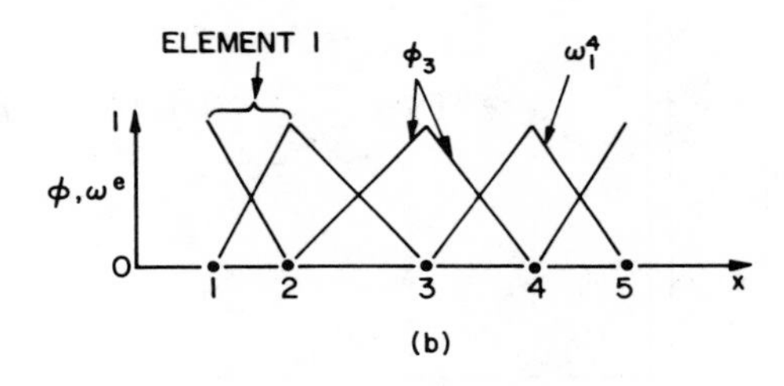


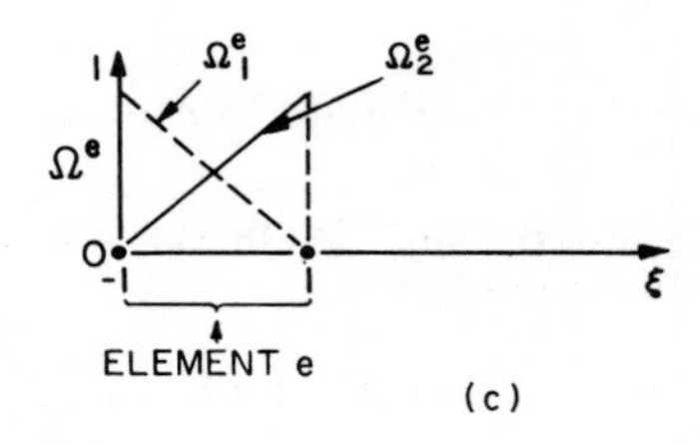


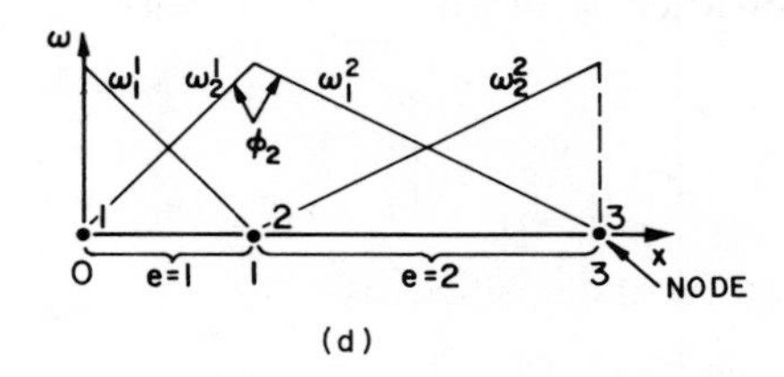


**Fig. 4.2.** The approximate function and corresponding coordinate functions.

The function $\hat{u}(x)$ may now be written for any element as

$$\hat{u} = \frac{x_{i+1}u_i - x_i u_{i+1}}{x_{i+1} - x_i} + \frac{u_{i+1} - u_i}{x_{i+1} - x_i} x$$

After rearrangement to obtain a form analogous to (4.1) this expression becomes

$$\hat{u} = u_i\left(\frac{x_{i+1} - x}{x_{i+1} - x_i}\right) + u_{i+1}\left(\frac{x - x_i}{x_{i+1} - x_i}\right) \tag{4.3}$$

Comparison of (4.3) and (4.1) demonstrates that the basis functions over the element are defined by

$$\phi_i = (x_{i+1} - x)/(x_{i+1} - x_i), \qquad \phi_{i+1} = (x - x_i)/(x_{i+1} - x_i)$$

In the formulation of these expressions for $\phi$, only one element has been considered. Thus there are two basis functions that have nonzero components over each element of this example. The linear $\phi$ basis function is often called the chapeau function because it is hat-shaped when defined for a given node, i.e.,

$$\phi_i(x) = \begin{cases} (x - x_{i-1})/(x_i - x_{i-1}) & (x_{i-1} \le x \le x_i) \\ (x_{i+1} - x)/(x_{i+1} - x_i) & (x_i \le x \le x_{i+1}) \end{cases} \tag{4.4}$$

Note that basis function $\phi_i(x)$ has a value of unity at node $i$ but goes to zero at all other nodes.

It should also be apparent at this point that the basis function $\phi_i$ has a component in each of two elements as illustrated in Fig. 4.2b. Basis functions, therefore, can be defined either by reference to the appropriate node or by reference to the element and the location within the element. For example, in element $e$, the two nonzero $\phi$ components may be denoted for the present case by

$$\phi_i(x_{i+1} \ge x \ge x_i) \equiv \omega_1^e(x) \equiv (x_{i+1} - x)/(x_{i+1} - x_i)$$

$$\phi_{i+1}(x_{i+1} \ge x \ge x_i) \equiv \omega_2^e(x) \equiv (x - x_i)/(x_{i+1} - x_i)$$

where the superscript $e$ designates element number. Thus an equivalent statement of Eq. (4.1) is

$$\hat{u} = \sum_{e=1}^{E} \sum_{j=1}^{n} u_j^e \omega_j^e(x) \tag{4.5}$$

where $E$ is the number of elements (four in the present example), and $n$ the number of nonzero basis functions per element (two in the present example).

### *4.2.2 Linear Basis Functions in Local Coordinates*

The basis functions are nearly always considered over one element at a time in practical application. Consequently, it is convenient to introduce a local coordinate system for each element to simplify integration. This is particularly true in the more complex element formulations where numerical integration is necessary. Linear basis functions defined in a local coordinate system are illustrated in Fig. 4.2c. Each element is defined such that $\xi$, the local coordinate, lies between $-1$ and 1. To distinguish basis functions defined in the local coordinate system from the basis functions in the global system, they are designated by $\Omega_i^e$. Thus, for the present case, the expression for $\hat{u}$ over element $e$ in local coordinates would be

$$\hat{u} = u_1^e \Omega_1^e + u_2^e \Omega_2^e \tag{4.6}$$

The appropriate expressions for the local basis functions are obtained as in the global coordinate system. A linear interpolation across the element is assumed, i.e.,

$$\hat{u}(\xi) = a + b\xi \tag{4.7}$$

and the constraints are imposed such that $\hat{u}(-1) = u_1^e$ and $\hat{u}(1) = u_2^e$. Solution of this equation for the constants $a$ and $b$ yields

$$\begin{Bmatrix} a \\ b \end{Bmatrix} = \frac{1}{2} \begin{bmatrix} 1 & 1 \\ -1 & 1 \end{bmatrix} \begin{Bmatrix} u_1^e \\ u_2^e \end{Bmatrix}$$

or

$$a = \tfrac{1}{2}(u_1^e + u_2^e), \qquad b = \tfrac{1}{2}(u_2^e - u_1^e)$$

Substitution of these expressions back into (4.7) and rearrangement yields

$$\hat{u}(\xi) = u_1^e[\tfrac{1}{2}(1 - \xi)] + u_2^e[\tfrac{1}{2}(1 + \xi)]$$

Comparison with Eq. (4.6) indicates that the linear basis functions in local coordinates are

$$\Omega_1^e = \tfrac{1}{2}(1 - \xi), \qquad \Omega_2^e = \tfrac{1}{2}(1 + \xi)$$

### *4.2.3 Transformation Functions*

To integrate the basis functions in the local coordinate system, it is necessary to develop a means of transforming from global to local coordinates. The required transformation functions are obtained in a manner analogous to the development of the basis functions. In many cases, in fact, the basis and transformation functions are identical.

The development and application of coordinate functions will be presented through an example problem. Consider the one-dimensional finite element of Fig. 4.2d. In both the local and global coordinate systems, values of $x$ and $\xi$ will increase in direct proportion to the distance along the axes. Thus the transformation function will be of the form

$$x = a + b\xi \tag{4.8}$$

In addition, there must be a one-to-one correspondence between the node locations in $x$ and in $\xi$. These constraints give rise to a set of equations analogous to those developed earlier for linear bases:

$$\begin{Bmatrix} x_1^e \\ x_2^e \end{Bmatrix} = \begin{bmatrix} 1 & -1 \\ 1 & 1 \end{bmatrix} \begin{Bmatrix} a \\ b \end{Bmatrix} \tag{4.9}$$

Solution of (4.9) and substitution of the resulting values of $a$ and $b$ into (4.8) yields

$$x = \tfrac{1}{2}(x_2^e + x_1^e) + \tfrac{1}{2}(x_2^e - x_1^e)\xi$$

which can be rearranged to give

$$x = x_1^e[\tfrac{1}{2}(1 - \xi)] + x_2^e[\tfrac{1}{2}(1 + \xi)] \tag{4.10}$$

It is apparent from (4.10) that the functions required for transformation from the global to the local coordinate system are indeed the basis functions defined earlier. Moreover, Eq. (4.10) can be written over an element as

$$x = \sum_{j=1}^{2} x_j^e \Omega_j^e$$

While there is little, if any, conservation of effort in utilizing local coordinates in one dimension, a simple problem will be considered to demonstrate the general procedure involved in utilizing the transformation functions, and formulating the finite element solution procedure.

**Example** Consider the following equations and the finite element net of Fig. 4.2d:

$$Lu \equiv d^2u(x)/dx^2 = 3$$

with boundary conditions $u(0) = 1$ and $du(3)/dx = 0$.

*Step 1* Formulate the approximate integral equations using the Galerkin method. From Chapter 3, we have

$$\langle L\hat{u} - f, \phi_i(x)\rangle = 0 \qquad (i = 1, 2, 3)$$

where, in this example, $f = 3$. Substitution of the governing differential equation yields

$$\langle (d^2\hat{u}/dx^2 - 3), \phi_i(x) \rangle = 0 \qquad (i = 1, 2, 3)$$

Application of integration by parts gives

$$\left\langle -\frac{d\hat{u}}{dx}, \frac{d\phi_i}{dx} \right\rangle - \langle 3, \phi_i \rangle + \frac{d\hat{u}}{dx}\phi_i \Bigg|_{x=0}^{x=3} = 0 \qquad (i = 1, 2, 3) \qquad (4.11)$$

*Step 2* Select basis functions for substitution into (4.11). Because the problem will be solved using linear bases, the appropriate expression for $\hat{u}$ in terms of these bases is

$$\hat{u}(x) = \sum_{j=1}^{3} u_j \phi_j(x) \qquad (4.12)$$

Substitution of (4.12) into (4.11) yields

$$\sum_{j=1}^{3} u_j \left\langle \frac{d\phi_j}{dx}, \frac{d\phi_i}{dx} \right\rangle + \langle 3, \phi_i \rangle - \frac{d\hat{u}}{dx}\phi_i \Bigg|_{x=0}^{x=3} = 0 \qquad (i = 1, 2, 3)$$

*Step 3* Formulate the system of algebraic equations. This is accomplished by writing in matrix form the equations generated successively by setting $i$ to 1, 2, and 3:

$$\begin{bmatrix} \left\langle \frac{d\phi_1}{dx}, \frac{d\phi_1}{dx} \right\rangle & \left\langle \frac{d\phi_2}{dx}, \frac{d\phi_1}{dx} \right\rangle & 0 \\ \left\langle \frac{d\phi_1}{dx}, \frac{d\phi_2}{dx} \right\rangle & \left\langle \frac{d\phi_2}{dx}, \frac{d\phi_2}{dx} \right\rangle & \left\langle \frac{d\phi_3}{dx}, \frac{d\phi_2}{dx} \right\rangle \\ 0 & \left\langle \frac{d\phi_2}{dx}, \frac{d\phi_3}{dx} \right\rangle & \left\langle \frac{d\phi_3}{dx}, \frac{d\phi_3}{dx} \right\rangle \end{bmatrix} \begin{Bmatrix} u_1 \\ u_2 \\ u_3 \end{Bmatrix}$$

$$= \begin{Bmatrix} -\langle 3, \phi_1 \rangle + \frac{d\hat{u}}{dx}\phi_1 \Big|_{x=0}^{x=3} \\ -\langle 3, \phi_2 \rangle + \frac{d\hat{u}}{dx}\phi_2 \Big|_{x=0}^{x=3} \\ -\langle 3, \phi_3 \rangle + \frac{d\hat{u}}{dx}\phi_3 \Big|_{x=0}^{x=3} \end{Bmatrix} \qquad (4.13)$$

Because $d\phi_1/dx$ and $d\phi_3/dx$ are never both nonzero in any part of the solution domain, zeros appear in the lower left and upper right corners of the coefficient matrix.

*Step 4* Apply the boundary conditions. Because of the way the basis functions are defined, $u_1 = u\ (x = 0)$. Thus, to apply the condition that $u(0) = 1$, it is only necessary to delete the first row of Eq. (4.13) and replace it with the boundary condition. This is done by putting 1 in the first column, zeros in column two and three, and 1 in the top row of the right-hand side vector. Note that the term in the second row of the right-side vector $\phi_2\, d\hat{u}/dx|_{x=0}^{x=3}$ will also be zero because $\phi_2 = 0$ at $x = 0$ and $x = 3$. For this problem, the term in the third row of the right-side vector, $\phi_3\, d\hat{u}/dx|_{x=0}^{x=3}$ will also be zero because $\phi_3 = 0$ at $x = 0$ and, by the boundary condition, $d\hat{u}/dx = 0$ at $x = 3$. Thus Eq. (4.13) becomes

$$\begin{bmatrix} 1 & 0 & 0 \\ \left\langle \frac{d\phi_1}{dx}, \frac{d\phi_2}{dx} \right\rangle & \left\langle \frac{d\phi_2}{dx}, \frac{d\phi_2}{dx} \right\rangle & \left\langle \frac{d\phi_3}{dx}, \frac{d\phi_2}{dx} \right\rangle \\ 0 & \left\langle \frac{d\phi_2}{dx}, \frac{d\phi_3}{dx} \right\rangle & \left\langle \frac{d\phi_3}{dx}, \frac{d\phi_3}{dx} \right\rangle \end{bmatrix} \begin{Bmatrix} u_1 \\ u_2 \\ u_3 \end{Bmatrix} = \begin{Bmatrix} 1 \\ -\langle 3, \phi_2 \rangle \\ -\langle 3, \phi_3 \rangle \end{Bmatrix} \tag{4.14}$$

Note that although the coefficient matrix in (4.13) was symmetric, the coefficient matrix in (4.14) is nonsymmetric. Because great savings in computer storage can be achieved when the coefficient matrix is symmetric, it is advantageous to further rearrange (4.14) to a symmetric form. Because $u_1$ is known to be unity from the boundary condition, the coefficients for $u_1$ in the second and third rows can be moved to the right side of the matrix equation and replaced by a zero. Thus (4.14) becomes

$$\begin{bmatrix} 1 & 0 & 0 \\ 0 & \left\langle \frac{d\phi_2}{dx}, \frac{d\phi_2}{dx} \right\rangle & \left\langle \frac{d\phi_3}{dx}, \frac{d\phi_2}{dx} \right\rangle \\ 0 & \left\langle \frac{d\phi_2}{dx}, \frac{d\phi_3}{dx} \right\rangle & \left\langle \frac{d\phi_3}{dx}, \frac{d\phi_3}{dx} \right\rangle \end{bmatrix} \begin{Bmatrix} u_1 \\ u_2 \\ u_3 \end{Bmatrix} = \begin{Bmatrix} 1 \\ -\langle 3, \phi_2 \rangle - \left\langle \frac{d\phi_1}{dx}, \frac{d\phi_2}{dx} \right\rangle \\ -\langle 3, \phi_3 \rangle \end{Bmatrix} \tag{4.15}$$

Because $u_1$ has now been completely uncoupled from the matrix problem, it is possible to condense the first row and first column out of the matrix and to solve the resulting $2 \times 2$ symmetric matrix problem enclosed in the dashed box.

*Step 5* Transform the integrals to local coordinates and evaluate them element by element. Some of the terms in (4.15) will be nonzero in both elements and therefore must be integrated over both elements. Thus (4.15) becomes

$$\begin{bmatrix} \left\langle \frac{d\Omega_2^e}{dx}, \frac{d\Omega_2^e}{dx} \right\rangle^1 + \left\langle \frac{d\Omega_1^e}{dx}, \frac{d\Omega_1^e}{dx} \right\rangle^2 & \left\langle \frac{d\Omega_2^e}{dx}, \frac{d\Omega_1^e}{dx} \right\rangle^2 \\ \left\langle \frac{d\Omega_1^e}{dx}, \frac{d\Omega_2^e}{dx} \right\rangle^2 & \left\langle \frac{d\Omega_2^e}{dx}, \frac{d\Omega_2^e}{dx} \right\rangle^2 \end{bmatrix} \begin{Bmatrix} u_2 \\ u_3 \end{Bmatrix}$$

$$= \begin{Bmatrix} -\langle 3, \Omega_2^e \rangle^1 - \langle 3, \Omega_1^e \rangle^2 - \left\langle \frac{d\Omega_1^e}{dx}, \frac{d\Omega_2^e}{dx} \right\rangle^1 \\ -\langle 3, \Omega_2^e \rangle^2 \end{Bmatrix} \tag{4.16}$$

where $\langle a, b \rangle^e$ denotes the integration of the product $ab$ over element $e$. Now perform the integrations over element 1. The term $\langle d\Omega_2^e/dx, d\Omega_2^e/dx \rangle^1$ is easily evaluated after being rewritten in $\xi$ coordinates:

$$\left\langle \frac{d\Omega_2^e}{dx}, \frac{d\Omega_2^e}{dx} \right\rangle^1 = \int_0^1 \frac{d\Omega_2^e}{dx} \frac{d\Omega_2^e}{dx} \, dx = \int_{-1}^1 \frac{d\Omega_2^e}{d\xi} \frac{d\xi}{dx} \frac{d\Omega_2^e}{d\xi} \frac{d\xi}{dx} \frac{dx}{d\xi} \, d\xi \tag{4.17}$$

Because $\Omega_2^e = \frac{1}{2}(1 + \xi)$, it is readily seen that $d\Omega_2^e/d\xi = \frac{1}{2}$. Derivatives of the form $dx/d\xi$ for element 1 can easily be obtained from Eq. (4.10):

$$x = x_1^1[\tfrac{1}{2}(1 - \xi)] + x_2^1[\tfrac{1}{2}(1 + \xi)]$$

Therefore

$$dx/d\xi = \tfrac{1}{2}(x_2^1 - x_1^1) = \tfrac{1}{2}(1 - 0) = \tfrac{1}{2}$$

Substitution back into (4.17) yields

$$\langle d\Omega_2^e/dx, d\Omega_2^e/dx \rangle^1 = \int_{-1}^1 \tfrac{1}{2}(2)\tfrac{1}{2}(2)\tfrac{1}{2} \, d\xi = 1 \tag{4.18a}$$

The other required integrals over element 1 are also easily obtained:

$$\langle 3, \Omega_2^e \rangle^1 = \int_{-1}^1 3[\tfrac{1}{2}(1 + \xi)]\tfrac{1}{2} \, d\xi = \tfrac{3}{2} \tag{4.18b}$$

$$\langle d\Omega_1^e/dx, d\Omega_2^e/dx \rangle^1 = \int_{-1}^1 -\tfrac{1}{2}(2)\tfrac{1}{2}(2)\tfrac{1}{2} \, d\xi = -1 \tag{4.18c}$$

Substitution of (4.18) back into (4.16) yields

$$\begin{bmatrix} 1 + \left\langle \frac{d\Omega_1^e}{dx}, \frac{d\Omega_1^e}{dx} \right\rangle^2 & \left\langle \frac{d\Omega_2^e}{dx}, \frac{d\Omega_1^e}{dx} \right\rangle^2 \\ \left\langle \frac{d\Omega_1^e}{dx}, \frac{d\Omega_2^e}{dx} \right\rangle^2 & \left\langle \frac{d\Omega_2^e}{dx}, \frac{d\Omega_2^e}{dx} \right\rangle^2 \end{bmatrix} \begin{Bmatrix} u_2 \\ u_3 \end{Bmatrix} = \begin{Bmatrix} -\frac{3}{2} - \langle 3, \Omega_1^e \rangle^2 + 1 \\ -\langle 3, \Omega_2^e \rangle^2 \end{Bmatrix} \tag{4.19}$$

The next step is to perform the integrations over element 2. For this element

$$x = x_1^2[\tfrac{1}{2}(1 - \xi)] + x_2^2[\tfrac{1}{2}(1 + \xi)]$$

and the transformation function is

$$dx/d\xi = \tfrac{1}{2}(x_2^2 - x_1^2) = \tfrac{1}{2}(3 - 1) = 1$$

Thus the integrals in (4.19) take the form

$$\left\langle \frac{d\Omega_1^e}{dx}, \frac{d\Omega_1^e}{dx} \right\rangle^2 = \int_{-1}^{1} \frac{d\Omega_1^e}{d\xi}\frac{d\xi}{dx}\frac{d\Omega_1^e}{d\xi}\frac{d\xi}{dx}\frac{dx}{d\xi}\, d\xi$$

$$= \int_{-1}^{1} \left(-\frac{1}{2}\right)(1)\left(-\frac{1}{2}\right)(1)(1)\, d\xi = \frac{1}{2} \tag{4.20a}$$

$$\left\langle \frac{d\Omega_1^e}{dx}, \frac{d\Omega_2^e}{dx} \right\rangle^2 = \int_{-1}^{1} \frac{d\Omega_1^e}{d\xi}\frac{d\xi}{dx}\frac{d\Omega_2^e}{d\xi}\frac{d\xi}{dx}\frac{dx}{d\xi}\, d\xi$$

$$= \int_{-1}^{1} \left(-\frac{1}{2}\right)(1)\left(\frac{1}{2}\right)(1)(1)\, d\xi = -\frac{1}{2} \tag{4.20b}$$

$$\left\langle \frac{d\Omega_2^e}{dx}, \frac{d\Omega_2^e}{dx} \right\rangle^2 = \int_{-1}^{1} \frac{d\Omega_2^e}{d\xi}\frac{d\xi}{dx}\frac{d\Omega_2^e}{d\xi}\frac{d\xi}{dx}\frac{dx}{d\xi}\, d\xi$$

$$= \int_{-1}^{1} \left(\frac{1}{2}\right)(1)\left(\frac{1}{2}\right)(1)(1)\, d\xi = \frac{1}{2} \tag{4.20c}$$

$$\langle 3, \Omega_1^e \rangle^2 = \int_{-1}^{1} 3\Omega_1^e\, dx/d\xi\, d\xi = \int_{-1}^{1} 3[\tfrac{1}{2}(1 - \xi)](1)\, d\xi = 3 \tag{4.20d}$$

$$\langle 3, \Omega_2^e \rangle^2 = \int_{-1}^{1} 3\Omega_2^e\, dx/d\xi\, d\xi = \int_{-1}^{1} 3[\tfrac{1}{2}(1 + \xi)](1)\, d\xi = 3 \tag{4.20e}$$

Substitution of the relations in (4.20) back into matrix equation (4.19) yields

$$\begin{bmatrix} \frac{3}{2} & -\frac{1}{2} \\ -\frac{1}{2} & \frac{1}{2} \end{bmatrix} \begin{Bmatrix} u_2 \\ u_3 \end{Bmatrix} = \begin{Bmatrix} -\frac{7}{2} \\ -3 \end{Bmatrix}$$

which has the solution $u_2 = -6.5$, $u_3 = -12.5$. It can be easily shown that the analytic solution to this problem is $u = \frac{3}{2}x^2 - 9x + 1$ and that the solutions obtained for $u_2$ and $u_3$ (as well as for $u_1$) are exact. The analytic solution is a quadratic function of $x$ while in this example, the approximate solution was assumed to be piecewise linear. Therefore, if values of $u$ are obtained by evaluating $\hat{u} = \sum_{j=1}^{3} u_j \phi_j$ at various locations on the $x$ axis, there will be some error between the nodes.

## 4.3 Higher-Degree Polynomial Basis Functions

Generally a quadratic polynomial fitted through three points will provide a better representation of the given function than two linear approximations between the same points. When this is the case, it is possible to define a three-node element such as shown in Fig. 4.3. The basis functions for this element are formulated by first assuming a quadratic function as the interpolating function $\hat{u}(x)$ and then imposing constraints analogous to those used for the linear case. The appropriate quadratic function defined in the local $\xi$ coordinates is

$$\hat{u} = a + b\xi + c\xi^2 \tag{4.21}$$

which corresponds to the element of Fig. 4.4a and generates the following

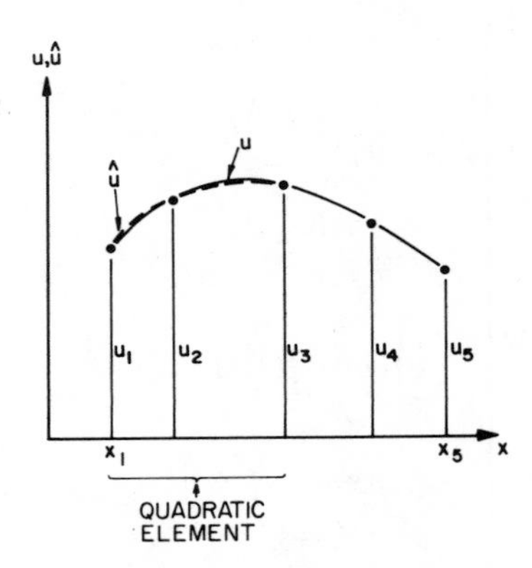


**Fig. 4.3.** One-dimensional quadratic finite element in global coordinates.

set of equations when the boundary nodes are located at $\xi_1 = -1$ and $\xi_3 = 1$ and the midelement node is at $\xi_2 = 0$:

$$\begin{Bmatrix} u_1^e \\ u_2^e \\ u_2^e \end{Bmatrix} = \begin{bmatrix} 1 & \xi_1 & \xi_1^2 \\ 1 & \xi_2 & \xi_2^2 \\ 1 & \xi_3 & \xi_3^2 \end{bmatrix} \begin{Bmatrix} a \\ b \\ c \end{Bmatrix}$$

Substitution of the values of $\xi_1$, $\xi_2$, and $\xi_3$ into this equation yields

$$\begin{Bmatrix} u_1^e \\ u_2^e \\ u_3^e \end{Bmatrix} = \begin{bmatrix} 1 & -1 & 1 \\ 1 & 0 & 0 \\ 1 & 1 & 1 \end{bmatrix} \begin{Bmatrix} a \\ b \\ c \end{Bmatrix}$$

which can be solved for $a$, $b$, and $c$:

$$\begin{Bmatrix} a \\ b \\ c \end{Bmatrix} = \begin{bmatrix} 0 & 1 & 0 \\ -\frac{1}{2} & 0 & \frac{1}{2} \\ \frac{1}{2} & -1 & \frac{1}{2} \end{bmatrix} \begin{Bmatrix} u_1^e \\ u_2^e \\ u_3^e \end{Bmatrix}$$

Substitution into Eq. (4.14) yields

$$\hat{u} = \{1, \xi, \xi^2\} \begin{Bmatrix} a \\ b \\ c \end{Bmatrix} = \{1, \xi, \xi^2\} \begin{bmatrix} 0 & 1 & 0 \\ -\frac{1}{2} & 0 & \frac{1}{2} \\ \frac{1}{2} & -1 & \frac{1}{2} \end{bmatrix} \begin{Bmatrix} u_1^e \\ u_2^e \\ u_3^e \end{Bmatrix}$$

which can be compared with the general expression for $\hat{u}$ over one element

$$\hat{u} = \sum_{j=1}^{3} u_j^e \Omega_j^e(\xi) = \{\Omega_1^e, \Omega_2^e, \Omega_3^e\} \begin{Bmatrix} u_1^e \\ u_2^e \\ u_3^e \end{Bmatrix}$$

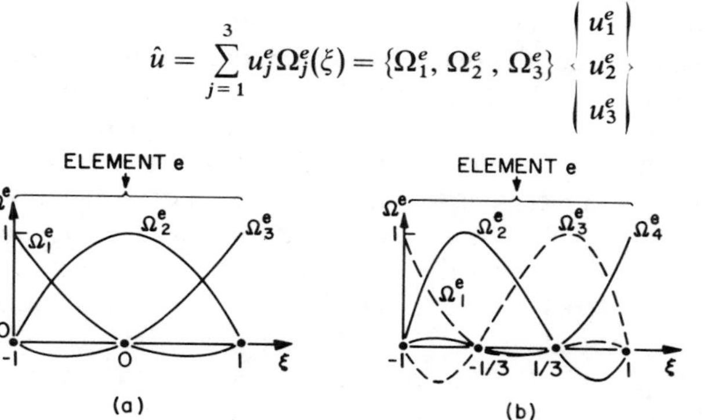


**Fig. 4.4.** Quadratic (a) and cubic (b) coordinate functions defined in local coordinates.

to obtain the basis functions for the quadratic element, namely,

$$\Omega_1^e = -\tfrac{1}{2}\xi + \tfrac{1}{2}\xi^2 = -\tfrac{1}{2}\xi(1 - \xi)$$

$$\Omega_2^e = 1 - \xi^2 = (1 + \xi)(1 - \xi)$$

$$\Omega_3^e = \tfrac{1}{2}\xi + \tfrac{1}{2}\xi^2 = \tfrac{1}{2}\xi(1 + \xi)$$

These functions can also be used as coordinate transformation functions and are presented graphically in Fig. 4.4a.

Cubic approximations can be formulated on elements containing four nodes. The appropriate basis functions are generated through a development analogous to that presented above for quadratic elements. In local coordinates, the nodes are located at $\xi_1 = -1, \xi_2 = -\frac{1}{3}, \xi_3 = \frac{1}{3}, \xi_4 = 1$. The resulting functions are shown in Fig. 4.4b and have the form

$$\Omega_1^e = \tfrac{1}{16}[-9\xi^3 + 9\xi^2 + \xi - 1] = -\tfrac{1}{16}(1 + 3\xi)(1 - 3\xi)(1 - \xi)$$
$$\Omega_2^e = \tfrac{9}{16}[3\xi^3 - \xi^2 - 3\xi + 1] = \tfrac{9}{16}(1 + \xi)(1 - 3\xi)(1 - \xi)$$
$$\Omega_3^e = \tfrac{9}{16}[-3\xi^3 - \xi^2 + 3\xi + 1] = \tfrac{9}{16}(1 + \xi)(1 + 3\xi)(1 - \xi)$$
$$\Omega_4^e = \tfrac{1}{16}[9\xi^3 + 9\xi^2 - \xi - 1] = -\tfrac{1}{16}(1 + \xi)(1 + 3\xi)(1 - 3\xi)$$

Extension to higher-degree polynomials is similar to that presented above.

## 4.4 Hermitian Polynomials

While problems in surface and subsurface hydrology generally require the determination of a function rather than its derivative, there are occasions when the specification of the derivative as a nodal parameter can be useful. To solve directly for nodal values of the derivative, first-derivative continuity between elements is required. Hermitian polynomials provide a vehicle for formulating the necessary trial functions directly [1]. In particular, we will consider Hermitian cubics, which consist of two kinds of functions defined over each element. One kind is associated with the interpolation of the approximating function while the other relates to the interpolation of its derivative. The two types of functions and their application are illustrated in Fig. 4.5. The bell-shaped curves $H_{0,i}$ in Fig. 4.5b are denoted as Hermitian cubics of the first kind [2]. They are similar to chapeau functions in their application, having one complete function per node with a value of unity at the node for which they are defined and dropping off to zero at adjacent nodes. A linear combination of two bell curves generates a cubic interpolation of nodal values of the function.

The second kind of Hermitian cubic $H_{1,i}$ in Fig. 4.5c is used to interpolate nonzero derivatives defined at the nodes. As indicated in the figure, this function has a value of zero and a slope of unity at the node with which it is associated. Both its value and its slope are zero at adjacent nodes. Because the slope of the bell curves is always zero at the node points, the coefficient of the unit-slope functions is the slope of the resulting approximation function at the node [2]. The Hermitian cubics are defined in local coordinates consistent with Fig. 4.5d, as

$$\lambda_{0,1}^e = \tfrac{1}{4}(\xi - 1)^2(\xi + 2)$$
$$\lambda_{0,2}^e = -\tfrac{1}{4}(\xi + 1)^2(\xi - 2), \qquad \xi = 2[(x - x_1^e)/L^e] - 1$$
$$\lambda_{1,1}^e = \tfrac{1}{8}L^e(\xi + 1)(\xi - 1)^2, \qquad L^e = x_2^e - x_1^e$$
$$\lambda_{1,2}^e = \tfrac{1}{8}L^e(\xi + 1)^2(\xi - 1)$$

$\xi$ ranges from $-1$ to 1.

The Hermite basis functions are obtained from the general cubic polynomial subject to somewhat different constraints from those imposed when deriving basis functions for which continuity of derivatives is not required. The first step in the development is to express the interpolating function in $\xi$ coordinates as

$$\hat{u} = u_1^e \lambda_{0,1}^e + u_2^e \lambda_{0,2}^e + (du_1^e/dx)\lambda_{1,1}^e + (du_2^e/dx)\lambda_{1,2}^e$$

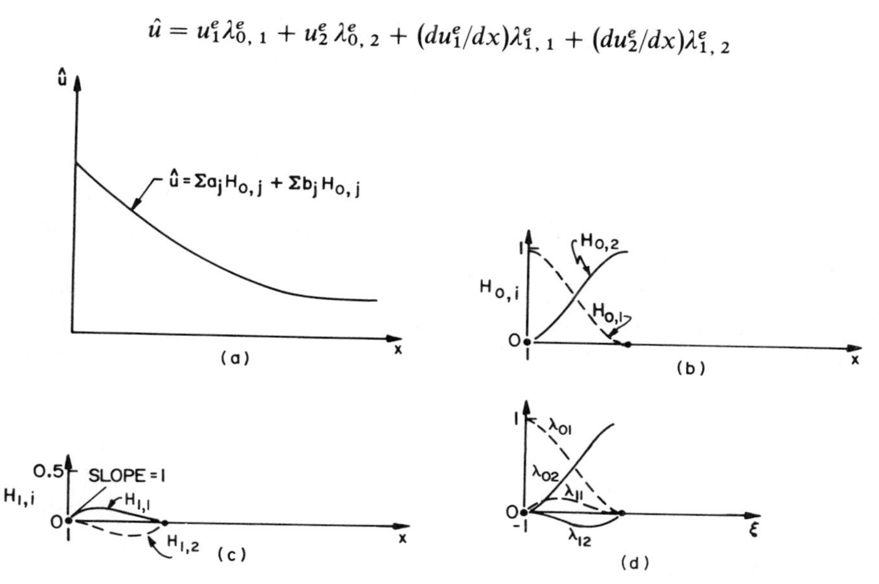


**Fig. 4.5.** Hermite cubics in global and local coordinates (modified from Doherty [2]).

or in matrix form

$$\hat{u} = \{\lambda_{0,1}^e, \lambda_{0,2}^e, \lambda_{1,1}^e, \lambda_{1,2}^e\} \begin{Bmatrix} u_1^e \\ u_2^e \\ du_1^e/dx \\ du_2^e/dx \end{Bmatrix} \tag{4.22}$$

where the lambda functions are the bases which have been defined a priori above and will be developed in this section. In contrast to earlier bases, then, the interpolating function uses the derivative of the unknown function at the node.

Let the interpolating functions $\hat{u}$ be expressed as the cubic polynomial

$$\hat{u} = a + b\xi + c\xi^2 + d\xi^3 \qquad \text{or} \qquad \hat{u} = \{1, \xi, \xi^2, \xi^3\} \begin{Bmatrix} a \\ b \\ c \\ d \end{Bmatrix} \tag{4.23}$$

Because the nodal values of $d\hat{u}/dx$ are also required, it is necessary to differentiate (4.23) to obtain

$$\frac{d\hat{u}}{dx} = \frac{d\hat{u}}{d\xi}\frac{d\xi}{dx} = (b + 2c\xi + 3d\xi^2)\frac{d\xi}{dx}$$

This equation can also be written in matrix form as

$$\frac{d\hat{u}}{dx} = \left\{0, \frac{d\xi}{dx}, 2\xi\frac{d\xi}{dx}, 3\xi^2\frac{d\xi}{dx}\right\}\begin{Bmatrix} a \\ b \\ c \\ d \end{Bmatrix}$$

To solve for the four constants $a$, $b$, $c$, and $d$, four conditions must be imposed. In earlier examples, additional nodes were added along the element sides to generate the necessary constraints. The Hermite element, however, has only two nodes and two conditions are imposed at each node. Both the function $\hat{u}_i$ and its first derivative $d\hat{u}_i/dx$ must be expressed as discrete values at the node $i$. Evaluated at the nodes, $d\xi/dx$ is merely the ratio of element lengths in local and global coordinates $2/L^e$. The following set of equations thereby arises:

$$\begin{Bmatrix} u_1^e \\ u_2^e \\ \dfrac{du_1^e}{dx} \\ \dfrac{du_2^e}{dx} \end{Bmatrix} = \begin{bmatrix} 1 & \xi_1 & \xi_1^2 & \xi_1^3 \\ 1 & \xi_2 & \xi_2^2 & \xi_2^3 \\ 0 & \dfrac{2}{L^e} & \dfrac{4\xi_1}{L^e} & \dfrac{6\xi_1^2}{L^e} \\ 0 & \dfrac{2}{L^e} & \dfrac{4\xi_2}{L^e} & \dfrac{6\xi_2^2}{L^e} \end{bmatrix}\begin{Bmatrix} a \\ b \\ c \\ d \end{Bmatrix} = \begin{bmatrix} 1 & -1 & 1 & -1 \\ 1 & 1 & 1 & 1 \\ 0 & \dfrac{2}{L^e} & \dfrac{-4}{L^e} & \dfrac{6}{L^e} \\ 0 & \dfrac{2}{L^e} & \dfrac{4}{L^e} & \dfrac{6}{L^e} \end{bmatrix}\begin{Bmatrix} a \\ b \\ c \\ d \end{Bmatrix}$$

$$\begin{Bmatrix} a \\ b \\ c \\ d \end{Bmatrix} = \frac{1}{8}\begin{bmatrix} 4 & 4 & L^e & -L^e \\ -6 & 6 & -L^e & -L^e \\ 0 & 0 & -L^e & L^e \\ 2 & -2 & L^e & L^e \end{bmatrix}\begin{Bmatrix} u_1^e \\ u_2^e \\ du_1^e/dx \\ du_2^e/dx \end{Bmatrix}$$

Equation (4.23) can now be written as

$$\hat{u} = \{1, \xi, \xi^2, \xi^3\}\frac{1}{8}\begin{bmatrix} 4 & 4 & L^e & -L^e \\ -6 & 6 & -L^e & -L^e \\ 0 & 0 & -L^e & L^e \\ 2 & -2 & L^e & L^e \end{bmatrix}\begin{Bmatrix} u_1^e \\ u_2^e \\ du_1^e/dx \\ du_2^e/dx \end{Bmatrix} \tag{4.24}$$

Comparison of (4.24) and (4.22) reveals that the required functions must have the form

$$\lambda^e_{0,1} = \tfrac{1}{4}(2 - 3\xi + \xi^3) = \tfrac{1}{4}(\xi - 1)^2(\xi + 2)$$
$$\lambda^e_{0,2} = \tfrac{1}{4}(2 + 3\xi - \xi^3) = -\tfrac{1}{4}(\xi + 1)^2(\xi - 2)$$
$$\lambda^e_{1,1} = \tfrac{1}{8}L^e(1 - \xi - \xi^2 + \xi^3) = \tfrac{1}{8}L^e(\xi + 1)(\xi - 1)^2$$
$$\lambda^e_{1,2} = \tfrac{1}{8}L^e(-1 - \xi + \xi^2 + \xi^3) = \tfrac{1}{8}L^e(\xi + 1)^2(\xi - 1)$$

which is identical to the definitions presented earlier. The above functions can be used as coordinate transformation functions as in the earlier examples or another set of appropriate functions may be selected.

It was previously mentioned that $dx/d\xi$ is equal to $L^e/2$ at the nodal points. The simplest way to satisfy this constraint is to use linear basis functions for the coordinate transformation over an element. Thus, over an element,

$$x = x^e_1[\tfrac{1}{2}(1 - \xi)] + x^e_2[\tfrac{1}{2}(1 + \xi)] \qquad \text{and} \qquad dx/d\xi = \tfrac{1}{2}(x^e_2 - x^e_1) = \tfrac{1}{2}L^e \tag{4.25}$$

For this case, the transformation function is $\frac{1}{2}L^e$ everywhere.

If the Hermitian cubic basis functions are to be used as the transformation functions, then

$$\begin{aligned} x = {} & x^e_1[\tfrac{1}{4}(\xi - 1)^2(\xi + 2)] - x^e_2[\tfrac{1}{4}(\xi + 1)^2(\xi - 2)] \\ & + (dx/dx)^e_1[\tfrac{1}{8}L^e(\xi + 1)(\xi - 1)^2] \\ & + (dx/dx)^e_2[\tfrac{1}{8}L^e(\xi + 1)^2(\xi - 1)] \end{aligned} \tag{4.26}$$

At both nodal locations, however, $dx/dx$ is simply unity and (4.26) becomes

$$x = x^e_1[\tfrac{1}{4}(\xi - 1)^2(\xi + 2)] - x^e_2[\tfrac{1}{4}(\xi + 1)^2(\xi - 2)] + \tfrac{1}{4}L^e\xi(\xi + 1)(\xi - 1)$$

For the case being considered here where the element is a straight line, $L^e = x^e_2 - x^e_1$. Use of this fact in the preceding equation and some rearrangement yields

$$x = x^e_1[\tfrac{1}{2}(1 - \xi)] + x^e_2[\tfrac{1}{2}(1 + \xi)]$$

which is the same relation as (4.25). Thus for an element using Hermitian cubics in one dimension, linear basis functions or the Hermitian cubic functions yield the same transformation function.

## 4.5 Transient Problem in One Space Variable

In a previous section an ordinary differential equation was used to introduce the concept of basis functions and coordinate transformation functions. In this section, we will consider an extension of this example to consider time dependence. While the space derivative will be approximated using finite elements, a finite difference scheme will be used for the time derivative. The system is described by a second-order, parabolic partial differential equation, an initial condition, and two boundary conditions

$$Lu \equiv \frac{\partial^2 u}{\partial x^2} - \frac{\partial u}{\partial t} = 0,$$

$$u(0, t) = 2, \qquad u(3, t) = 1$$

$$u(x, 0) = 0 \qquad (0 < x < 3)$$

The approximate integral equations are developed directly from the Galerkin requirement that

$$\langle L\hat{u}(x, t), \phi_i(x)\rangle = 0 \qquad (i = 1, 2, \ldots, M)$$

Expansion of the operator $L$ and integration by parts yields

$$-\left\langle \frac{\partial \hat{u}}{\partial x}(x, t), \frac{d\phi_i}{dx}(x)\right\rangle - \left\langle \frac{\partial \hat{u}}{\partial t}(x, t), \phi_i(x)\right\rangle + \frac{\partial \hat{u}}{\partial x}(x, t)\phi_i(x)\Bigg|_{x=0}^{x=3} = 0 \qquad (i = 1, 2, 3) \tag{4.27}$$

The trial functions and finite element grid (Fig. 4.2d) are similar to the previous case with the exception of the undetermined parameters which are now time dependent:

$$\hat{u}(x, t) = \sum_{j=1}^{3} a_j(t)\phi_j(x) = \sum_{e=1}^{2} \sum_{j=1}^{2} a_j^e(t)\omega_j^e(x) \tag{4.28}$$

Combining (4.27) with (4.28), we obtain the required system of algebraic equations:

$$-\sum_{j=1}^{3} a_j(t)\left\langle \frac{d\phi_j(x)}{dx}, \frac{d\phi_i(x)}{dx}\right\rangle - \sum_{j=1}^{3} \frac{da_j(t)}{dt}\langle \phi_j(x), \phi_i(x)\rangle + \frac{\partial \hat{u}}{\partial x}\phi_i(x)\Bigg|_0^3 = 0 \qquad (i = 1, 2, 3) \tag{4.29}$$

With $u$ specified at the two ends of the domain, $u(0, t) = a_1 = 2$ and

$u(3, t) = a_3 = 1$. Because of the Dirichlet boundary conditions, there is no need to form the finite element equations by weighting with respect to the basis functions associated with the end nodes. The problem thus reduces to solving one equation for one unknown:

$$-\sum_{j=1}^{3} a_j(t)\left\langle \frac{d\phi_j}{dx}, \frac{d\phi_2}{dx} \right\rangle - \sum_{j=1}^{3} \frac{da_j}{dt}\langle \phi_j, \phi_2 \rangle + \frac{\partial \hat{u}}{\partial x}\phi_2 \Bigg|_{x=0}^{x=3} = 0 \quad (4.30)$$

Because $\phi_2$ is zero at $x = 3$ and at $x = 0$, the last term on the left side of (4.30) may be dropped. In matrix form, this equation becomes

$$\left\{\left\langle \frac{d\phi_1}{dx}, \frac{d\phi_2}{dx} \right\rangle, \left\langle \frac{d\phi_2}{dx}, \frac{d\phi_2}{dx} \right\rangle, \left\langle \frac{d\phi_3}{dx}, \frac{d\phi_2}{dx} \right\rangle\right\} \begin{Bmatrix} 2 \\ a_2 \\ 1 \end{Bmatrix}$$

$$+ \{\langle \phi_1, \phi_2 \rangle, \langle \phi_2, \phi_2 \rangle, \langle \phi_3, \phi_2 \rangle\} \begin{Bmatrix} 0 \\ da_2/dt \\ 0 \end{Bmatrix} = 0 \quad (4.31)$$

Similar to Eq. (4.15), the terms in the coefficient matrices can be transferred to local coordinates and evaluated element by element to obtain

$$\{-1, \tfrac{3}{2}, -\tfrac{1}{2}\} \begin{Bmatrix} 2 \\ a_2 \\ 1 \end{Bmatrix} + \{\tfrac{1}{6}, 1, \tfrac{1}{3}\} \begin{Bmatrix} 0 \\ da_2/dt \\ 0 \end{Bmatrix} = 0$$

or

$$\tfrac{3}{2}a_2 + (da_2/dt) = \tfrac{5}{2} \quad (4.32)$$

Because the problem being solved involves only three nodes, Eq. (4.32) is of a very simple form. If a larger number of nodes were considered, a matrix equation of the form

$$[G]\{A\} + [B]\{dA/dt\} = \{F\}$$

would have resulted. It can be seen that (4.32) is a $1 \times 1$ degeneration of this form.

Now that all the integrals in space have been evaluated, the finite difference discretization of the time derivative must be considered. In Section 2.3.1, three weighted average schemes that are applicable to this type of problem were discussed: explicit, implicit, and Crank–Nicholson or centered implicit methods. If $da_2/dt$ is finite differenced, (4.32) can be written as

$$\tfrac{3}{2}\varepsilon a_{2_{t+\Delta t}} + \tfrac{3}{2}(1 - \varepsilon)a_{2_t} + (a_{2_{t+\Delta t}} - a_{2_t})/\Delta t = \tfrac{5}{2} \quad (4.33)$$

where $\varepsilon = 0$ corresponds to an explicit scheme, $\varepsilon = \frac{1}{2}$ is the centered implicit scheme, and $\varepsilon = 1$ is the implicit scheme. As an example, the first time step is computed with $\Delta t = \frac{1}{4}$ and the initial condition $a_{2_t} = u(1, 0) = 0$. Thus (4.33) becomes

$$\tfrac{3}{2}\varepsilon a_2 + \tfrac{3}{2}(1 - \varepsilon)0 + 4(a_2 - 0) = \tfrac{5}{2} \qquad \text{or} \qquad \tfrac{3}{2}\varepsilon a_2 + 4a_2 = \tfrac{5}{2} \tag{4.33a}$$

The solution is $a_2 = 5/(3\varepsilon + 8)$, which is $\hat{u}(1, \frac{1}{4})$. Thus, at $t = \frac{1}{4}$, for the explicit case, $a_2 = 0.625$; for the centered case, $a_2 = 0.526$; and for the fully implicit case, $a_2 = 0.455$. These answers should be compared to the analytical solution $u(1, \frac{1}{4}) = 0.319$. The first-order correct backward difference implicit scheme gives a better answer in this problem than the second-order centered scheme because the errors associated with the time and space discretization are cancelling out.

## 4.6 Finite Elements in Two Space Dimensions

### 4.6.1 *Rectangular Net in Two Space Dimensions*

The one-dimensional techniques can be readily extended to two space dimensions when rectangular finite elements are used. To demonstrate the formulation and application of rectangular finite elements consider the system of equations

$$\begin{gathered} Lu \equiv (\partial^2 u/\partial x^2) + (\partial^2 u/\partial y^2) = Q \\ u(x, 2) = 1, \quad u(0, y) = 1, \quad \partial u/\partial y(x, 0) = 0, \quad \partial u/\partial x(2, y) = 0 \\ Q(x, y) = Q_w(1, 1)\, \delta(x - 1, y - 1) \end{gathered} \tag{4.34}$$

where $Q(x, y)$ is a sink function, $Q_w(1, 1)$ the magnitude of the sink, and $\delta(x - 1, y - 1)$ the Dirac delta function. We will use the net indicated in Fig. 4.6, which, for simplicity, has linear sides and $\Delta x = \Delta y = 1$.

The Galerkin method of approximation generates the following set of integral equations:

$$\left\langle \frac{\partial^2 \hat{u}}{\partial x^2} + \frac{\partial^2 \hat{u}}{\partial y^2}, \phi_i(x, y) \right\rangle - \langle Q, \phi_i(x, y) \rangle = 0 \qquad (i = 1, 2, \ldots, 9) \tag{4.35}$$

where, because the elements are rectangles, the trial functions are obtained by taking the product of two linear basis functions, one in $x$ and the other in $y$, i.e.,

$$\hat{u} = \sum_{k=1}^{9} a_k \phi_k(x, y) = \sum_{i=-1}^{1} \sum_{j=-1}^{1} a_{ij} \pi_i(x) \pi_j(y) \tag{4.36}$$

where the bases $\pi_i(x)$ and $\pi_j(y)$ are chapeau functions defined as before:

$$\pi_i(x) = \begin{cases} \dfrac{x - x_{i-1}}{x_i - x_{i-1}} & (x_{i-1} \le x \le x_i), \\[2ex] \dfrac{x_{i+1} - x}{x_{i+1} - x_i} & (x_i \le x \le x_{i+1}), \end{cases}$$

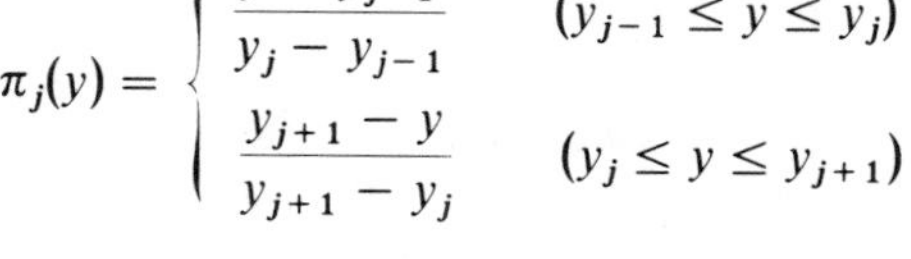

$$\pi_j(y) = \begin{cases} \dfrac{y - y_{j-1}}{y_j - y_{j-1}} & (y_{j-1} \le y \le y_j) \\[2ex] \dfrac{y_{j+1} - y}{y_{j+1} - y_j} & (y_j \le y \le y_{j+1}) \end{cases}$$

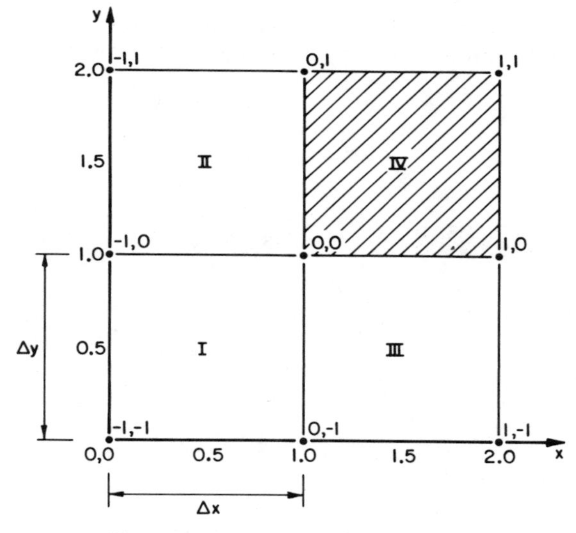


**Fig. 4.6.** Rectangular finite element net.

These functions are illustrated in Fig. 4.7; it is important to recognize that while the basis functions are linear along the element edges, they are, in general, of second degree on the interior.

The second derivatives in (4.35) are eliminated using the two-dimensional form of integration by parts generally known as Green's theorem:

$$-\left\langle \frac{\partial \hat{u}}{\partial x}, \frac{\partial \phi_i}{\partial x} \right\rangle - \left\langle \frac{\partial \hat{u}}{\partial y}, \frac{\partial \phi_i}{\partial y} \right\rangle + \int_\Gamma \left( \frac{\partial \hat{u}}{\partial x} l_x + \frac{\partial \hat{u}}{\partial y} l_y \right) \phi_i \, ds - \langle Q, \phi_i \rangle = 0 \qquad (i = 1, 2, \ldots, 9) \quad (4.37)$$

where $l_x$ and $l_y$ are direction cosines which, when multiplied by $\partial u/\partial x$ and $\partial u/\partial y$, respectively, provide the normal gradient of the function to the surface $\Gamma$. Equations (4.36) and (4.37) are now combined to provide a set of linear algebraic equations in $a_k$ which can be written in matrix form as

$$[B]\{A\} - \{F\} = 0$$

where a typical element of $[B]$ is

$$b_{ij} = -\left\langle \frac{\partial \phi_i}{\partial x}, \frac{\partial \phi_j}{\partial x} \right\rangle - \left\langle \frac{\partial \phi_i}{\partial y}, \frac{\partial \phi_j}{\partial y} \right\rangle$$

By using the symmetry of the finite element net, the required number of integrations can be minimized for this problem. Moreover, the integrals can

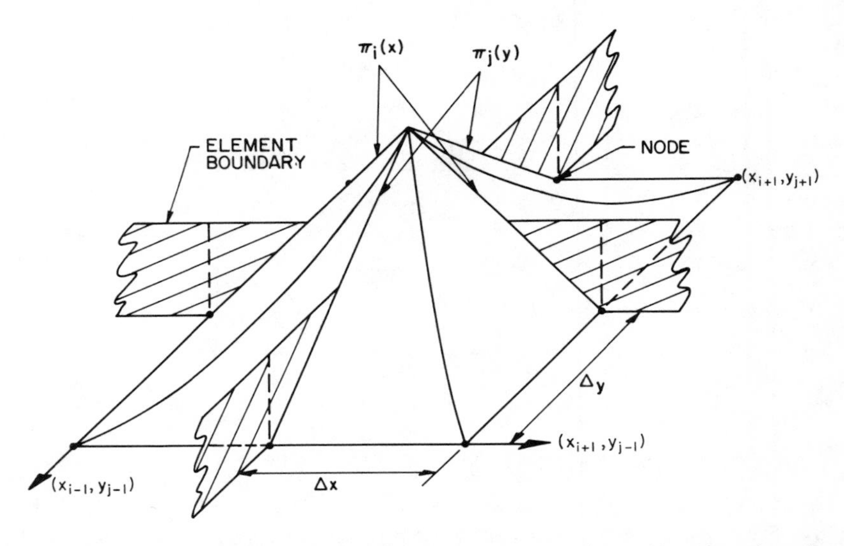


**Fig. 4.7.** Coordinate function for rectangular elements with linear sides. Note above function for central node spans four elements.

be computed elementwise using the two-dimensional analog of the $\omega_i^e(x)$ functions introduced earlier and illustrated in Fig. 4.2. A typical element of matrix $\{F\}$ is written as

$$f_i = -\int_\Gamma \left( \frac{\partial \hat{u}}{\partial x} l_x + \frac{\partial \hat{u}}{\partial y} l_y \right) \phi_i(x, y)\, ds + Q_{wi}$$

Evaluating the elements of $[B]$ and $\{F\}$, we obtain

$$
\begin{bmatrix}
\frac{2}{3} & -\frac{1}{6} & 0 & -\frac{1}{6} & -\frac{1}{3} & 0 & 0 & 0 & 0 \\
-\frac{1}{6} & \frac{4}{3} & -\frac{1}{6} & -\frac{1}{3} & -\frac{1}{3} & -\frac{1}{3} & 0 & 0 & 0 \\
0 & -\frac{1}{6} & \frac{2}{3} & 0 & -\frac{1}{3} & -\frac{1}{6} & 0 & 0 & 0 \\
-\frac{1}{6} & -\frac{1}{3} & 0 & \frac{4}{3} & -\frac{1}{3} & 0 & -\frac{1}{6} & -\frac{1}{3} & 0 \\
-\frac{1}{3} & -\frac{1}{3} & -\frac{1}{3} & -\frac{1}{3} & \frac{8}{3} & -\frac{1}{3} & -\frac{1}{3} & -\frac{1}{3} & -\frac{1}{3} \\
0 & -\frac{1}{3} & -\frac{1}{6} & 0 & -\frac{1}{3} & \frac{4}{3} & 0 & -\frac{1}{3} & -\frac{1}{6} \\
0 & 0 & 0 & -\frac{1}{6} & -\frac{1}{3} & 0 & \frac{2}{3} & -\frac{1}{6} & 0 \\
0 & 0 & 0 & -\frac{1}{3} & -\frac{1}{3} & -\frac{1}{3} & -\frac{1}{6} & \frac{4}{3} & -\frac{1}{6} \\
0 & 0 & 0 & 0 & -\frac{1}{3} & -\frac{1}{6} & 0 & -\frac{1}{6} & \frac{2}{3}
\end{bmatrix}
\begin{Bmatrix} a_{-1,-1} \\ a_{-1,0} \\ a_{-1,1} \\ a_{0,-1} \\ a_{0,0} \\ a_{0,1} \\ a_{1,-1} \\ a_{1,0} \\ a_{1,1} \end{Bmatrix}
-
\begin{Bmatrix} q^*_{-1,-1} \\ q^*_{-1,0} \\ q^*_{-1,1} \\ 0 \\ Q_w \\ q^*_{0,1} \\ 0 \\ 0 \\ q^*_{1,1} \end{Bmatrix}
= 0
$$

where

$$
q^*_{k,l} \equiv -\int_\Gamma \left( \frac{\partial \hat{u}}{\partial x} l_x + \frac{\partial \hat{u}}{\partial y} l_y \right) \pi_k(x)\pi_l(y)\, ds
$$

Equations written for Dirichlet nodes are partitioned from the matrix, and boundary condition information involving these nodes is transferred to the known matrix. The reduced system of equations can be written as

$$
\begin{bmatrix}
\frac{4}{3} & -\frac{1}{3} & -\frac{1}{6} & -\frac{1}{3} \\
-\frac{1}{3} & \frac{8}{3} & -\frac{1}{3} & -\frac{1}{3} \\
-\frac{1}{6} & -\frac{1}{3} & \frac{2}{3} & -\frac{1}{6} \\
-\frac{1}{3} & -\frac{1}{3} & -\frac{1}{6} & \frac{4}{3}
\end{bmatrix}
\begin{Bmatrix} a_{0,-1} \\ a_{0,0} \\ a_{1,-1} \\ a_{1,0} \end{Bmatrix}
-
\begin{Bmatrix} \frac{1}{2} \\ -Q_w + \frac{5}{3} \\ 0 \\ \frac{1}{2} \end{Bmatrix}
= 0
$$

Assuming $Q_{wi}$ to be unity and taking the inverse of $[B]$ the matrix of unknown coefficients is obtained directly as

$$
\{A\} = [B]^{-1}\{F\}
$$

or

$$
\begin{Bmatrix} a_{0,-1} \\ a_{0,0} \\ a_{1,-1} \\ a_{1,0} \end{Bmatrix}
=
\begin{bmatrix}
0.942 & 0.214 & 0.429 & 0.342 \\
0.214 & 0.471 & 0.342 & 0.214 \\
0.429 & 0.342 & 1.88 & 0.429 \\
0.342 & 0.214 & 0.429 & 0.942
\end{bmatrix}
\begin{Bmatrix} \frac{1}{2} \\ \frac{2}{3} \\ 0 \\ \frac{1}{2} \end{Bmatrix}
=
\begin{Bmatrix} 0.783 \\ 0.525 \\ 0.655 \\ 0.783 \end{Bmatrix}
$$

Because the unknown coefficient $a_{i,j}$ is identical to the trial function at that point, the solution to the system of equations (4.34) is

$$
\begin{aligned}
&\{\hat{u}(0, 0), \hat{u}(0, 1), \hat{u}(0, 2), \hat{u}(1, 0), \hat{u}(1, 1), \hat{u}(1, 2), \hat{u}(2, 0), \hat{u}(2, 1), \hat{u}(2, 2)\} \\
&\quad = \{1.0, 1.0, 1.0, 0.783, 0.525, 1.0, 0.655, 0.783, 1.0\}
\end{aligned}
$$

### 4.6.2 *Rectangular Net in One Space Dimension and Time*

Although one generally uses a finite difference scheme for time discretization, the Galerkin method may also be employed directly. To demonstrate the use of finite elements in time let us consider once again the system of equations

$$Lu \equiv \frac{\partial^2 u}{\partial x^2} - \frac{\partial u}{\partial t} = 0$$

$$u(0, t) = 2, \qquad u(3, t) = 1, \qquad u(x, 0) = 0 \quad (0 < x < 3)$$

Application of Galerkin's method results in the set of equations

$$\langle \partial^2 \hat{u}/\partial x^2, \phi_i(x, t)\rangle - \langle \partial \hat{u}/\partial t, \phi_i(x, t)\rangle = 0 \qquad (i = 1, 2, 3) \tag{4.38}$$

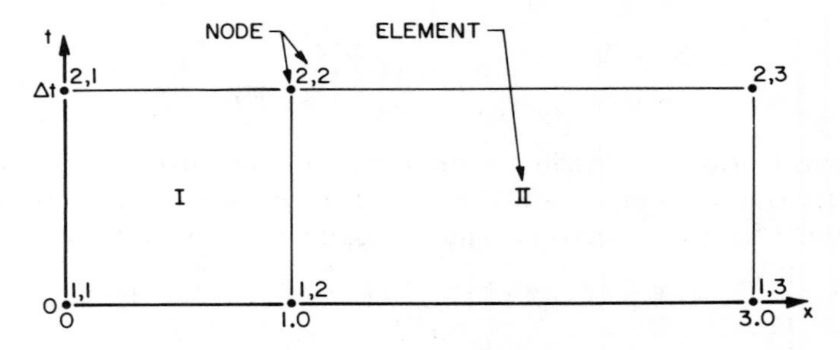


**Fig. 4.8.** Rectangular finite element net for transient problem in one space dimension.

In contrast to the earlier example where the time dependence of the trial functions was introduced through the undetermined parameter $a(t)$, consider now a new trial function analogous to (4.36):

$$\hat{u} = \sum_{k=1}^{6} a_k \phi_k(x, t) = \sum_{i=1}^{2} \sum_{j=1}^{3} a_{i,j} \pi_i(t) \pi_j(x)$$

where $k = 3i + j - 3$ can be defined for this problem. The finite element net is illustrated in Fig. 4.8.

Because this is an initial value problem the time domain is "open ended." Consequently the solution is propagated through time one element at a time; the solution for time level $i$ becomes the initial condition for time level $i + 1$. While it is possible to consider several elements in time simultaneously, this would only increase the computational effort without improving the accuracy of the solution. If, however, elements using higher-degree polynomials in the time domain are used, the increased accuracy one would expect to achieve might justify the additional work required for a solution.

This trade-off will be examined later when two space dimensions and time are considered.

Returning to the example problem, one can apply integration by parts to (4.38) to obtain

$$-\left\langle \frac{\partial \hat{u}}{\partial x}, \frac{\partial \phi_i}{\partial x} \right\rangle - \left\langle \frac{\partial \hat{u}}{\partial t}, \phi_i \right\rangle + \phi_i \frac{\partial \hat{u}}{\partial x}\bigg|_{x=0}^{x=3} = 0 \qquad (i = 1, 2, 3, \ldots, 6)$$

Combination of the approximate integral equations and the definition of $\hat{u}$ provides the system of equations

$$-\sum_{k=1}^{6} a_k \left\langle \frac{\partial \phi_k(x, t)}{\partial x}, \frac{\partial \phi_l(x, t)}{\partial x} \right\rangle - \sum_{k=1}^{6} a_k \left\langle \frac{\partial \phi_k(x, t)}{\partial t}, \phi_l(x, t) \right\rangle$$
$$+ \frac{\partial \hat{u}(x, t)}{\partial x} \phi_l(x, t)\bigg|_{x=0}^{x=3} = 0 \qquad (l = 1, \ldots, 6)$$

This equation can be written as a $6 \times 6$ matrix. However, because $a_j$ is known either by the boundary conditions or the initial conditions at every node except (2, 2), it is really only necessary to consider the row of the matrix found by weighting with respect to the basis function associated with (2, 2). After transforming to local coordinates and performing the integration, this equation becomes

$$-\left(\tfrac{1}{12} + \tfrac{1}{6}\,\Delta t\right)a_{1,1} - \left(\tfrac{1}{2} - \tfrac{1}{4}\,\Delta t\right)a_{1,2} - \left(\tfrac{1}{6} + \tfrac{1}{12}\,\Delta t\right)a_{1,3}$$
$$+ \left(\tfrac{1}{12} - \tfrac{1}{3}\,\Delta t\right)a_{2,1} + \left(\tfrac{1}{2} + \tfrac{1}{2}\,\Delta t\right)a_{2,2} + \left(\tfrac{1}{6} - \tfrac{1}{6}\,\Delta t\right)a_{2,3} = 0$$

which can be rearranged to give the equivalent equation

$$\frac{1}{6}\left(\frac{a_{2,1} - a_{1,1}}{\Delta t}\right) + \left(\frac{a_{2,2} - a_{1,2}}{\Delta t}\right) + \frac{1}{3}\left(\frac{a_{2,3} - a_{1,3}}{\Delta t}\right)$$
$$= \tfrac{1}{3}\left(a_{1,1} - \tfrac{3}{2}a_{1,2} + \tfrac{1}{2}a_{1,3}\right) + \tfrac{2}{3}\left(a_{2,1} - \tfrac{3}{2}a_{2,2} + \tfrac{1}{2}a_{2,3}\right) \qquad (4.39)$$

The right-hand side of (4.39) is simply the finite element spatial approximation evaluated at $\varepsilon = \frac{2}{3}$. The interpretation of the left-hand side of (4.39) is less obvious, but apparently represents a weighted average of time derivatives at the three spatial nodes, discretized using first-order correct finite differences. It will be shown later that this weighted average actually represents a spatial integration of the time derivative approximation.

Finally, it can be seen that after substituting the values of the known coefficients in (4.39), the solution $a_{2,2} = 0.500$ is identical to that obtained from (4.33a) using the finite difference method in time with $\varepsilon = \frac{2}{3}$.

**Table 4.1**

Elements of the Lagrangian and Serendipity Families

| Elements | Coordinate basis functions (let $\xi_0 = \xi\xi_i$, $\eta_0 = \eta\eta_i$) |
|---|---|
| | Lagrangian elements |
| $\eta=1$<br>$\xi=-1$ $\eta$ $\xi$ $\xi=1$<br>$\eta=0$<br>LINEAR | All Nodes<br>$\Omega_i^e = \frac{1}{4}(1+\xi_0)(1+\eta_0)$ |
| QUADRATIC | Corner Nodes<br>$\Omega_i^e = [\frac{1}{2}\xi_0(1+\xi_0)][\frac{1}{2}\eta_0(1+\eta_0)]$<br>Midside Nodes<br>$\xi_i = 0, \quad \Omega_i^e = (1-\xi^2)[\frac{1}{2}\eta_0(1+\eta_0)]$<br>$\eta_i = 0, \quad \Omega_i^e = (1-\eta^2)[\frac{1}{2}\xi_0(1+\xi_0)]$<br>Midelement Nodes<br>$\xi_i = 0, \quad \eta_i = 0, \quad \Omega_i^e = (1-\xi^2)(1-\eta^2)$ |
| CUBIC | Corner Nodes<br>$\Omega_i^e = [\frac{1}{16}(1+\xi_0)(9\xi^2-1)][\frac{1}{16}(1+\eta_0)(9\eta^2-1)]$<br>Midside Nodes<br>$\xi_i = \pm 1, \quad \eta_i = \pm\frac{1}{3}, \quad \Omega_i^e = [\frac{1}{16}(1+\xi_0)(9\xi^2-1)][\frac{9}{16}(1-\eta^2)(9\eta_0+1)]$<br>$\xi_i = \pm\frac{1}{3}, \quad \eta_i = \pm 1, \quad \Omega_i^e = [\frac{9}{16}(1-\xi^2)(9\xi_0+1)][\frac{1}{16}(1+\eta_0)(9\eta^2-1)]$<br>Midelement Nodes<br>$\eta_i = \pm\frac{1}{3}, \quad \xi_i = \pm\frac{1}{3}, \quad \Omega_i^e = [\frac{9}{16}(1-\xi^2)(9\xi_0+1)][\frac{9}{16}(1-\eta^2)(9\eta_0+1)]$ |
| | Serendipity elements |
| LINEAR | All Nodes<br>$\Omega_i^e = \frac{1}{4}(1+\xi_0)(1+\eta_0)$ |
| QUADRATIC | Corner Nodes<br>$\Omega_i^e = \frac{1}{4}(1+\xi_0)(1+\eta_0)(\xi_0+\eta_0-1)$<br>Midside Nodes<br>$\xi_i = 0, \quad \Omega_i^e = \frac{1}{2}(1-\xi^2)(1+\eta_0)$<br>$\eta_i = 0, \quad \Omega_i^e = \frac{1}{2}(1+\xi_0)(1-\eta^2)$ |
| CUBIC | Corner Nodes<br>$\Omega_i^e = \frac{1}{32}(1+\xi_0)(1+\eta_0)[-10+9(\xi^2+\eta^2)]$<br>Midside Nodes<br>$\xi_i = \pm 1, \quad \eta_i = \pm\frac{1}{3}, \quad \Omega_i^e = \frac{9}{32}(1+\xi_0)(1-\eta^2)(1+9\eta_0)$<br>$\xi_i = \pm\frac{1}{3}, \quad \eta_i = \pm 1, \quad \Omega_i^e = \frac{9}{32}(1+\eta_0)(1-\xi^2)(1+9\xi_0)$ |

### *4.6.3 Rectangular Finite Elements Based on Higher-Degree Polynomials*

Heretofore we have focused on elements that use linear coordinate functions along the element sides. One can also design a suite of elements employing higher-degree polynomial bases along the sides. When the general procedure outlined above for linear bases is used in conjunction with higher-degree polynomials, one generates elements that contain nodes on the interior as well as on the perimeter of the element. Because the basis functions used on this type of grid are members of the Lagrangian family of polynomials, these elements are denoted as Lagrangian elements.

In practice, the Lagrangian element is usually replaced by a serendipity element developed from the same degree polynomial. The serendipity element is characterized by a lack of nodal points in the interior of higher degree elements. Basis functions for elements of this family are most easily obtained through a combination of ingenuity and trial and error; one examines plausible combinations of functions until a set fulfilling the requirements of coordinate functions is obtained. It is, therefore, appropriate that this family of elements derives its name from the famous princes of Serendip who were noted for their chance discoveries [3]. It should be emphasized that these elements can be derived rigorously; it is just less tedious to derive them using reasonable guesses. As is demonstrated in the next section, they are very nearly as accurate as Lagrangian elements but require fewer nodes. Both types of elements up to cubic degree are illustrated in Table 4.1.

## 4.7 Relationship between the Finite Element and Finite Difference Methods

Before leaving this discussion of rectangular finite elements let us examine the possibility of gaining further insight into the finite element method using concepts generally associated with finite difference schemes. In this section two important principles will be demonstrated:

(1) Whereas in finite difference approximations of a differential equation the discrete representation of the equation is applied at a point, in the finite element method the discrete equations are applicable over a region.

(2) The finite element representation of the $x$ and $z$ derivatives at a node $i$ can be interpreted as simple finite difference formulas integrated over the $z$ and $x$ directions, respectively.

To illustrate these two concepts only two of several possible element configurations will be considered. Because first-order spatial derivatives will be encountered later in the chapters considering applications, a slightly more general equation than examined so far, the convective diffusion equation, is introduced at this point. This equation is written in two space dimensions as

$$Lc \equiv \frac{\partial c}{\partial t} + v\frac{\partial c}{\partial x} + w\frac{\partial c}{\partial z} - D\frac{\partial^2 c}{\partial x^2} - D\frac{\partial^2 c}{\partial z^2} + kc = 0 \qquad (4.40)$$

where $v$ and $w$ are the velocity components in the $x$ and $z$ directions, respectively, $D$ the diffusion coefficient, and $k$ the chemical rate constant. The physical significance of this equation will be discussed in a later section on simulating mass transport.

By applying Galerkin's method to the spatial derivatives one obtains the integral equations

$$\langle Lc, \phi_i(x, z)\rangle = 0 \qquad (i = 1, 2, \ldots, N) \qquad (4.41)$$

Combining (4.41) with the trial function

$$c(x, z, t) \simeq \sum_{j=1}^{N} C_j(t)\phi_j(x, z)$$

and applying Green's theorem to the resulting integral equations, we obtain

$$\sum_{j=1}^{N} \frac{dC_j}{dt}\langle\phi_j, \phi_i\rangle + \sum_{j=1}^{N} C_j\left[u\left\langle\frac{\partial\phi_j}{\partial x}, \phi_i\right\rangle + w\left\langle\frac{\partial\phi_j}{\partial z}, \phi_i\right\rangle + D\left\langle\frac{\partial\phi_j}{\partial x}, \frac{\partial\phi_i}{\partial x}\right\rangle + D\left\langle\frac{\partial\phi_j}{\partial z}, \frac{\partial\phi_i}{\partial z}\right\rangle + k\langle\phi_j, \phi_i\rangle\right] = 0 \qquad (4.42)$$

$$(i = 1, 2, \ldots, N)$$

where node $i$ is assumed to lie on the interior of the mesh.† This equation is analogous to those considered in earlier examples with the exception of the reaction term and the convection terms (terms involving velocity).

### 4.7.1 *Regular Bilinear Rectangular Grid*

As the first and simplest example, consider the approximating algebraic system obtained by combining (4.42) with the net illustrated in Fig. 4.6.‡.By suitably grouping elements of the resulting coefficient matrix as shown in Table 4.2, formulas containing classical finite difference approximations can be generated. Each difference expression is tabulated immediately below the

† Note that this assumption in effect removes surface integrals associated with natural boundary conditions from the equation.

‡ Note we have interchanged the symbols $z$ and $y$ in this example.

term it approximates. Because the time derivative has not been discretized a difference expression does not appear in the first column. When numerical differentiation does arise, the appropriate formulas resulting from classical Taylor series expansions are provided on the third row of the table. Thus one can determine a truncation error for the spatial approximation in the manner outlined in Chapter 2.

Careful examination of row one of Table 4.2 reveals that the coefficient multiplying each derivative, or its difference representation, can be interpreted as an integration formula. The numerical integration formulas represented by these weighting coefficients and their associated truncation error are shown in the third row of the table. Apparently the finite element approximation may be interpreted as a finite difference discretization in one spatial dimension integrated over the other. Moreover, the time derivative is integrated over both spatial dimensions.

The truncation error for each approximating scheme can be obtained by examining the error associated with differentiation and integration. In Table 4.2, for example, the truncation error is second order for the discretization and fourth order for integration resulting in a second-order accurate scheme, $O(\Delta x^2) + O(\Delta z^2)$. Thus the finite element formulation on a regular grid with linear basis functions appears to be of the same order of accuracy as the analogous finite difference approximation. However, examination of the overall truncation error associated with this finite element representation indicates that the second-order error associated with the convective term is cancelled out by the truncation error associated with the integration of the time derivative over space. Thus the finite element method gives a higher-order (fourth-order) approximation to the convective term than the finite difference method, but the error associated with the diffusive term is second order in both schemes.

### *4.7.2 Irregular, Bilinear, Rectangular Grid*

Analysis of the finite element approximation of the convective–diffusion equation on an irregular rectangular grid is summarized in Table 4.3. The truncation error for this grid configuration and linear basis functions is seen to be first order, $O(\Delta x) + O(\Delta z)$, which is of the same order as the finite difference truncation error on the same grid.

Whereas the two element configurations considered thus far involved only one nodal type, namely the corner node, the analysis of elements using higher-degree basis functions involves the introduction of additional types of nodes. The complexity of the analysis increases with the introduction of these nodes and is presented in detail by Gray and Pinder [4]. It may be

**Table 4.3**

Irregular Bilinear Rectangular Grid

| TERM | $\frac{\partial C}{\partial t}$ | $u\frac{\partial C}{\partial x}$ | $w\frac{\partial C}{\partial z}$ |
|---|---|---|---|
| Finite Element Representation | $\frac{\Delta x \Delta z}{36}\left\{\left[\frac{dC_{-1,1}}{dt} + 4\frac{dC_{0,1}}{dt} + \frac{dC_{1,1}}{dt}\right]\right.$<br>$+ 4\left[\frac{dC_{-1,0}}{dt} + 4\frac{dC_{0,0}}{dt} + \frac{dC_{1,0}}{dt}\right]$<br>$\left. + \left[\frac{dC_{-1,-1}}{dt} + 4\frac{dC_{0,-1}}{dt} + \frac{dC_{1,-1}}{dt}\right]\right\}$ | $u \cdot \frac{\Delta x \Delta z}{6}\left\{\left[\frac{C_{1,1}-C_{-1,1}}{2\Delta x}\right]\right.$<br>$+ 4\left[\frac{C_{1,0}-C_{-1,0}}{2\Delta x}\right]$<br>$\left. + \left[\frac{C_{1,-1}-C_{-1,-1}}{2\Delta x}\right]\right\}$ | $w\,\frac{\Delta x \Delta z}{6}\left\{\left[\frac{C_{1,1}-C_{1,-1}}{2\Delta z}\right]\right.$<br>$+ 4\left[\frac{C_{0,1}-C_{0,-1}}{2\Delta z}\right]$<br>$\left. + \left[\frac{C_{-1,1}-C_{-1,-1}}{2\Delta z}\right]\right\}$ |
| Numerical Differentiation Formulae Used | — | $\frac{\partial C_{0,k}}{\partial x} = \frac{C_{1,k}-C_{-1,k}}{2\Delta x} - \frac{1}{6}\Delta x^2 \frac{\partial^3 C_{0,k}}{\partial x^3}$<br>$k = -1,0,1$<br>$2^{nd}$ order accurate | $\frac{\partial C_{j,0}}{\partial z} = \frac{C_{j,1}-C_{j,-1}}{2\Delta z} - \frac{1}{6}\Delta z^2 \frac{{}^3C_{j,0}}{\Delta z^3}$<br>$j = -1,0,1$<br>$2^{nd}$ order accurate |
| Numerical Integration Formulae Used | $\int \frac{\partial C}{\partial t}\,dxdz = \frac{\Delta x \Delta z}{9}\left\{\left[\frac{\partial C_{-1,1}}{\partial t} + 4\frac{\partial C_{0,1}}{\partial t} + \frac{\partial C_{1,1}}{\partial t}\right]\right.$<br>$+ 4\left[\frac{\partial C_{-1,0}}{\partial t} + 4\frac{\partial C_{0,0}}{\partial t} + \frac{\partial C_{1,0}}{\partial t}\right]$<br>$\left. + \left[\frac{\partial C_{-1,-1}}{\partial t} + 4\frac{\partial C_{0,-1}}{\partial t} + \frac{\partial C_{1,-1}}{\partial t}\right]\right\}$<br>$- \frac{\Delta x \Delta z^5}{45}\frac{\partial^4}{\partial z^4}\left(\frac{\partial C}{\partial t}\right) - \frac{\Delta z \Delta x^5}{45}\frac{\partial^4}{\partial x^4}\left(\frac{\partial C}{\partial t}\right)$<br>$4^{th}$ order accurate<br>(2-dimensional Simpson's rule) | $\int \frac{\partial C_{0,k}}{\partial x}\,dz = \frac{\Delta z}{3}\left\{\frac{\partial C_{0,-1}}{\partial x}\right.$<br>$\left. + 4\frac{\partial C_{0,0}}{\partial x} + \frac{\partial C_{0,1}}{\partial x}\right\}$<br>$- \frac{\Delta z^5}{90}\frac{\partial^5 C}{\partial x \partial z^4}$<br>$4^{th}$ order accurate<br>(1-dimensional Simpson's rule) | $\int \frac{\partial^2 C}{\partial z^2}\,dx = \frac{\Delta x}{3}\left\{\frac{\partial^2 C_{-1,0}}{\partial z^2} + 4\frac{\partial^2 C_{0,0}}{\partial z^2}\right.$<br>$\left. + \frac{\partial^2 C_{1,0}}{\partial z^2}\right\} - \frac{\Delta x^5}{90}\frac{\partial^6 C}{\partial x^4 \partial z^2}$<br>$4^{th}$ order accurate<br>(1-dimensional Simpson's rule) |

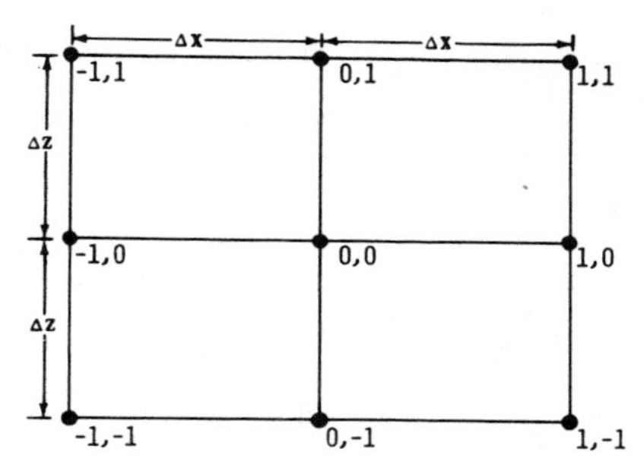


| $-D\frac{\partial^2 C}{\partial x^2}$ | $D\frac{\partial^2 C}{\partial z^2}$ | $kC$ |
|---|---|---|
| $-D\frac{\Delta x \Delta z}{6}\left\{\left[\frac{C_{1,1}-2C_{0,1}+C_{-1,1}}{\Delta x^2}\right]\right.$<br>$+4\left[\frac{C_{1,0}-2C_{0,0}+C_{-1,0}}{\Delta x^2}\right]$<br>$\left.+\left[\frac{C_{1,-1}-2C_{0,-1}+C_{-1,-1}}{\Delta x^2}\right]\right\}$ | $-D\frac{\Delta x \Delta z}{6}\left\{\left[\frac{C_{1,1}-2C_{1,0}+C_{1,-1}}{\Delta z^2}\right]\right.$<br>$+4\left[\frac{C_{0,1}-2C_{0,0}+C_{0,-1}}{\Delta z^2}\right]$<br>$\left.+\left[\frac{C_{-1,1}-2C_{-1,0}+C_{-1,-1}}{\Delta z^2}\right]\right\}$ | $k\frac{\Delta x \Delta z}{36}\left\{\left[C_{-1,1}+4C_{0,1}+C_{1,1}\right]\right.$<br>$+4\left[C_{-1,0}+4C_{0,0}+C_{1,0}\right]$<br>$\left.+\left[C_{-1,-1}+4C_{0,-1}+C_{1,-1}\right]\right\}$ |
| $\frac{\partial^2 C_{0,k}}{\partial x^2}=\frac{C_{1,k}-2C_{0,k}+C_{-1,k}}{\Delta x^2}-\frac{1}{12}\Delta x^2\frac{\partial^4 C}{\partial x^4}$<br>$k=-1,0,1$<br>$2^{nd}$ order accurate | $\frac{\partial^2 C_{j,0}}{\partial z^2}=\frac{C_{j,1}-2C_{j,0}+C_{j,-1}}{\Delta z^2}-\frac{1}{12}\Delta z^2\frac{\partial^4 C}{\partial z^4}$<br>$j=-1,0,1$<br>$2^{nd}$ order accurate | |
| $\int\frac{\partial^2 C_{0,k}}{\partial x^2}dz=\frac{\Delta z}{3}\left\{\frac{\partial^2 C_{0,-1}}{\partial x^2}+4\frac{\partial^2 C_{0,0}}{\partial x^2}\right.$<br>$\left.+\frac{\partial^2 C_{0,1}}{\partial x^2}\right\}-\frac{\Delta z^5}{90}\frac{\partial^6 C}{\partial x^2\partial z^4}$<br>$4^{th}$ order accurate<br>(1-dimensional Simpson's rule) | $\int\frac{\partial C_{j,0}}{\partial z}dx=\frac{\Delta x}{3}\left\{\frac{\partial C_{-1,0}}{\partial z}+4\frac{\partial C_{0,0}}{\partial z}\right.$<br>$\left.+\frac{\partial C_{1,0}}{\partial z}\right\}-\frac{\Delta x^5}{90}\frac{\partial^5 C}{\partial z\partial x^4}$<br>$4^{th}$ order accurate<br>(1-dimensional Simpson's rule) | $\int C\,dx\,dz=\frac{\Delta x\Delta z}{9}\left\{\left[C_{-1,1}+4C_{0,1}+C_{1,1}\right]\right.$<br>$+4\left[C_{-1,0}+4C_{0,0}+C_{1,0}\right]$<br>$\left.+\left[C_{-1,-1}+4C_{0,-1}+C_{1,-1}\right]\right\}$<br>$-\frac{\Delta x\Delta z^5}{45}\frac{\partial^4 C}{\partial z^4}-\frac{\Delta z\Delta x^5}{45}\frac{\partial^4 C}{\partial x^4}$ |

## Table 4.2

### Regular Bilinear Rectangular Grid

| Term | $\frac{\partial C}{\partial t}$ | $u\,\frac{\partial C}{\partial x}$ | $w\,\frac{\partial C}{\partial z}$ |
|---|---|---|---|
| Finite Element Representation | $\frac{\Delta x \Delta z}{36}\left\{\left[\frac{dC_{-1,1}}{dt} + 2(1+\beta)\frac{dC_{0,1}}{dt} + \beta\frac{dC_{1,1}}{dt}\right]\right.$<br>$+ 2(1+\alpha)\left[\frac{dC_{-1,0}}{dt} + 2(1+\beta)\frac{dC_{0,0}}{dt} + \beta\frac{dC_{1,0}}{dt}\right]$<br>$\left. + \alpha\left[\frac{dC_{-1,-1}}{dt} + 2(1+\beta)\frac{dC_{0,-1}}{dt} + \beta\frac{dC_{1,-1}}{dt}\right]\right\}$ | $u\,\frac{(\beta+1)\Delta x \Delta z}{12}\left\{\left[\frac{C_{1,1}-C_{-1,1}}{(\beta+1)\Delta x}\right]\right.$<br>$+ 2(\alpha+1)\left[\frac{C_{1,0}-C_{-1,0}}{(\beta+1)\Delta x}\right]$<br>$\left. + \alpha\left[\frac{C_{1,-1}-C_{-1,-1}}{(\beta+1)\Delta x}\right]\right\}$ | $w\,\frac{(\alpha+1)\Delta x \Delta z}{12}\left\{\beta\left[\frac{C_{1,1}-C_{1,-1}}{(\alpha+1)\Delta z}\right]\right.$<br>$+ 2(\beta+1)\left[\frac{C_{0,1}-C_{0,-1}}{(\alpha+1)\Delta z}\right]$<br>$\left. + \left[\frac{C_{-1,1}-C_{-1,-1}}{(\alpha+1)\Delta z}\right]\right\}$ |
| Numerical Differentiation Formulae Used | — | $\frac{\partial C_{0,k}}{\partial x} = \frac{C_{1,k}-C_{-1,k}}{(\beta+1)\Delta x}$<br>$- \frac{\partial^2 C_{0,k}}{\partial x^2}\frac{(\beta-1)\Delta x}{2}$<br>$k = -1, 0, 1$<br>1st order accurate | $\frac{\partial C_{j,0}}{\partial z} = \frac{C_{j,1}-C_{j,-1}}{(\alpha+1)\Delta z}$<br>$- \frac{\partial^2 C_{j,0}}{\partial z^2}\frac{(1-\alpha)\Delta x}{2}$<br>$j = -1, 0, 1$<br>1st order accurate |
| Numerical Integration Formulae Used | $\iint \frac{\partial C}{\partial t}\,dxdz = \frac{\Delta x \Delta z}{9}\left\{\left[\frac{\partial C_{-1,1}}{\partial t} + 2(1+\beta)\frac{\partial C_{0,1}}{\partial t} + \beta\frac{\partial C_{1,1}}{\partial t}\right]\right.$<br>$2(1+\alpha)\left[\frac{\partial C_{-1,0}}{\partial t} + 2(1+\beta)\frac{\partial C_{0,0}}{\partial t} + \beta\frac{\partial C_{1,0}}{\partial t}\right]$<br>$\left. + \alpha\left[\frac{\partial C_{-1,-1}}{\partial t} + 2(1+\beta)\frac{\partial C_{0,-1}}{\partial t} + \beta\frac{\partial C_{1,-1}}{\partial t}\right]\right\}$<br>$+ \frac{(1-\alpha^2)(1+\beta)\Delta x \Delta z^2}{6}\frac{\partial}{\partial z}\left(\frac{\partial C}{\partial t}\right)$<br>$+ \frac{(\beta^2-1)(1+\alpha)\Delta x^2 \Delta z}{6}\frac{\partial}{\partial x}\left(\frac{\partial C}{\partial t}\right)$<br>1st order accurate | $\int \frac{\partial C_{0,k}}{\partial x}\,dz = \frac{\Delta z}{3}\left\{\alpha\frac{\partial C_{0,-1}}{\partial x}\right.$<br>$\left. + 2(\alpha+1)\frac{\partial C_{0,0}}{\partial x} + \frac{\partial C_{0,1}}{\partial x}\right\}$<br>$+ \frac{(1-\alpha^2)\Delta z^2}{6}\frac{\partial^2 C}{\partial z \partial x}$<br>1st order accurate | $\int \frac{\partial C_{j,0}}{\partial z}\,dx = \frac{\Delta x}{3}\left\{\frac{\partial C_{-1,0}}{\partial z}\right.$<br>$\left. + 2(\beta+1)\frac{\partial C_{0,0}}{\partial z} + \beta\frac{\partial C_{1,0}}{\partial z}\right\}$<br>$+ \frac{(\beta^2-1)\Delta x^2}{6}\frac{\partial^2 C}{\partial x \partial z}$<br>1st order accurate |

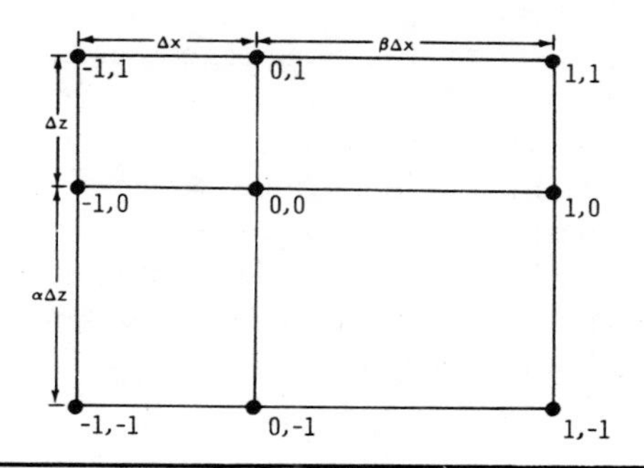


| $-D\frac{\partial^2 C}{\partial x^2}$ | $-D\frac{\partial^2 C}{\partial z^2}$ | $kC$ |
|---|---|---|
| $-D\frac{(\beta+1)\Delta x\Delta z}{12}\left\{\left[\frac{\frac{C_{1,1}-C_{0,1}}{\Delta x}-\frac{C_{0,1}-C_{-1,1}}{\beta\Delta x}}{(\beta+1)\Delta x/2}\right] + 2(\alpha+1)\left[\frac{\frac{C_{1,0}-C_{0,0}}{\Delta x}-\frac{C_{0,0}-C_{-1,0}}{\beta\Delta x}}{(\beta+1)\Delta x/2}\right] + \alpha\left[\frac{\frac{C_{1,-1}-C_{0,-1}}{\Delta x}-\frac{C_{0,-1}-C_{-1,-1}}{\beta\Delta x}}{(\beta+1)\Delta x/2}\right]\right\}$ | $-D\frac{(\alpha+1)\Delta x\Delta z}{12}\left\{\beta\left[\frac{\frac{C_{1,1}-C_{1,0}}{\Delta z}-\frac{C_{1,0}-C_{1,-1}}{\alpha\Delta z}}{(\alpha+1)\Delta z/2}\right] + 2(\beta+1)\left[\frac{\frac{C_{0,1}-C_{0,0}}{\Delta z}-\frac{C_{0,0}-C_{0,-1}}{\alpha\Delta z}}{(\alpha+1)\Delta z/2}\right] + \left[\frac{\frac{C_{-1,1}-C_{-1,0}}{\Delta z}-\frac{C_{-1,0}-C_{-1,-1}}{\alpha\Delta z}}{(\alpha+1)\Delta z/2}\right]\right\}$ | $\frac{k\Delta x\Delta z}{36}\left\{\left[C_{-1,1}+2(1+\beta)C_{0,1}+\beta C_{1,1}\right] + 2(1+\alpha)\left[C_{-1,0}+2(1+\beta)C_{0,0}+\beta C_{1,0}\right] + \alpha\left[C_{-1,-1}+2(1+\beta)C_{0,-1}+\beta C_{1,-1}\right]\right\}$ |
| $\frac{\partial^2 C_{0,k}}{\partial x^2} = \frac{\frac{C_{1,k}-C_{0,k}}{\Delta x}-\frac{C_{0,k}-C_{-1,k}}{\beta\Delta x}}{(\beta+1)\Delta x/2} - \frac{\partial^3 C_{0,k}}{\partial x^3}\frac{(\beta-1)\Delta x}{3}$ <br> $k = -1,0,1$ <br> 1st order accurate | $\frac{\partial^2 C_{j,0}}{\partial z^2} = \frac{\frac{C_{j,1}-C_{j,0}}{\Delta z}-\frac{C_{j,0}-C_{j,-1}}{\alpha\Delta z}}{(\alpha+1)\Delta z/2} - \frac{\partial^3 C_{j,0}}{\partial x^3}\frac{(\beta-1)\Delta x}{3}$ <br> $j = -1,0,1$ <br> 1st order accurate | — |
| $\int \frac{\partial^2 C_{0,k}}{\partial x^2}dz = \frac{\Delta z}{3}\left\{\alpha\frac{\partial^2 C_{0,-1}}{\partial x^2} + 2(\alpha+1)\frac{\partial^2 C_{0,0}}{\partial x^2} + \frac{\partial^2 C_{0,1}}{\partial x^2}\right\} + \frac{(1-\alpha^2)\Delta z^2}{6}\frac{\partial^3 C}{\partial z\partial x^2}$ <br> 1st order accurate | $\int \frac{\partial^2 C_{j,0}}{\partial z^2}dx = \frac{\Delta x}{3}\left\{\frac{\partial^2 C_{-1,0}}{\partial z^2} + 2(\beta+1)\frac{\partial^2 C_{0,0}}{\partial z^2} + \beta\frac{\partial^2 C_{1,0}}{\partial z^2}\right\} + \frac{(\beta^2-1)\Delta x^2}{6}\frac{\partial^3 C}{\partial z^2\partial x}$ <br> 1st order accurate | $\iint C\,dx\,dz = \frac{\Delta x\Delta z}{9}\left\{\left[C_{-1,1}+2(1+\beta)C_{0,1}+\beta C_{1,1}\right] + 2(1+\alpha)\left[C_{-1,0}+2(1+\beta)C_{0,0}+\beta C_{1,0}\right] + \alpha\left[C_{-1,-1}+2(1+\beta)C_{0,-1}+\beta C_{1,-1}\right]\right\} + \frac{(1-\alpha^2)(1+\beta)\Delta x\Delta z^2}{6}\frac{\partial C}{\partial z} + \frac{(\beta^2-1)(1-\alpha)\Delta x^2\Delta z}{6}\frac{\partial C}{\partial x}$ <br> 1st order accurate |

of particular interest to cite the following general conclusions which can be drawn from this analysis. While in most finite difference procedures the order of accuracy of the solution to a differential equation is the same at all nodes in the domain of the solution, a finite element solution generated on a grid with midside nodes will be more accurate at the corner nodes than at the midside nodes. Furthermore, the addition of midelement nodes to element grids that make use of second-order basis functions improves the accuracy of the solution only slightly.

## 4.8 Triangular Finite Elements in Two Space Dimensions

While rectangular finite elements are a natural extension of the one-dimensional formulation, they are not readily applicable to problems of irregular geometry. One approach to overcoming this difficulty is the development of a triangular element such as the one illustrated in Fig. 4.9. It

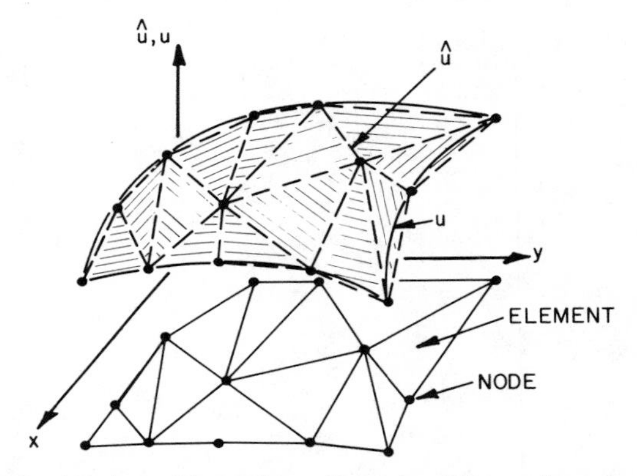


**Fig. 4.9.** Assemblage of two-dimensional linear elements.

is apparent that the triangular element can be used effectively not only to represent irregular boundaries, but also to concentrate coordinate functions in those regions of the domain where a rapidly varying solution is anticipated.

### *4.8.1 Basis Functions*

This section is concerned with the important concept of basis functions defined over triangles. The appropriate functions will be formulated using two avenues of approach. In the first approach the functions will be developed in the global $xy$ coordinates. While this development is attractive because of its conceptual simplicity, a second formulation using local coordinates is also presented to facilitate element integrations.

First consider the more general problem of generating basis functions over the element of Fig. 4.10. These functions must fulfill the requirements that they are unity at the node for which they are defined and zero at all other nodes. Moreover, in the case of linear triangles considered here, they must describe a plane over the triangular element. The general expression for such a plane is

$$\phi_i = a_i + b_i x + c_i y \tag{4.43}$$

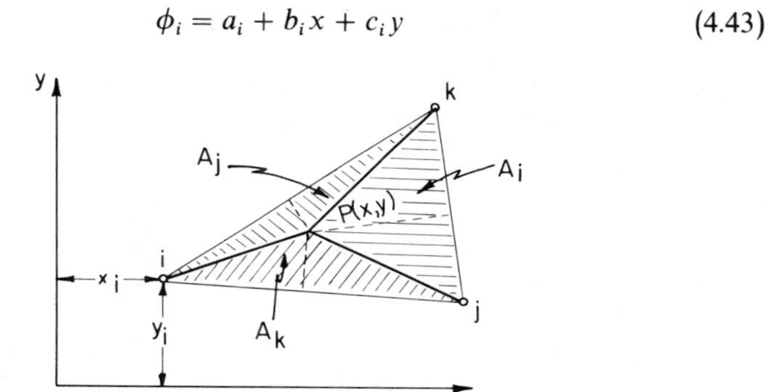


**Fig. 4.10.** Element illustrating area coordinates.

where $a_i$, $b_i$, and $c_i$ are constants identified with the $i$th basis function. From the definition and the indicated constraints, the following set of equations for the $i$th basis function may be written

$$\begin{Bmatrix} \phi_i(x_i, y_i) \\ \phi_i(x_j, y_j) \\ \phi_i(x_k, y_k) \end{Bmatrix} = \begin{Bmatrix} 1 \\ 0 \\ 0 \end{Bmatrix} = \begin{bmatrix} 1 & x_i & y_i \\ 1 & x_j & y_j \\ 1 & x_k & y_k \end{bmatrix} \begin{Bmatrix} a_i \\ b_i \\ c_i \end{Bmatrix} \tag{4.44}$$

where $i$, $j$, and $k$ refer to the element nodes in counterclockwise order, and the $3 \times 3$ matrix will be referred to as the matrix $[C]$. Solution of this equation yields

$$a_i = (x_j y_k - x_k y_j)/2A \tag{4.45a}$$

$$b_i = (y_j - y_k)/2A \tag{4.45b}$$

$$c_i = (x_k - x_j)/2A \tag{4.45c}$$

where $2A = \det[C]$ and is twice the area of the triangular element. This solution is easily verified through substitution in (4.44). The basis function sought for node $i$ thus becomes, from (4.43) and (4.45),

$$\phi_i = [(x_j y_k - x_k y_j) + (y_j - y_k)x + (x_k - x_j)y]/2A \tag{4.46a}$$

Analogous expressions for nodes $j$ and $k$ are readily obtained by cyclically permuting the indices $i$, $j$, and $k$ in (4.46a) for those nodes. The resulting expressions are

$$\phi_j = [(x_k y_i - x_i y_k) + (y_k - y_i)x + (x_i - x_k)y]/2A \tag{4.46b}$$

and

$$\phi_k = [(x_i y_j - x_j y_i) + (y_i - y_j)x + (x_j - x_i)y]/2A \tag{4.46c}$$

From these definitions it is apparent that another property of functions that are admissible as bases is also satisfied, namely

$$\phi_i + \phi_j + \phi_k = 1 \tag{4.47}$$

It should be pointed out that these functions are not independent since, given any two, the third is automatically specified through (4.47).

Given the bases defined by (4.46) there remains the formidable task of integrating products of these functions and their derivatives over the triangular finite element. To achieve this objective the bases of Eq. (4.46) will be rederived using area coordinates. In this local coordinate system the integrations required in the finite element method are carried out easily. The following development follows in part the formulation by Connor and Will [5].

Consider the triangular element illustrated in Fig. 4.10. Let $A$ denote the area of this triangle and $A_i$, $A_j$, $A_k$ the areas of the subtriangles such that $A = A_i + A_j + A_k$. The "triangular" area coordinates of point $P(x, y)$ are defined as $L_i \equiv A_i/A$, $L_j \equiv A_j/A$, and $L_k \equiv A_k/A$, and consequently $L_i + L_j + L_k = 1$. Moreover these functions possess the properties that $L_i = 0$ at nodes $i$ and $k$, $L_i = 1$ at node $i$, and $L_j = L_k = 0$ at node $i$. Thus $L$ fulfills the requirements outlined earlier for basis functions. To show that $L_i$, $L_j$, and $L_k$ are, in fact, identical to $\phi_i$, $\phi_j$, and $\phi_k$ in (4.46), respectively, the area of the triangular element is obtained using the vector cross product

$$2A = [(x_j - x_i)\hat{\mathbf{i}} + (y_j - y_i)\hat{\mathbf{j}}] \times [(x_k - x_i)\hat{\mathbf{i}} + (y_k - y_i)\hat{\mathbf{j}}] \cdot \hat{\mathbf{k}}$$

where $\hat{\mathbf{k}}$ is the unit vector normal to the triangle surface. This can also be written as the determinant of a $3 \times 3$ matrix:

$$2A = \begin{vmatrix} 1 & x_i & y_i \\ 1 & x_j & y_j \\ 1 & x_k & y_k \end{vmatrix}$$

Consider now the area $A_k$ which is written as

$$2A_k = \begin{vmatrix} 1 & x & y \\ 1 & x_i & y_i \\ 1 & x_j & y_j \end{vmatrix}$$

Expansion of the determinant yields

$$2A_k = x_i y_j - x_j y_i + (y_i - y_j)x + (x_j - x_i)y$$

The triangular coordinate $L_k$ thus becomes

$$L_k = A_k/A = [(x_i y_j - x_j y_i) + (y_i - y_j)x + (x_j - x_i)y]/2A \quad (4.48)$$

which is identical to $\phi_k$ as defined by (4.46c).

The question naturally arises as to why one would go to the effort of demonstrating that area coordinates are suitable basis functions for triangular elements. The answer lies in the ease with which functions defined in area coordinates can be integrated. An integration of the form

$$\int_A f(\phi_i, \phi_j, \phi_k)\, dA = \int_A f(L_i, L_j, L_k)\, dA$$

can be readily performed if the differential element $dA$ is expressed in local $(L_i, L_j, L_k)$ coordinates. It is well known from differential calculus that

$$dL_i\, dL_j = \det[J]\, dA$$

where $[J]$ is the Jacobian matrix defined as

$$[J] = \begin{bmatrix} \partial L_i/\partial x & \partial L_j/\partial x \\ \partial L_i/\partial y & \partial L_j/\partial y \end{bmatrix} = \begin{bmatrix} (y_j - y_k)/2A & (y_k - y_i)/2A \\ (x_k - x_j)/2A & (x_i - x_k)/2A \end{bmatrix}$$

and the determinant of $[J]$ is

$$\det[J] = (1/4A^2)[(y_j - y_k)(x_i - x_k) - (x_k - x_j)(y_k - y_i)] = 2A/4A^2 = 1/2A$$

Thus the transformation from a global to local integration element is

$$dA = 2A\, dL_i\, dL_j$$

It should be pointed out that only two area coordinates appear in the Jacobian because the third can always be expressed in terms of the other two. One can easily show that the choice of coordinates to use in defining $[J]$ has no influence on the form of this transformation relationship.

The typical integration indicated earlier can now be restated as

$$\int_A f(\phi_i, \phi_j)\, dA = \int_{L_i}\int_{L_j} f(\phi_i, \phi_j) 2A\, dL_i\, dL_j$$

$$= 2A \int_{L_j=0}^{L_j=1} \left( \int_{L_i=0}^{L_i=(1-L_j)} f(L_i, L_j)\, dL_i \right) dL_j$$

In this form integration is easily performed. By performing these integrations, some of the most common integrals can be shown to have the values

$$\int_A L_l\, dA = \tfrac{1}{3}A \qquad (l = i, j, k)$$

$$\int_A L_l^2\, dA = \tfrac{1}{6}A \qquad (l = i, j, k)$$

$$\int_A L_l L_m\, dA = \tfrac{1}{12}A \qquad (l = i, j, k; \quad m = i, j, k; \quad l \neq m)$$

$$\int_A \frac{\partial L_i}{\partial x}\frac{\partial L_j}{\partial x}\, dA = \frac{1}{4A}(y_j - y_k)(y_k - y_i)$$

$$\int_A \frac{\partial L_i}{\partial y}\frac{\partial L_j}{\partial y}\, dA = \frac{1}{4A}(x_k - x_j)(x_i - x_k)$$

In fact it is possible to integrate any polynomial in area coordinates using the simple relationship

$$\int_A L_l^{n_1} L_m^{n_2} L_p^{n_3}\, dA = 2A \frac{n_1!\, n_2!\, n_3!}{(n_1 + n_2 + n_3 + 2)!}$$

where $n_1$, $n_2$, $n_3$ are positive integers. Consider for example the integrals evaluated above:

$$\int_A L_l^1\, dA = 2A \frac{(1)!}{(3)!} = \frac{A}{3}$$

$$\int_A L_l^2\, dA = 2A \frac{(2)!}{(4)!} = \frac{A}{6}$$

$$\int_A L_l L_m\, dA = 2A \frac{(1)!\,(1)!}{(1 + 1 + 2)!} = \frac{A}{12}$$

Line integrals are easily evaluated using a natural coordinate system. For example, a function $f$ can be integrated along the side $\overline{ij}$ of the element illustrated in Fig. 4.10 in the following way. Along this side $f = f(L_j)$ [or equivalently $f = f(L_i)$] and the integral is of the form

$$\int_{(x_i, y_i)}^{(x_j, y_j)} f(L_j)\, ds = \int_0^1 f(L_j)\, ds/dL_j\, dL_j = \int_0^1 f(L_j) l_{ij}\, dL_j$$

$$= l_{ij} \int_0^1 f(L_j)\, dL_j$$

where $l_{ij} = [(x_i - x_j)^2 + (y_i - y_j)^2]^{1/2}$. Let the function $f(L_j)$ be defined such that $f = f_i L_i + f_j L_j$, where $f$ is specified at the nodes and varies linearly along the side $ij$. The line integral becomes

$$l_{ij} \int_0^1 (f_i L_i + f_j L_j)\, dL_j = l_{ij} f_i \int_0^1 (1 - L_j)\, dL_j + l_{ij} f_j \int_0^1 L_j\, dL_j$$

$$= l_{ij} f_i(\tfrac{1}{2}) + l_{ij} f_j(\tfrac{1}{2})$$

### 4.8.2 *A Triangular Finite Element Example*

To illustrate the application of triangular finite elements we will consider the system of equations

$$Lu \equiv \frac{\partial^2 u}{\partial x^2} + \frac{\partial^2 u}{\partial y^2} - Q = 0$$

$$u(0, y) = 1, \quad \frac{\partial u}{\partial x}(2, y) = 0, \quad \frac{\partial u}{\partial y}(x, 0) = 0, \quad \frac{\partial u}{\partial y}(x, 2) = 0$$

$$Q(x, y) = Q_w(1, 1)\, \delta(x - 1, y - 1)$$

where the sink $Q_w$ is of magnitude 1. The finite element net generated to solve this problem is illustrated in Fig. 4.11. The network consists of five nodes and four elements with the point sink located at the central node.

The integral equations for the problem are identical to those of the rectangular element example except for the number of equations which has been reduced from nine to five, i.e.,

$$\left\langle \frac{\partial^2 u}{\partial x^2} + \frac{\partial^2 u}{\partial y^2}, \phi_i(x, y) \right\rangle - \langle Q, \phi_i(x, y) \rangle = 0 \qquad (i = 1, 2, \ldots, 5) \quad (4.49)$$

Application of Green's theorem and substitution of the trial solution

$$u \simeq \hat{u}(x, y) = \sum_{j=1}^{5} a_j \phi_j$$

yield

$$\sum_{j=1}^{5}\left\{-a_j\left\langle\frac{\partial\phi_j}{\partial x}, \frac{\partial\phi_i}{\partial x}\right\rangle - a_j\left\langle\frac{\partial\phi_j}{\partial y}, \frac{\partial\phi_i}{\partial y}\right\rangle\right\}$$

$$+\int_\Gamma \frac{\partial\hat{u}}{\partial n}\phi_i\, ds - \langle Q, \phi_i\rangle = 0 \qquad (i = 1, \ldots, 5) \qquad (4.50)$$

where $\Gamma$ is the boundary of the domain and $\partial\hat{u}/\partial n$ the outward normal derivative to $\Gamma$.

The integrals that appear in (4.50) are most easily evaluated in the local coordinate system. The transformations are made elementwise, the integrations are performed over an element, and the results of these local integrations are assembled into the global coefficient matrix at the appropriate

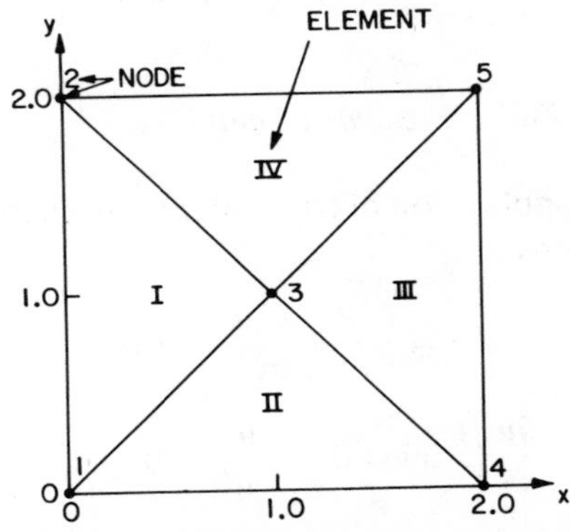


**Fig. 4.11.** Triangular finite element net for example problem.

locations. The information obtained from integration over one triangular element is initially stored in an element coefficient matrix. Because there are three nodes per element, each matrix will be of order three. A typical element coefficient matrix will be of the form

$$\begin{bmatrix}
\left\langle\frac{\partial\omega_1^e}{\partial x}, \frac{\partial\omega_1^e}{\partial x}\right\rangle + \left\langle\frac{\partial\omega_1^e}{\partial y}, \frac{\partial\omega_1^e}{\partial y}\right\rangle & \left\langle\frac{\partial\omega_2^e}{\partial x}, \frac{\partial\omega_1^e}{\partial x}\right\rangle + \left\langle\frac{\partial\omega_2^e}{\partial y}, \frac{\partial\omega_1^e}{\partial y}\right\rangle & \left\langle\frac{\partial\omega_3^e}{\partial x}, \frac{\partial\omega_1^e}{\partial x}\right\rangle + \left\langle\frac{\partial\omega_3^e}{\partial y}, \frac{\partial\omega_1^e}{\partial y}\right\rangle \\
\left\langle\frac{\partial\omega_1^e}{\partial x}, \frac{\partial\omega_2^e}{\partial x}\right\rangle + \left\langle\frac{\partial\omega_1^e}{\partial y}, \frac{\partial\omega_2^e}{\partial y}\right\rangle & \left\langle\frac{\partial\omega_2^e}{\partial x}, \frac{\partial\omega_2^e}{\partial x}\right\rangle + \left\langle\frac{\partial\omega_2^e}{\partial y}, \frac{\partial\omega_2^e}{\partial y}\right\rangle & \left\langle\frac{\partial\omega_3^e}{\partial x}, \frac{\partial\omega_2^e}{\partial x}\right\rangle + \left\langle\frac{\partial\omega_3^e}{\partial y}, \frac{\partial\omega_2^e}{\partial y}\right\rangle \\
\left\langle\frac{\partial\omega_1^e}{\partial x}, \frac{\partial\omega_3^e}{\partial x}\right\rangle + \left\langle\frac{\partial\omega_1^e}{\partial y}, \frac{\partial\omega_3^e}{\partial y}\right\rangle & \left\langle\frac{\partial\omega_2^e}{\partial x}, \frac{\partial\omega_3^e}{\partial x}\right\rangle + \left\langle\frac{\partial\omega_2^e}{\partial y}, \frac{\partial\omega_3^e}{\partial y}\right\rangle & \left\langle\frac{\partial\omega_3^e}{\partial x}, \frac{\partial\omega_3^e}{\partial x}\right\rangle + \left\langle\frac{\partial\omega_3^e}{\partial y}, \frac{\partial\omega_3^e}{\partial y}\right\rangle
\end{bmatrix}$$

Notice that the indices are local, in the sense that they pertain to nodes numbered counterclockwise on the triangle and do not exceed three. This matrix now contains all the information pertinent to element $e$ and, provided the global nodal numbers of the nodes of this element are known, it is straightforward to retrieve this information later for assembly of the global coefficient matrix.

Now consider the problem of how to obtain the element coefficient matrices for the example problem. Assuming linear basis functions, employ-

**Table 4.4**

| Matrix | Node number: Global | Node number: Element |
|---|---|---|
| Element I | | |
| $\begin{bmatrix} (\frac{1}{4}+\frac{1}{4}) & (-\frac{1}{2}+0) & (\frac{1}{4}-\frac{1}{4}) \\ (-\frac{1}{2}+0) & (1+0) & (-\frac{1}{2}+0) \\ (\frac{1}{4}-\frac{1}{4}) & (-\frac{1}{2}+0) & (\frac{1}{4}+\frac{1}{4}) \end{bmatrix}$ | 1<br>3<br>2 | 1<br>2<br>3 |
| Element II | | |
| $\begin{bmatrix} (\frac{1}{4}+\frac{1}{4}) & (-\frac{1}{4}+\frac{1}{4}) & (0-\frac{1}{2}) \\ (-\frac{1}{4}+\frac{1}{4}) & (\frac{1}{4}+\frac{1}{4}) & (0-\frac{1}{2}) \\ (0-\frac{1}{2}) & (0-\frac{1}{2}) & (0+1) \end{bmatrix}$ | 1<br>4<br>3 | 1<br>2<br>3 |
| Element III | | |
| $\begin{bmatrix} (\frac{1}{4}+\frac{1}{4}) & (\frac{1}{4}-\frac{1}{4}) & (-\frac{1}{2}+0) \\ (\frac{1}{4}-\frac{1}{4}) & (\frac{1}{4}+\frac{1}{4}) & (-\frac{1}{2}+0) \\ (-\frac{1}{2}+0) & (-\frac{1}{2}+0) & (1+0) \end{bmatrix}$ | 4<br>5<br>3 | 1<br>2<br>3 |
| Element IV | | |
| $\begin{bmatrix} (\frac{1}{4}+\frac{1}{4}) & (-\frac{1}{4}+\frac{1}{4}) & (0-\frac{1}{2}) \\ (-\frac{1}{4}+\frac{1}{4}) & (\frac{1}{4}+\frac{1}{4}) & (0-\frac{1}{2}) \\ (0-\frac{1}{2}) & (0-\frac{1}{2}) & (0+1) \end{bmatrix}$ | 5<br>2<br>3 | 1<br>2<br>3 |

ing the integration formulas, and at the same time recalling Fig. 4.11, one obtains the matrices shown in Table 4.4. The two main features common to these matrices are the symmetry and the fact that each row sums to zero. One can use this information to halve the computational effort and also to check the accuracy of the integrations. Unfortunately element coefficient matrices arising from other differential equations may not possess these characteristics.

The global coefficient matrix is obtained by summing, for a given global node, the contributions to that node from each element coefficient matrix. This task can be performed by systematically examining the global node numbers for each element matrix and cataloging the contribution from the element matrix in the global matrix. Following this operational procedure one obtains

$$\begin{bmatrix} (\frac{1}{2}+\frac{1}{2}) & 0 & (-\frac{1}{2}-\frac{1}{2}) & 0 & 0 \\ 0 & (\frac{1}{2}+\frac{1}{2}) & (-\frac{1}{2}-\frac{1}{2}) & 0 & 0 \\ (-\frac{1}{2}-\frac{1}{2}) & (-\frac{1}{2}-\frac{1}{2}) & (1+1+1+1) & (-\frac{1}{2}-\frac{1}{2}) & (-\frac{1}{2}-\frac{1}{2}) \\ 0 & 0 & (-\frac{1}{2}-\frac{1}{2}) & (\frac{1}{2}+\frac{1}{2}) & 0 \\ 0 & 0 & (-\frac{1}{2}-\frac{1}{2}) & 0 & (\frac{1}{2}+\frac{1}{2}) \end{bmatrix}$$

It is apparent that in this case the global coefficient matrix exhibits characteristics analogous to the element coefficient matrix: once again, however, this is not always the case.

The boundary conditions imposed on this system of equations lead to the matrix equation

$$\left[\begin{array}{cc|ccc} 1 & 0 & -1 & 0 & 0 \\ 0 & 1 & -1 & 0 & 0 \\ \hline -1 & -1 & 4 & -1 & -1 \\ 0 & 0 & -1 & 1 & 0 \\ 0 & 0 & -1 & 0 & 1 \end{array}\right] \begin{Bmatrix} a_1 \\ a_2 \\ a_3 \\ a_4 \\ a_5 \end{Bmatrix} - \left\{\begin{array}{c} \int (\partial\hat{u}/\partial n)\phi_1 \, ds \\ \int (\partial\hat{u}/\partial n)\phi_2 \, ds \\ \hline -Q_w \\ 0 \\ 0 \end{array}\right\} = 0$$

which, when partitioned for Dirichlet nodes, becomes

$$\begin{bmatrix} 4 & -1 & -1 \\ -1 & 1 & 0 \\ -1 & 0 & 1 \end{bmatrix} \begin{Bmatrix} a_3 \\ a_4 \\ a_5 \end{Bmatrix} = \begin{Bmatrix} -Q_w + a_1 + a_2 \\ 0 \\ 0 \end{Bmatrix} = \begin{Bmatrix} 1 \\ 0 \\ 0 \end{Bmatrix}$$

Because the coefficient matrix is small, the inverse of this matrix is used to obtain

$$\begin{Bmatrix} a_3 \\ a_4 \\ a_5 \end{Bmatrix} = \frac{1}{2} \begin{bmatrix} 1 & 1 & 1 \\ 1 & 3 & 1 \\ 1 & 1 & 3 \end{bmatrix} \begin{Bmatrix} 1 \\ 0 \\ 0 \end{Bmatrix} = \begin{Bmatrix} \frac{1}{2} \\ \frac{1}{2} \\ \frac{1}{2} \end{Bmatrix}$$

and the solution is

$$\{\hat{u}(0, 0), \hat{u}(0, 2), \hat{u}(1, 1), \hat{u}(2, 0), \hat{u}(2, 2)\} = \{1, 1, \tfrac{1}{2}, \tfrac{1}{2}, \tfrac{1}{2}\}$$

## 4.9 Use of Triangular Finite Elements in Two Space Dimensions for Transient Problems

### 4.9.1 *Finite Difference Representations of the Time Derivative*

The time-dependent analog of the previous example is described by the system of equations

$$Lu \equiv \frac{\partial^2 u}{\partial x^2} + \frac{\partial^2 u}{\partial y^2} - \frac{\partial u}{\partial t} - Q = 0$$

$$u(0, y, t) = 1, \quad \frac{\partial u}{\partial x}(2, y, t) = 0, \quad \frac{\partial u}{\partial y}(x, 0, t) = 0, \quad \frac{\partial u}{\partial y}(x, 2, t) = 0$$

$$u(x, y, 0) = 1 \quad (x > 0), \qquad Q(x, y) = Q_w(1, 1)\,\delta(x - 1, y - 1), \quad Q_w = 1$$

Applying Galerkin's method in conjunction with the trial function

$$u \simeq \hat{u}(x, y, t) = \sum_{j=1}^{5} a_j(t)\phi_j(x, y)$$

we obtain the matrix equation

$$[B]\{A\} + [C]\{dA/dt\} = \{F\}$$

where $[B]$ and $\{F\}$ are identical to the coefficient and right-hand side matrices obtained in the previous example and a typical element of $[C]$ would be

$$c_{i,j} = \langle \phi_i, \phi_j \rangle$$

which can be written in terms of element basis functions for element $e$ as

$$\begin{bmatrix} \langle \omega_1^e, \omega_1^e \rangle & \langle \omega_1^e, \omega_2^e \rangle & \langle \omega_1^e, \omega_3^e \rangle \\ \langle \omega_2^e, \omega_1^e \rangle & \langle \omega_2^e, \omega_2^e \rangle & \langle \omega_2^e, \omega_3^e \rangle \\ \langle \omega_3^e, \omega_1^e \rangle & \langle \omega_3^e, \omega_2^e \rangle & \langle \omega_3^e, \omega_3^e \rangle \end{bmatrix}$$

Use of the same relationship between global and element nodal numbers as in the previous section yields the following element coefficient matrix for $[C]$

which, because all elements have the same geometry and size is identical for all elements:

$$\begin{bmatrix} \frac{1}{6} & \frac{1}{12} & \frac{1}{12} \\ \frac{1}{12} & \frac{1}{6} & \frac{1}{12} \\ \frac{1}{12} & \frac{1}{12} & \frac{1}{6} \end{bmatrix}$$

| | Nodal number | | | | |
|---|---|---|---|---|---|
| | Global | | | | Local |
| Element | 1 | 2 | 3 | 4 | All |
| Node numbers | 1 | 1 | 4 | 5 | 1 |
| | 3 | 4 | 5 | 2 | 2 |
| | 2 | 3 | 3 | 3 | 3 |

The global coefficient matrix $[C]$ can now be assembled to give

$$\begin{bmatrix} (\frac{1}{6}+\frac{1}{6}) & \frac{1}{12} & (\frac{1}{12}+\frac{1}{12}) & \frac{1}{12} & 0 \\ \frac{1}{12} & (\frac{1}{6}+\frac{1}{6}) & (\frac{1}{12}+\frac{1}{12}) & 0 & \frac{1}{12} \\ (\frac{1}{12}+\frac{1}{12}) & (\frac{1}{12}+\frac{1}{12}) & (\frac{1}{6}+\frac{1}{6}+\frac{1}{6}+\frac{1}{6}) & (\frac{1}{12}+\frac{1}{12}) & (\frac{1}{12}+\frac{1}{12}) \\ \frac{1}{12} & 0 & (\frac{1}{12}+\frac{1}{12}) & (\frac{1}{6}+\frac{1}{6}) & \frac{1}{12} \\ 0 & \frac{1}{12} & (\frac{1}{12}+\frac{1}{12}) & \frac{1}{12} & (\frac{1}{6}+\frac{1}{6}) \end{bmatrix} = [C]$$

If an implicit finite difference approximation ($\varepsilon = 1$) is used in the time domain, substitution for $[B]$, $[C]$, and $\{F\}$ yields

$$\left[\begin{array}{cc|ccc} 1 & 0 & -1 & 0 & 0 \\ 0 & 1 & -1 & 0 & 0 \\ \hline -1 & -1 & 4 & -1 & -1 \\ 0 & 0 & -1 & 1 & 0 \\ 0 & 0 & -1 & 0 & 1 \end{array}\right] \left\{\begin{array}{c} a_{1,1} \\ a_{2,1} \\ \hline a_{3,1} \\ a_{4,1} \\ a_{5,1} \end{array}\right\}$$

$$+\frac{1}{12\,\Delta t}\left[\begin{array}{cc|ccc} 4 & 1 & 2 & 1 & 0 \\ 1 & 4 & 2 & 0 & 1 \\ \hline 2 & 2 & 8 & 2 & 2 \\ 1 & 0 & 2 & 4 & 1 \\ 0 & 1 & 2 & 1 & 4 \end{array}\right] \left\{\begin{array}{c} a_{1,1}-a_{1,0} \\ a_{2,1}-a_{2,0} \\ \hline a_{3,1}-a_{3,0} \\ a_{4,1}-a_{4,0} \\ a_{5,1}-a_{5,0} \end{array}\right\} = \left\{\begin{array}{c} \int (\partial\hat{u}/\partial n)\phi_1\, ds \\ \int (\partial\hat{u}/\partial n)\phi_2\, ds \\ \hline -Q_w \\ 0 \\ 0 \end{array}\right\} \tag{4.51}$$

where the second subscript on $a$ denotes the time level with "0" and "1" denoting the solution at the known and unknown time levels, respectively.

Introduction of the initial conditions $a_{i,0}$ and the boundary conditions $a_{1,1}$ and $a_{2,1}$, and selection of a time step $\Delta t = \frac{1}{12}$ yields

$$\begin{bmatrix} 12 & 1 & 1 \\ 1 & 5 & 1 \\ 1 & 1 & 5 \end{bmatrix} \begin{Bmatrix} a_{3,1} \\ a_{4,1} \\ a_{5,1} \end{Bmatrix} = \begin{bmatrix} 8 & 2 & 2 \\ 2 & 4 & 1 \\ 2 & 1 & 4 \end{bmatrix} \begin{Bmatrix} 1 \\ 1 \\ 1 \end{Bmatrix} + \begin{Bmatrix} 1 \\ 0 \\ 0 \end{Bmatrix}$$

where the matrices have been partitioned to exclude the known value of $a$ at time level 1. Calculation of the inverse matrix leads to the solution at $t = \Delta t = \frac{1}{12}$:

$$\begin{Bmatrix} a_{3,1} \\ a_{4,1} \\ a_{5,1} \end{Bmatrix} = \frac{1}{280} \begin{bmatrix} 24 & -4 & -4 \\ -4 & 59 & -11 \\ -4 & -11 & 59 \end{bmatrix} \begin{Bmatrix} 13 \\ 7 \\ 7 \end{Bmatrix} = \begin{Bmatrix} 0.914 \\ 1.01 \\ 1.01 \end{Bmatrix} = \begin{Bmatrix} \hat{u}(1, 1, \frac{1}{12}) \\ \hat{u}(2, 0, \frac{1}{12}) \\ \hat{u}(2, 2, \frac{1}{12}) \end{Bmatrix}$$

To calculate the second time step, i.e., $t = 2\,\Delta t = \frac{1}{6}$, the computed values at $t = \frac{1}{12}$ are substituted for the initial conditions in (4.51) and one obtains

$$\begin{Bmatrix} a_{3,1} \\ a_{4,1} \\ a_{5,1} \end{Bmatrix} = \frac{1}{280} \begin{bmatrix} 24 & -4 & -4 \\ -4 & 59 & -11 \\ -4 & -11 & 59 \end{bmatrix} \begin{Bmatrix} 12.34 \\ 6.88 \\ 6.88 \end{Bmatrix} = \begin{Bmatrix} 0.861 \\ 1.003 \\ 1.003 \end{Bmatrix} = \begin{Bmatrix} \hat{u}(1, 1, \frac{1}{6}) \\ \hat{u}(2, 0, \frac{1}{6}) \\ \hat{u}(2, 2, \frac{1}{6}) \end{Bmatrix}$$

Note that the second subscript does not denote the time level but rather distinguishes between the known and unknown levels. This solution may be extended in a stepwise fashion until the period of analysis has been simulated.

### 4.9.2 *Finite Element Representation of the Time Derivative*

The preceding example can be solved using a finite element, rather than finite difference, representation of the time derivative. As in the transient, one space dimension example presented earlier (Section 4.5.2), we assume a trial solution of the form

$$u(x, y, t) \simeq \hat{u}(x, y, t) = \sum_{j=1}^{5} a_j \phi_j(x, y, t)$$

The difficulty, however, is the formulation of a basis function which will provide an efficient numerical scheme. One successful approach is based on a columnar element with triangular cross section [6]. The time axis runs the length of the column and is subdivided into elements which may be linear, quadratic, or cubic. These elements are illustrated in Figs. 4.12a, 4.12b, and 4.12c, respectively.

The basis function can be considered as consisting of two parts, a spatial part and a temporal part. The spatial part is the well-known basis function for linear triangles presented earlier; the temporal part can be expressed as a Lagrangian polynomial in one dimension. For the $l$th time level this polynomial has the form

$$\beta_l = \prod_{r=1,\, r \neq l}^{m} (t_r - t)/(t_r - t_l) \qquad (m \leq 4) \tag{4.52}$$

where the $m$ possible subscripts for $t$ refer to the $m$ time positions in the

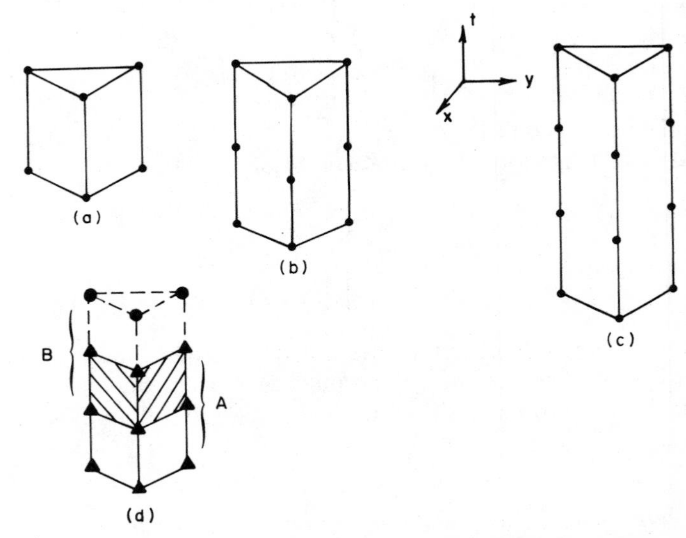


**Fig. 4.12.** Three-dimensional finite elements for solving transient problems in two space dimensions. The elements are (a) linear, (b) quadratic, and (c) cubic in time. Part (d) is an overlapping quadratic (quadratic lag 1), where element $A$ generates the solution at time $T$ and $B$ generates the solution at time $t + \Delta t$.

element (larger subscripts refer to larger times). The basis function for an element with $m$ different time locations becomes

$$\phi_j(x, y, t) = \Phi_j(x, y) \cdot \beta_j(t)$$

where $\beta_j$ is given by Eq. (4.52) and $\Phi_j(x, y)$ is the linear basis function for a triangle.

By formulating the basis functions in this way the integral equations can be simplified as in the earlier transient one-dimensional problem.

Specifically one can write the approximating equations as

$$\sum_{j=1}^{5} a_j \left[\left\{\left\langle \frac{\partial \Phi_j}{\partial x}(x, y), \frac{\partial \Phi_i}{\partial x}(x, y)\right\rangle + \left\langle \frac{\partial \Phi_j}{\partial y}(x, y), \frac{\partial \Phi_i}{\partial y}(x, y)\right\rangle\right\} \cdot \left\{\int_{D_t} \beta_i \beta_j \, dt\right\}\right]$$
$$+ \sum_{j=1}^{5} a_j \left[\{\langle \Phi_j(x, y), \Phi_i(x, y)\rangle\} \cdot \left\{\int_{D_t} \beta_i \frac{d\beta_j}{dt} dt\right\}\right]$$
$$+ \left[\{\langle Q, \Phi_i(x, y)\rangle\} + \left\{\int_{\Gamma} \Phi_i(x, y) \frac{\partial \hat{u}}{\partial n} ds\right\}\right] \cdot \left[\int_{D_t} \beta_i \, dt\right] = 0$$
$$(i = 1, 2, 3, 4, 5)$$

where $D_t$ is the time domain, and $Q$ and $\partial \hat{u}/\partial n$ are assumed to be independent of time. The integrals are all simple polynomials and therefore can be obtained readily as functions of the polynomial coefficients without resorting to numerical integration.†

There are several methods available for marching through the time domain. The most obvious is the case where $m = 2$ and linear basis functions in time are used. This is analogous to the procedure presented earlier for one space variable which was shown to be equivalent to a finite difference time discretization with $\varepsilon = 0.67$. The conceptual models for quadratic and cubic time approximations are direct extensions of the linear case. One should keep in mind, however, that quadratic time elements require the simultaneous solution of two time levels (the earliest level is assumed known) and consequently twice as many equations are involved. Similarly, three time levels must be considered simultaneously using cubic elements. The resulting larger systems of algebraic equations require, not only greater computer storage (assuming peripheral storage is not utilized), but also increased computational effort.

In an attempt to circumvent the disadvantages of solving for multiple time levels simultaneously, an overlapping element procedure has been developed [6]. This scheme involves the use of known time levels, not only at the base of the time element, but also at other levels heretofore assumed unknown. A typical element formulated in this way is illustrated in Fig. 4.12d. The elements effectively overlap as the simulation proceeds stepwise through time. This scheme provides a higher-order approximation, in a sense, while minimizing the number of unknown time levels and accordingly

† Numerical integration will be discussed in the next section.

the computational effort. This is denoted as a lagged finite element in time method.

Unfortunately, numerical experiments have shown that, in general, the optimal choice among finite difference, finite element, and lagged finite element approximations to the time derivative is problem and time-step dependent. It appears that none of the methods tested performed significantly better than the centered finite difference procedure. This observation along with the simplicity in application and theoretical development lead to the conclusion that the finite difference scheme in time is the best overall choice in the majority of transient finite element analyses.

## 4.10 Curved Isoparametric Elements

The discussion of finite elements has proceeded from the simple, one-dimensional elements to more complex two-dimensional formulations. It has been shown that the rectangular element is a direct extension of the line element, but is limited in application by the requirement of a regular net similar to classical finite difference schemes. By formulating a triangular element, it was possible to utilize an irregular net which not only provided an improved representation of irregular boundaries, but also facilitated the distribution of nodes within the domain of solution in a manner consistent with known information concerning the physical system. In this section, elements will be considered which, in addition to being formulated on an irregular grid, can also have curved sides.

The basic idea is to generate a simple element in one coordinate system, say a rectangle or triangle, and then distort this element to a more convenient shape in another curvilinear coordinate system. The general procedure is similar to changing from global to local coordinates to simplify integration—in fact it is precisely this in a slightly more generalized form. The principal obstacle is the formulation of a mapping procedure to transform from curvilinear to Cartesian coordinates.

### 4.10.1 *Curvilinear Coordinates*

Consider two coordinate systems, a local $(\xi, \eta)$ system and a global Cartesian system $(x, y)$. Visualize an element such as one of those illustrated in Fig. 4.13, that is irregularly shaped in $(x, y)$ but reduces to a square in $(\xi, \eta)$. Furthermore, let the square in $(\xi, \eta)$ be such that its corners are located at $\xi = \pm 1, \eta = \pm 1$. Similarly to the one-dimensional local coordinates introduced earlier, the relationship between the global and local coordinates can be established by introducing a general expression of the form [7]

$$x = \Omega_1^e x_1^e + \Omega_2^e x_2^e + \Omega_3^e x_3^e + \cdots = \sum_{j=1}^{N} \Omega_j^e x_j^e$$

$$y = \Omega_1^e y_1^e + \Omega_2^e y_2^e + \Omega_3^e y_3^e + \cdots = \sum_{j=1}^{N} \Omega_j^e y_j^e \tag{4.53}$$

where $\Omega_j^e$ are some as yet undetermined functions of $\eta$ and $\xi$ and $x_j^e$, $y_j^e$ are nodal coordinates in $x$ and $y$. Because adjacent elements must fit together, their sides must be uniquely determined by their common points. The simplest element, Fig. 4.13a, will have straight sides defined by a linear function,

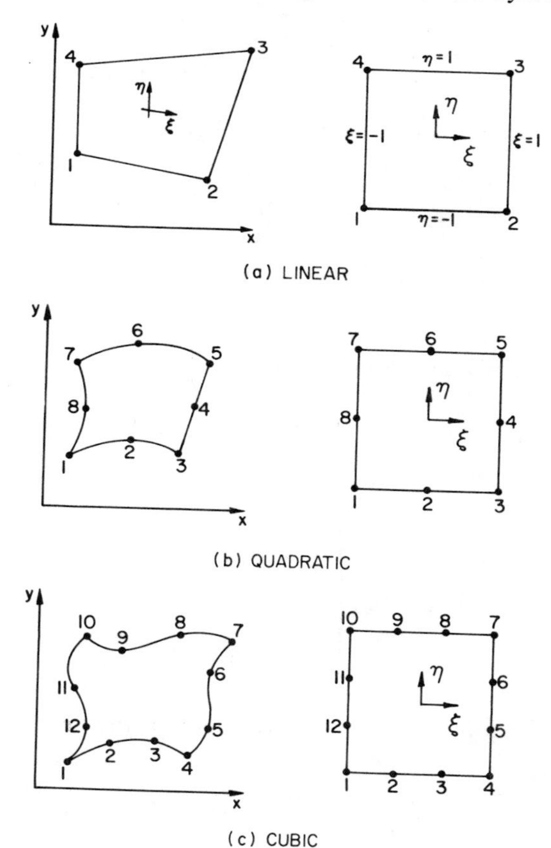


**Fig. 4.13.** (a) Linear, (b) quadratic and (c) cubic isoparametric elements in local and global coordinates.

while the elements with curved sides will require quadratic or cubic transformation functions in $\xi$ and $\eta$.

To obtain values for $\Omega_j^e$ that will provide the required variation along element sides, we write a general polynomial for each element type. For example, a linear element such as appears in Fig. 4.13a would require a polynomial of the form

$$x = a + b\xi + c\eta + d\xi\eta \tag{4.54}$$

which is linear in $\xi$ when $\eta = \pm 1$ and linear in $\eta$ when $\xi = \pm 1$. It is apparent from Fig. 4.13 that the following requirements must be met:

$$\begin{aligned} x &= x_1 && \text{when} \quad \xi = \eta = -1 \\ x &= x_2 && \text{when} \quad \xi = 1, \quad \eta = -1 \\ x &= x_3 && \text{when} \quad \xi = \eta = 1 \\ x &= x_4 && \text{when} \quad \xi = -1, \quad \eta = 1 \end{aligned}$$

Substitution of these constraints into (4.54) yields

$$\begin{Bmatrix} x_1 \\ x_2 \\ x_3 \\ x_4 \end{Bmatrix} = \begin{bmatrix} 1 & \xi_1 & \eta_1 & \xi_1\eta_1 \\ 1 & \xi_2 & \eta_2 & \xi_2\eta_2 \\ 1 & \xi_3 & \eta_3 & \xi_3\eta_3 \\ 1 & \xi_4 & \eta_4 & \xi_4\eta_4 \end{bmatrix} \begin{Bmatrix} a \\ b \\ c \\ d \end{Bmatrix} = \begin{bmatrix} 1 & -1 & -1 & 1 \\ 1 & 1 & -1 & -1 \\ 1 & 1 & 1 & 1 \\ 1 & -1 & 1 & -1 \end{bmatrix} \begin{Bmatrix} a \\ b \\ c \\ d \end{Bmatrix} = [A]\{\gamma\} \tag{4.55}$$

which may be readily solved for $\{\gamma\}$ as

$$\{\gamma\} = [A]^{-1}\{x\} \tag{4.56}$$

where

$$[A]^{-1} = \frac{1}{4} \begin{bmatrix} 1 & 1 & 1 & 1 \\ -1 & 1 & 1 & -1 \\ -1 & -1 & 1 & 1 \\ 1 & -1 & 1 & -1 \end{bmatrix} \tag{4.57}$$

Comparison of Eq. (4.53) and (4.54) reveals that the functions $\Omega_j^e$ must be specified by

$$\{\Omega^e\}^{\mathrm{T}}\{x\} = \{1, \xi, \eta, \xi\eta\}\{\gamma\} = \{1, \xi, \eta, \xi\eta\}[A]^{-1}\{x\}$$

which requires that

$$\{\Omega^e\}^{\mathrm{T}} = \{1, \xi, \eta, \xi\eta\}[A]^{-1}$$

Through expansion of the right-hand side, this expression becomes

$$\begin{aligned}
\Omega_1^e &= \tfrac{1}{4}(1 - \xi - \eta + \xi\eta) = \tfrac{1}{4}(1 - \xi)(1 - \eta) \\
\Omega_2^e &= \tfrac{1}{4}(1 + \xi - \eta - \xi\eta) = \tfrac{1}{4}(1 + \xi)(1 - \eta) \\
\Omega_3^e &= \tfrac{1}{4}(1 + \xi + \eta + \xi\eta) = \tfrac{1}{4}(1 + \xi)(1 + \eta) \\
\Omega_4^e &= \tfrac{1}{4}(1 - \xi + \eta - \xi\eta) = \tfrac{1}{4}(1 - \xi)(1 + \eta)
\end{aligned} \tag{4.58}$$

Equations (4.58) are identical to the basis functions defined for a rectangular element as presented in Table 4.1. Thus the polynomials used to define the coordinate transformation also fulfill the requirements of bases defined in local coordinates. It can be shown [3, 7] that bases defined in this way satisfy the two criteria necessary for convergence:

(1) that any required derivative of the unknown function can be adequately reproduced on an element (that is analogous to requiring $\sum \Omega_j^e = 1$) and

(2) that the function be continuous between adjacent elements.

Appropriate transformation and basis functions can be generated for quadratic and higher-degree elements using procedures analogous to those presented above. The resulting functions for either Lagrangian or serendipity elements are as shown in Table 4.1.

### 4.10.2 *Mixed, Curved, Isoparametric Elements*

Although the isoparametric element provides considerable flexibility, one is still required to design a mesh that uses elements of only one kind. To allow the transition from, say, linear elements in one region to quadratic elsewhere, a transition element is required. The mixed, isoparametric element is just such a mechanism [7a]. This element characteristically has different sides described by polynomials of different degrees such as illustrated in Fig. 4.14. The basis functions are somewhat more complex than those

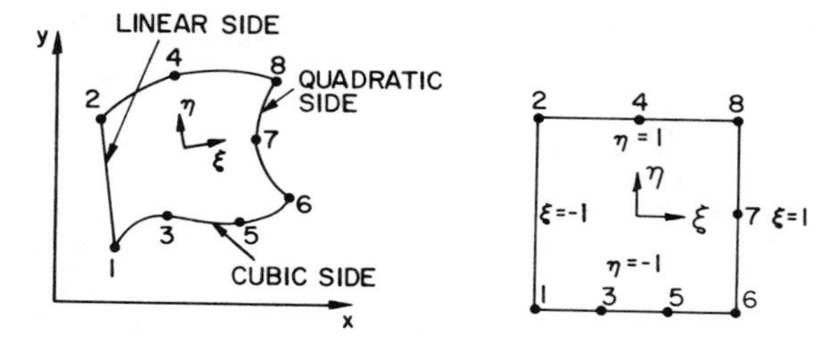


**Fig. 4.14.** Mixed, curved isoparametric elements.

specified thus far. Formulas for the side nodes are the same as given in Table 4.1 but the corner nodes require the bases of adjacent sides to be of different degrees. Accordingly, basis functions and coordinate transformation functions at corner nodes are defined as

$$\Omega_i^e(\xi, \eta) = \beta_i \sigma_i$$

where $\beta_i = \frac{1}{4}(1 + \xi_0)(1 + \eta_0)$, $\xi_0 = \xi\xi_i$, $\eta_0 = \eta\eta_i$, and $\sigma_i = \sigma_\xi + \sigma_\eta$. Values of $\sigma_\xi$ and $\sigma_\eta$ are determined by the order of the element sides adjacent to the corner node. These parameters are given in Table 4.5. Thus for node 1 in Fig. 4.14, $\Omega_1^e = \frac{1}{4}(1 - \xi)(1 - \eta)(\frac{9}{8}\xi^2 - \frac{5}{8} + \frac{1}{2})$.

**Table 4.5**

Parameters of $\sigma_\xi$ and $\sigma_\eta$

| Side | $\sigma_\xi$ | $\sigma_\eta$ |
|---|---|---|
| Linear | $\frac{1}{2}$ | $\frac{1}{2}$ |
| Quadratic | $\xi_0 - \frac{1}{2}$ | $\eta_0 - \frac{1}{2}$ |
| Cubic | $\frac{9}{8}\xi^2 - \frac{5}{8}$ | $\frac{9}{8}\eta^2 - \frac{5}{8}$ |

### *4.10.3 Jacobian Matrix*

To perform the integrations over isoparametric elements, it is necessary to transform derivatives in global coordinates to local coordinates. This is achieved through the use of the Jacobian matrix which relates derivatives in the two coordinate systems, i.e.,

$$\begin{Bmatrix} \partial\Omega_j^e/\partial\xi \\ \partial\Omega_j^e/\partial\eta \end{Bmatrix} = \begin{bmatrix} \partial x/\partial\xi & \partial y/\partial\xi \\ \partial x/\partial\eta & \partial y/\partial\eta \end{bmatrix} \begin{Bmatrix} \partial\Omega_j^e/\partial x \\ \partial\Omega_j^e/\partial y \end{Bmatrix} = [J] \begin{Bmatrix} \partial\Omega_j^e/\partial x \\ \partial\Omega_j^e/\partial y \end{Bmatrix}$$

where $[J]$ is the Jacobian matrix. Differentiation of (4.53) with respect to the local coordinates reveals that $[J]$ can be easily evaluated numerically from the relationship

$$[J] = \begin{bmatrix} \partial\Omega_1^e/\partial\xi & \partial\Omega_2^e/\partial\xi & \cdots & \partial\Omega_M^e/\partial\xi \\ \partial\Omega_1^e/\partial\eta & \partial\Omega_2^e/\partial\eta & \cdots & \partial\Omega_M^e/\partial\eta \end{bmatrix} \begin{bmatrix} x_1 & y_1 \\ x_2 & y_2 \\ \vdots & \vdots \\ x_M & y_M \end{bmatrix}$$

where $M$ is the number of nodes in the element.

In addition to transforming the derivatives from $(x, y)$ to $(\xi, \eta)$ the area of the element must be changed using

$$dx\,dy = \det[J]\,d\eta\,d\xi$$

and also the limits of integration over each element must be modified to $-1$ and 1. Having made the above substitutions it is now possible to perform the integration required for the coefficient matrix. Fortunately this can be easily accomplished using Gaussian quadrature.

### *4.10.4 Numerical Integration by Gaussian Quadrature*

In the Gaussian quadrature technique a polynomial of degree $(2n - 1)$ is integrated exactly using a weighted mean of $n$ particular values of the polynomial evaluated at particular points. Specifically the definite integral

**Table 4.6**

Tabulated Truncated Values of Gauss Points and Weighting Factors

$$\int_{-1}^{1} f(x)\,dx = \sum_{j=1}^{n} H_j f(x_j)$$

| $\pm x$ | n | H |
|---|---|---|
| 0.577350269189626 | 2 | 1.00 |
| 0.774596669241483 | 3 | 0.555555555555556 |
| 0.000000000000000 | | 0.888888888888889 |
| 0.861136311594053 | 4 | 0.347854845137454 |
| 0.339981043584856 | | 0.652145154862546 |

$\int_{-1}^{1} f(x)\,dx$ is replaced by a finite series $\sum_{j=1}^{n} H_j f(x_j)$, where $H_j$ are the weighting coefficients, $f(x_j)$ the value of the function at the specified point $x_j$, and $n$ the number of Gauss points used for integration.

Consider, for example, the integral

$$\int_{-1}^{1} f(x)\,dx = \int_{-1}^{1} (x^3 + x^2 + x + 1)\,dx \tag{4.59}$$

This integral may be evaluated in conjunction with the tabulated Gauss point locations and weighting factors for $n = 2$ in Table 4.6. With

$x_1 = -0.577$, $x_2 = 0.577$, and $H = 1.00$, integral (4.59) becomes

$$\int_{-1}^{1} f(x)\,dx = \sum_{j=1}^{2} H_j f(x_j)$$
$$= (1.00)f(-0.577) + (1.00)f(0.577)$$
$$= 0.564 + 2.10 = 2.66$$

which is exact to three significant figures. An exact integration should have been anticipated because two Gauss points will exactly integrate a third-order polynomial.

The two-dimensional analog of this procedure requires the evaluation of an integral of the form

$$I = \int_{-1}^{1} \int_{-1}^{1} f(x, y)\,dx\,dy$$

The integration proceeds stepwise by first considering the inner integral, holding $y$ constant,

$$\int_{-1}^{1} f(x, y)\,dx = \sum_{j=1}^{n} H_j f(x_j, y) = \psi(y)$$

The outer integral can now be evaluated as

$$\int_{-1}^{1} \psi(y)\,dy = \sum_{i=1}^{n} H_i \psi(y_i) = \sum_{i=1}^{n} H_i \sum_{j=1}^{n} H_j f(x_j, y_i)$$
$$= \sum_{i=1}^{n} \sum_{j=1}^{n} H_j H_i f(x_j, y_i)$$

where for $n = 3$, a polynomial up to the fifth power in both $x$ and $y$ can be integrated exactly.

It is apparent that the computational effort in numerical integration is proportional, in two dimensions, to $n^2$ and, in three dimensions, to $n^3$. Accordingly, the accuracy of the numerical scheme required to assure convergence of the finite element method is of considerable importance. Convergence is assured if the integration is sufficient to evaluate exactly the area or volume of the element [3]. Because the volume integration requires evaluation of the Jacobian determinant the required minimum number of Gauss points can be determined by examining the derivatives in the determinant. In a two-dimensional quadratic element, for example, the determinant will be a cubic expression which requires a two-point Gaussian formulation as a minimum. Our experience, however, indicates that while this criterion may guarantee convergence in the limit there are many problems that require more exact integration of all functions to obtain acceptable accuracy using a reasonable number of nodes. Clearly, when only a few elements are used to approximate a region, a more accurate integration is justified than when a large number of smaller elements are used.

### *4.10.5 Curved Triangular Elements*

While arbitrarily shaped quadrilaterals are sometimes considered to be better mesh units than triangles, triangles with curved sides are also used, due to the simplification of the mesh generation [8]. A typical curved triangular element is illustrated in Fig. 4.15 in both global $(x, y)$ and local $(\zeta_i)$ space.

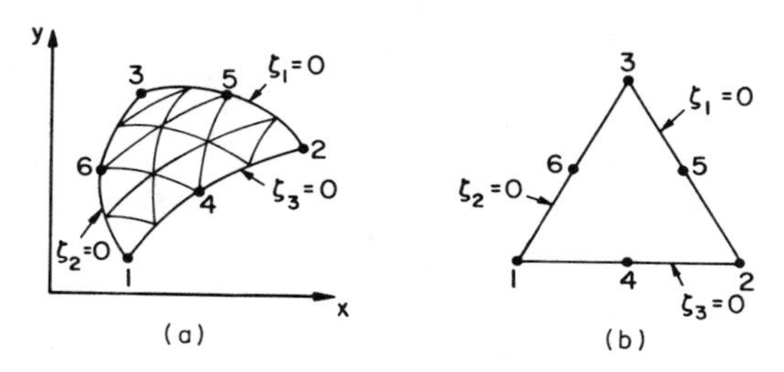


**Fig. 4.15.** General curved quadratic element in (a) global and (b) local space.

In area coordinates, the basis functions for the quadratic triangular element are of the form $\phi_k = L_k(2L_k - 1)$ for corner nodes and $\phi_k = 4L_{k-1}L_{k+1}$ for side nodes, where $k$ refers to the local element indices numbered sequentially in a counterclockwise order around the element. Thus, for regular element geometry, the simple relationship for integration, given earlier for linear triangles, may be used. For isoparametric elements, however, numerical integration is required.

Whereas numerical integration for quadrilaterals is relatively straightforward, several formulas have been derived for integration over a triangular area. The advantages and disadvantages of each are discussed by Cowper [9], who corrects formulas reported by Felippa [10] and Zienkiewicz [3], and introduces Gaussian quadrature formulas of his own. The formulas are applied in a fashion analogous to our earlier development, but using an equilateral triangle for the local integration. An arbitrary triangle can be mapped onto the equilateral triangle by a linear transformation. The appropriate Gauss points and weighting functions are tabulated by Cowper [9] and selected values appear in Table 4.7.

**Table 4.7**

Tabulated Truncated Values of Gauss Points and Weighting Factors for Triangles[a]

$$(1/A) \iint_A f\,dA = \sum_{j=1}^{N} H_j f(L_{1j}, L_{2j}, L_{3j})$$

| Number of Gauss points | Coordinates | | | Weighting factor $H$ |
|---|---|---|---|---|
| | $L_1$ | $L_2$ | $L_3$ | |
| 3 | 0.66666667 | 0.16666667 | 0.16666667 | 0.33333333 |
| | 0.16666667 | 0.66666667 | 0.16666667 | 0.33333333 |
| | 0.16666667 | 0.16666667 | 0.66666667 | 0.33333333 |
| 4 | 0.60000000 | 0.20000000 | 0.20000000 | 0.52083333 |
| | 0.20000000 | 0.60000000 | 0.20000000 | 0.52083333 |
| | 0.20000000 | 0.20000000 | 0.60000000 | 0.52083333 |
| | 0.33333333 | 0.33333333 | 0.33333333 | −0.56250000 |
| 6 | 0.81684757 | 0.09157621 | 0.09157621 | 0.10995174 |
| | 0.09157621 | 0.81684757 | 0.09157621 | 0.10995174 |
| | 0.09157621 | 0.09157621 | 0.81684757 | 0.10995174 |
| | 0.10810302 | 0.44594849 | 0.44594849 | 0.22338159 |
| | 0.44594849 | 0.10810302 | 0.44594849 | 0.22338159 |
| | 0.44594849 | 0.44594849 | 0.10810302 | 0.22338159 |

[a] From Cowper [9].

### 4.10.6 *Isoparametric Hermitian Finite Elements*

Hermitian polynomials were introduced in Section 4.3 to solve a problem in one space dimension. The same general approach can be used in two dimensions and has recently been extended to the consideration of isoparametric elements [11] (see Fig. 4.16). In the most general case for rectangles, the bases and transformation functions are generated as the product of the well-known one-dimensional expressions given in Section 4.3. This formulation results in four unknowns per node $i$; $u_i$, $\partial u_i/\partial x$, $\partial u_i/\partial y$, $\partial^2 u_i/\partial x\, \partial y$. It is also possible, however, to obtain highly accurate basis functions which are also useful with isoparametric rectangles that require only three unknowns per node, i.e. $u_i$, $\partial u_i/\partial x$, $\partial u_i/\partial y$. These bases still preserve many of the attractive features of the Hermitian approximation. Because these functions are formulated in a manner analogous to the serendipity isoparametric elements they are denoted as serendipity-type Hermitian basis functions. These functions are presented in Figs. 4.17a–c.

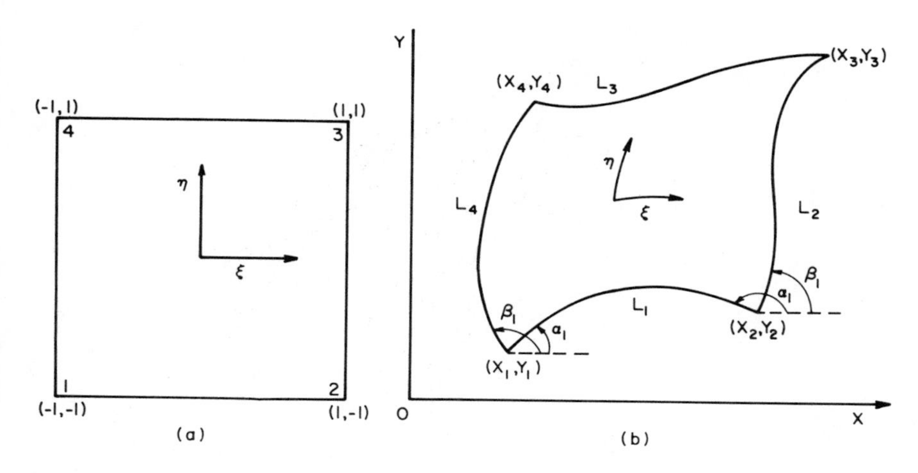


**Fig. 4.16.** Deformed isoparametric Hermitian elements in (a) local and (b) global coordinates.

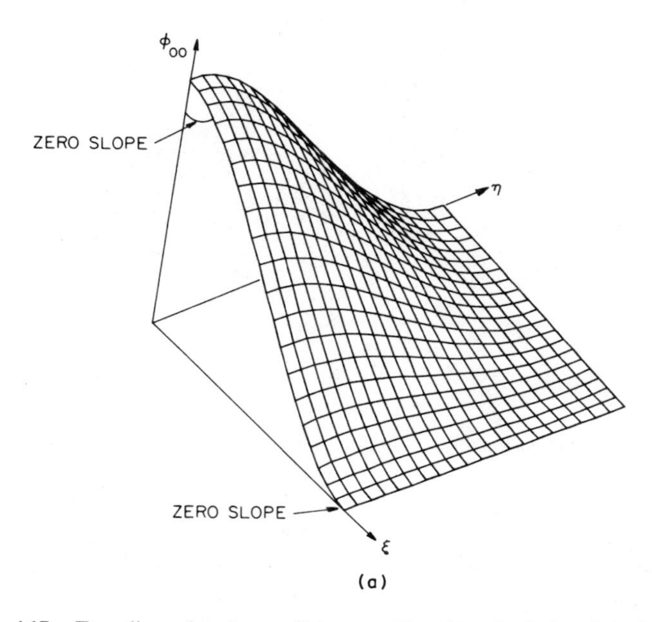


**Fig. 4.17a.** Two-dimensional serendipity-type Hermitian basis function: (a) $\phi_{00}$.

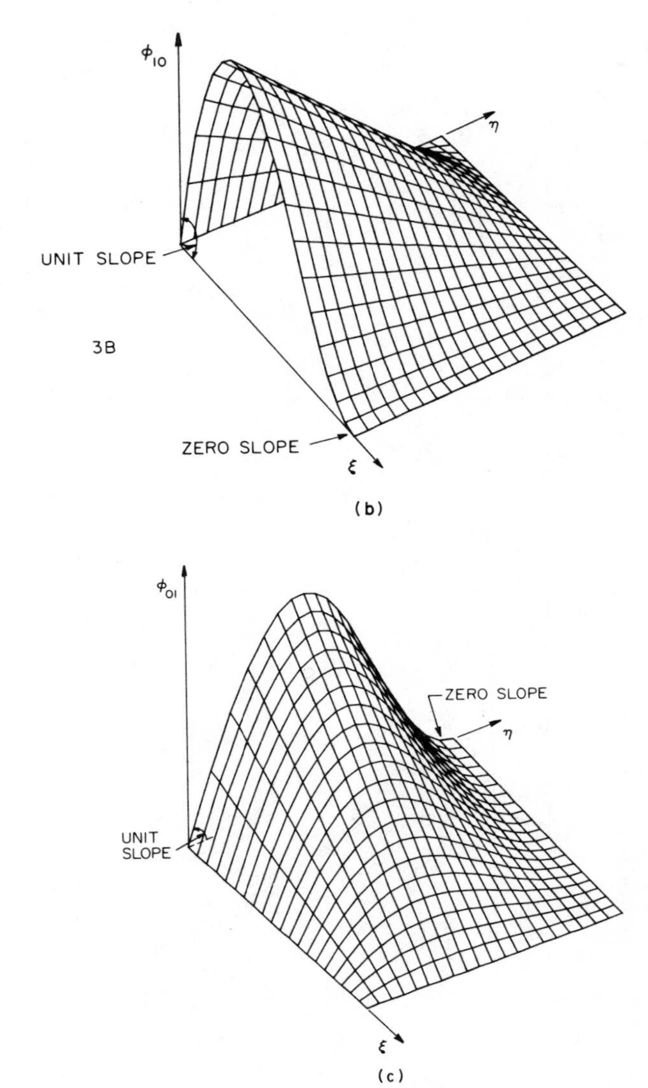


**Fig. 4.17b, c.** Two-dimensional serendipity-type Hermitian basis function: (b) $\phi_{10}$; (c) $\phi_{01}$.

In numerical experiments the Hermitian scheme was found to require 25–40% fewer degrees of freedom to obtain the same accuracy as the zero-order continuous cubic elements in general use. It should be pointed out, however, that whenever an improved approximation of a continuous normal gradient across interelement boundaries is desired, the fourth unknown

$\partial^2 u_i/\partial x\, \partial y$ should be represented. In this case the Hermitian element is somewhat less attractive requiring only on the order of 25% fewer degrees of freedom as compared to the zero order continuous cubic.

The reader interested in utilizing the Hermitian isoparametric approach should refer to the paper by Van Genuchten *et al.* [11] for details regarding the transformation procedures.

## 4.11 Three-Dimensional Elements

The formulation and application of three-dimensional elements is a direct extension of the two-dimensional case. In fact, one type of three-dimensional element has already been considered in the discussion of the finite element approximation to the time derivative. This element, formulated for two space dimensions and time, is a prism (see Fig. 4.12).

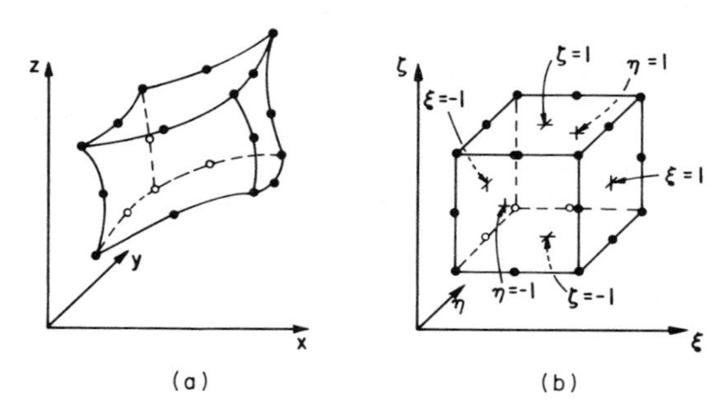


**Fig. 4.18.** Three-dimensional isoparametric element in (a) global and (b) local space.

The three-dimensional equivalent to the isoparametric quadrilateral is illustrated in Fig. 4.18. The basis functions in three dimensions are analogous to their two-dimensional counterpart and are given in [3] Table 4.8. It is also possible to apply the mixed element formulation in three dimensions.

The three-dimensional equivalent of a triangle is a tetrahedron, which is illustrated in Fig. 4.19. The concept of local coordinates, as applied to the tetrahedral element, is a direct extension of the scheme introduced for triangles. The basis functions are formulated using volume coordinates (the analog of area coordinates in two dimensions) and the appropriate integration formula is [12]

$$\int_{\forall} \zeta_1^p \zeta_2^q \zeta_3^r \zeta_4^s \, d\forall = 6\forall \frac{p!\,q!\,r!\,s!}{(p+q+r+s+3)!}$$

**Table 4.8**

Basis Functions for Three-Dimensional Isoparametric Elements

| Type | Corner node | Side node |
|---|---|---|
| Linear | $\Omega_i^e = \frac{1}{8}(1+\xi_0)(1+\eta_0)(1+\zeta_0)$ | |
| Quadratic | $\Omega_i^e = \frac{1}{8}(1+\xi_0)(1+\eta_0)(1+\zeta_0)$ $\cdot(\xi_0+\eta_0+\zeta_0-2)$ | $\xi_i = 0,\quad \eta_i = \pm 1,\quad \zeta_i = \pm 1$ $\Omega_i^e = \frac{1}{4}(1-\xi^2)(1+\eta_0)(1+\zeta_0)$ |
| Cubic | $\Omega_i^e = \frac{1}{64}(1+\xi_0)(1+\eta_0)(1+\zeta_0)$ $\cdot[9(\xi^2+\eta^2+\zeta^2)-19]$ | $\xi_i = \pm\frac{1}{3},\quad \eta_i = \pm 1,\quad \zeta_i = \pm 1$ $\Omega_i^e = \frac{9}{64}(1-\xi^2)(1+9\xi_0)(1+\eta_0)(1+\zeta_0)$ |

$\xi_0 = \xi\xi_i, \qquad \eta_0 = \eta\eta_i, \qquad \zeta_0 = \zeta\zeta_i$

Cartesian coordinates are related to volume coordinates by

$$\begin{Bmatrix} 1 \\ x \\ y \\ z \end{Bmatrix} = \begin{bmatrix} 1 & 1 & 1 & 1 \\ x_1 & x_2 & x_3 & x_4 \\ y_1 & y_2 & y_3 & y_4 \\ z_1 & z_2 & z_3 & z_4 \end{bmatrix} \begin{Bmatrix} \zeta_1 \\ \zeta_2 \\ \zeta_3 \\ \zeta_4 \end{Bmatrix} \tag{4.60}$$

and the volume of the tetrahedron is given by one-sixth of the determinant of the matrix in (4.60).

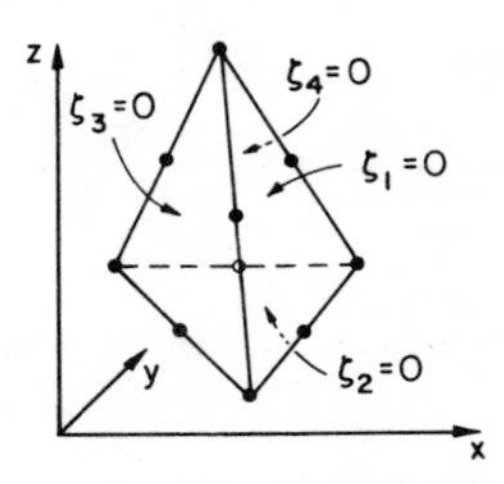


**Fig. 4.19.** Tetrahedral element.

## 4.12 Odds and Ends

### *4.12.1 Subparametric and Superparametric Elements*

In the isoparametric finite element concept the element geometry and the unknown function are approximated in terms of the same bases. In other words the geometry of the element is described by the same degree polynomial as that used in the trial solution. Elements using the same bases for both purposes are denoted as isoparametric elements. There are occasions, however, when complex geometry dictates the use of polynomials of higher degree than is required in the trial solution. Moreover, when solving systems of differential equations it may be advantageous to expand different variables in basis functions of different degrees. When a lower-degree polynomial is used to describe the geometry than the unknown function, the resulting element is termed subparametric. Elements that use higher-degree polynomials to define the geometry than the function are called superparametric (see Fig. 4.20). Convergence can be demonstrated for all subparametric elements and for some forms of superparametric elements [12, 13].

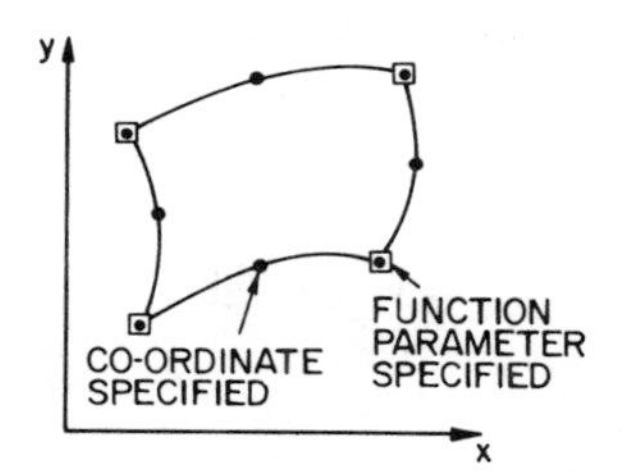


**Fig. 4.20.** Superparametric element.

### 4.12.2 *Treatment of Neumann Boundary Conditions*

Neumann boundary conditions generally appear in the equations as the surface integral resulting from the application of Green's theorem to second order equations. In two dimensions this integral is of the form

$$\int_\Gamma \left(\frac{\partial \hat{u}}{\partial x} l_x + \frac{\partial \hat{u}}{\partial y} l_y\right) \phi_i \, ds \qquad \text{or} \qquad \int_\Gamma \frac{\partial \hat{u}}{\partial n} \phi_i \, ds$$

where $l_x$ and $l_y$ are the direction cosines and $\Gamma$ the boundary of the solution domain. When the boundary condition is specified per unit length of $\Gamma$ it is convenient to use the dimensionless coordinate system $\bar{s} = s/l$, where $l$ is the side length. The surface integral becomes

$$I = l \int_0^1 \frac{\partial \hat{u}}{\partial n} \Omega_i^e \, d\bar{s} \tag{4.61}$$

If, for example, a quadratic basis function is assumed along $\Gamma$ as illustrated in Fig. 4.21b, there are three equations of the form of (4.61), one for each node along the side of the element:

$$I_1 = l \int_0^1 \bar{q}(1 - 3\bar{s} + 2\bar{s}^2)\, d\bar{s} = \tfrac{1}{6} l\bar{q}$$

$$I_2 = l \int_0^1 \bar{q}(4\bar{s} - 4\bar{s}^2)\, d\bar{s} = \tfrac{2}{3} l\bar{q}$$

$$I_3 = l \int_0^1 \bar{q}(-\bar{s} + 2\bar{s}^2)\, d\bar{s} = \tfrac{1}{6} l\bar{q}$$

where $\bar{q}$ is defined as the average value of $\partial\hat{u}/\partial n$ applied along the side. Similar integrals arise for linear and cubic sides and their appropriate dis-

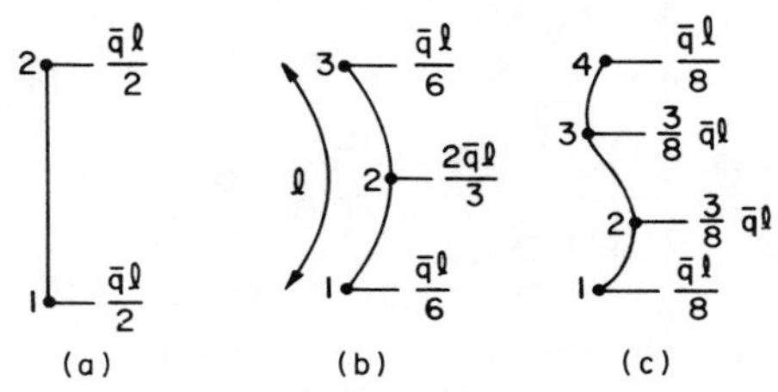


**Fig. 4.21.** Nodal fluxes corresponding to uniform boundary flux $q$ on a Neumann boundary: (a) linear, (b) quadratic, and (c) cubic sides.

tributions are indicated on Fig. 4.21a and c, respectively. If the value of $q$ is known at each node, it is possible to generate a polynomial of the form

$$q(s) = \sum_{j=1}^{3} q_j \Omega_j(\bar{s})$$

and the necessary integrations performed [5]. When an element with curved sides is used or the nodes are not positioned evenly along the sides, it may be difficult to perform the integration. In such cases one can use one-dimensional Gaussian quadrature to evaluate easily the integrals

$$I_i = \int_{-1}^{1} \bar{q}\Omega_i[(\partial x/\partial\xi)^2 + (\partial y/\partial\xi)^2]^{1/2}\eta \, d\xi \qquad (\eta = \pm 1)$$

$$I_i = \int_{-1}^{1} \bar{q}\Omega_i[(\partial x/\partial\eta)^2 + (\partial y/\partial\eta)^2]^{1/2}\xi \, d\eta \qquad (\xi = \pm 1)$$

When the total stress along a side, i.e., $\bar{q}l$, is given, then the allocation is simply given directly by Fig. 4.21.

### 4.12.3 *Treatment of Anisotropy*

When a parameter in a differential equation is a tensor, it is sometimes advantageous to generate a coordinate system colinear with the principal components of the tensor. When the principal components change direction over the region, one can use local coordinates to maintain the diagonal form of the tensor. The local coordinates are specified for each element so as to be colinear with the principal components. Given the angle between the local and global systems, one can transform the element coordinates from the global to the local system so that the appropriate integrations can be performed and the coefficient matrices generated. This procedure is outlined for anisotropic seepage by Zienkiewicz *et al.* [14].

## References

1. J. C. Cavendish, H. S. Price, and R. S. Varga, Galerkin methods for the numerical solution of boundary value problems, *Soc. Pet. Eng. J.* **9** **(2)**, 204 (1969).
2. P. C. Doherty, Unsaturated Darcian Flow by the Galerkin Method. U.S. Geolog. Surv. Comput. Contribution, Program C938 (1972).
3. O. C. Zienkiewicz, "The Finite Element Method in Engineering Science." McGraw-Hill, New York, 1971.
4. W. G. Gray and G. F. Pinder, On the relationship between the finite element and finite difference methods, *Int. J. Numer. Methods Eng.* **10**, 843 (1976).
5. J. Connor and G. Will, Computer-Aided Teaching of the Finite Element Displacement Method. Dept. of Civil Eng. Res., Vol. 69-23, Massachusetts Inst. Technol. (1969).
6. W. G. Gray and G. F. Pinder, Galerkin approximation of the time derivative in the finite element analysis of groundwater flow, *Water Resources Res.* **10**, 821–828 (1974).
7. I. Ergatoudis, B. M. Irons, and O. C. Zienkiewicz, Curved isoparametric "quadrilateral" elements for finite element analysis, *Int. J. Solids Struct.* **4**, 31–42 (1968).

7a. G. F. Pinder and E. O. Frind, Application of Galerkin's procedure to aquifer analysis, *Water Resources Res.* **8**, (1), 108–120 (1972).

8. C. A. Felippa and R. W. Clough, The finite element method in solid mechanics, *Numer. Solut. Field Probl. Continuum Phys. SIAM-AMS Proc.* **11**, 210–251 (1970).
9. G. R. Cowper, Gaussian quadrature formulas for triangles, *Int. J. Numer. Methods Eng.* **7** **(3)**, 405–408 (1973).
10. C. A. Felippa, Refined Finite Element Analysis of Linear and Non-Linear Two-Dimensional Structure, Rep. 66-22, Dept. of Civil Eng., Univ. California, Berkeley, California (1966).
11. Th. M. Van Genuchten, G. F. Pinder, and E. O. Frind, Simulation of two-dimensional contaminant transport with isoparametric Hermitian finite elements, *Water Resources Res.* (in press).
12. C. S. Desai and J. F. Abel, "Introduction to the Finite-Element Method." Van Nostrand-Reinhold, Princeton, New Jersey, 1972.
13. O. C. Zienkiewicz, B. M. Irons, J. Ergatoudis, S. Ahmand, and F. C. Scott, Isoparametric and associated element families for two and three dimensional analysis, FEM Tapir (1969). From Desai and Abel [12].
14. O. C. Zienkiewicz, P. Mayer, and Y. K. Cheung, Solution of anisotropic seepage by finite elements, *J. Eng. Mech. Div. ASCE* **92**, No. EM1, 111–120 (1966).

# Chapter 5 Finite Element Method in Subsurface Hydrology

## 5.1 Introduction

The governing equations describing the transport of mass, energy, and momentum in porous media form the foundation for the development of simulation methods in subsurface hydrology. In combination with various constitutive relationships, we formulate systems of equations to describe flow, mass transport, and energy transport. We encounter linear and nonlinear governing equations and boundary conditions. The focus is primarily on solution methods rather than development of the differential equations. The discussion proceeds from the specific to the more general because this also leads us from the relatively simple to the more complex problems.

## 5.2 Flow of Homogeneous Fluids in Saturated Porous Media

### *5.2.1 Governing Equations*

The flow of a nonhomogeneous fluid through a saturated porous medium is described by the conservation of mass equation

$$\partial(\theta\rho)/\partial t + \nabla \cdot (\rho \mathbf{q}) = 0 \tag{5.1}$$

where $\theta$ is porosity, $\mathbf{q}$ specific discharge, and $\rho$ fluid density. The specific discharge is given by a simplified form of the momentum balance equation known as Darcy's law

$$\mathbf{q} = -(\mathbf{k}/\mu) \cdot (\nabla p + \rho g\, \nabla Z) \tag{5.2}$$

where $\mathbf{k}$ is permeability, $\mu$ dynamic viscosity, $p$ fluid pressure, $g$ gravitational acceleration, and $Z$ elevation above some reference datum. When the fluid is assumed to be homogeneous, a potential function $\phi^*$ may be introduced such that

$$\phi^* = gZ + \int_{p_0}^{p} 1/[\rho(p)]\, dp + \tfrac{1}{2}v^2 \tag{5.3}$$

where $\mathbf{v}$ is the fluid velocity and $\phi^*$ Hubbert's potential [1]. It is important to note that density must be a function of pressure only for the potential to be unique. When the kinetic energy of the fluid is assumed to be small by virtue of the low velocities generally encountered in subsurface flow, combination of (5.3), (5.2), and (5.1) yields

$$\partial(\theta\rho)/\partial t - \nabla \cdot [\rho(\mathbf{k}/\mu) \cdot (\rho g\, \nabla h)] = 0 \tag{5.4}$$

where $h \equiv \phi^*/g$, and is called the piezometric head. Moreover, if the hydraulic conductivity is defined by $\mathbf{K} \equiv \rho g \mathbf{k}/\mu$, Eq. (5.4) reduces to

$$\partial(\theta\rho)/\partial t - \nabla \cdot (\rho \mathbf{K} \cdot \nabla h) = 0 \tag{5.5}$$

Introduction of additional constitutive relationships regarding aquifer compressibility and fluid compressibility as well as an equation of state [2] allows one to modify Eq. (5.5) to

$$\rho S_s \frac{\partial h}{\partial t} - \nabla \cdot (\rho \mathbf{K} \cdot \nabla h) = 0 \tag{5.6}$$

where $S_s$ denotes specific storage and is defined as $S_s \equiv \rho_0 g(\alpha + \theta\beta)$, where $\rho_0$ is a reference fluid density, $\alpha$ the aquifer compressibility, and $\beta$ the fluid compressibility. Because of its relative simplicity, simulation of flow through a cross section might seem to be the logical point of departure for subsequent discussion. However, the presence of the free surface makes the problem more complex than one might anticipate. Therefore, the discussion of this case is postponed to a later section and we initially consider two-dimensional flow in the horizontal or areal plane.

### 5.2.2 *Two-Dimensional Horizontal Flow*

#### *Confined Aquifer*

The two-dimensional flow problem to be considered is illustrated in Fig. 5.1. The aquifer is assumed confined above and below by layers of material which have a hydraulic conductivity much lower than that of the aquifer. It is further assumed that only vertical flow exists in the confining bed and occurs in response to a pressure drop in the aquifer. This is typical of the aquifer systems encountered in evaluating regional groundwater resources and their response to exploitation.

The general equation (5.6) is reduced to two-dimensional form by integration over the aquifer thickness. With reference to Fig. 5.1, this requires integration from $z_1$ to $z_2$ which yields†

$$\bar{\rho} S_s \int_{z_1}^{z_2} \frac{\partial h}{\partial t}\, dz + \int_{z_1}^{z_2} \left[ \frac{\partial}{\partial x}(\rho q_x) + \frac{\partial}{\partial y}(\rho q_y) + \frac{\partial}{\partial z}(\rho q_z) \right] dz = 0 \qquad (5.7)$$

where $\bar{\rho}$ is the vertical mean density of the fluid in the aquifer, and $z_1$ and $z_2$

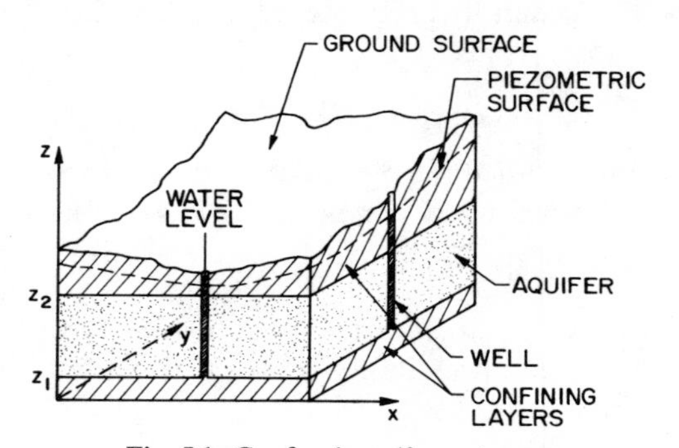


**Fig. 5.1.** Confined aquifer system.

are functions of $x$, $y$, and $t$. Applying Leibnitz' rule for differentiation of an integral to each term, we obtain

$$\bar{\rho} S_s \int_{z_1}^{z_2} \frac{\partial h}{\partial t}\, dz = \bar{\rho} S_s \frac{\partial}{\partial t} \int_{z_1}^{z_2} h\, dz - \bar{\rho} S_s h \bigg|_{z_2} \frac{\partial z_2}{\partial t} + \bar{\rho} S_s h \bigg|_{z_1} \frac{\partial z_1}{\partial t}$$

$$\int_{z_1}^{z_2} \frac{\partial}{\partial x}(\rho q_x)\, dz = \frac{\partial}{\partial x} \int_{z_1}^{z_2} \rho q_x\, dz - \rho q_x \bigg|_{z_2} \frac{\partial z_2}{\partial x} + \rho q_x \bigg|_{z_1} \frac{\partial z_1}{\partial x}$$

$$\int_{z_1}^{z_2} \frac{\partial}{\partial y}(\rho q_y)\, dz = \frac{\partial}{\partial y} \int_{z_1}^{z_2} \rho q_y\, dz - \rho q_y \bigg|_{z_2} \frac{\partial z_2}{\partial y} + \rho q_y \bigg|_{z_1} \frac{\partial z_1}{\partial y}$$

$$\int_{z_1}^{z_2} \frac{\partial}{\partial z}(\rho q_z)\, dz = \rho q_z \bigg|_{z_2} - \rho q_z \bigg|_{z_1}$$

Substitution of these equalities into (5.7) yields a two-dimensional equation

† Notice that the subscript notation refers to a vector component and not differentiation.

involving vertically integrated parameters

$$\bar{\rho}S_s \frac{\partial}{\partial t}(\bar{h}l) + \frac{\partial}{\partial x}(\bar{\rho}\bar{q}_x l) + \frac{\partial}{\partial y}(\bar{\rho}\bar{q}_y l)$$

$$- \left[ \rho q_x \bigg|_{z_2} \frac{\partial z_2}{\partial x} - \rho q_x \bigg|_{z_1} \frac{\partial z_1}{\partial x} + \rho q_y \bigg|_{z_2} \frac{\partial z_2}{\partial y} - \rho q_y \bigg|_{z_1} \frac{\partial z_1}{\partial y} \right.$$

$$\left. - \rho q_z \bigg|_{z_2} + \rho q_z \bigg|_{z_1} + \bar{\rho}S_s \left( h \bigg|_{z_2} \frac{\partial z_2}{\partial t} - h \bigg|_{z_1} \frac{\partial z_1}{\partial t} \right) \right] = 0 \qquad (5.8)$$

where $l \equiv z_2 - z_1$ and the overbar notation indicates a vertically averaged value. In writing $\bar{\rho}\bar{q}_x$ in lieu of $\overline{\rho q_x}$, the products involving small deviations from mean values have been neglected. If vertical flow in the aquifer is assumed negligible, then

$$h \bigg|_{z_2} \simeq h \bigg|_{z_1} \simeq \bar{h}$$

and the last term in (5.8) becomes

$$\rho S_s \left( h \bigg|_{z_2} \frac{\partial z_2}{\partial t} - h \bigg|_{z_1} \frac{\partial z_1}{\partial t} \right) \simeq \rho S_s \bar{h} \frac{\partial l}{\partial t} \qquad (5.9)$$

Expanding the time derivative in the first term of (5.8), and introducing (5.9) yields

$$\bar{\rho}S_s l \frac{\partial \bar{h}}{\partial t} + \frac{\partial}{\partial x}(\bar{\rho}\bar{q}_x l) + \frac{\partial}{\partial y}(\bar{\rho}\bar{q}_y l)$$

$$+ \rho q_z \bigg|_{z_2} - \rho q_z \bigg|_{z_1} - \rho \mathbf{q} \bigg|_{z_2} \cdot \nabla_{xy} z_2 + \rho \mathbf{q} \bigg|_{z_1} \cdot \nabla_{xy} z_1 = 0 \qquad (5.10)$$

and

$$\nabla_{xy} \equiv \frac{\partial}{\partial x}\mathbf{i} + \frac{\partial}{\partial y}\mathbf{j}$$

In general, one assumes that $z_1$ and $z_2$, the confining bed elevations, are spatially invariant whereupon the last two terms vanish except at a well. The well bore can be considered as a $z_2$ surface discontinuity giving rise to a Dirac delta function. Thus the discharging well at an arbitrary location $x_k$, $y_k$ can be described as

$$\left( \mathbf{q} \bigg|_{z_2} \cdot \nabla_{xy} z_2 \right) \bigg|_{x_k,\, y_k} \equiv Q_w(x_k, y_k)\, \delta(x - x_k, y - y_k)$$

and $Q = \sum_{k=1}^{N} Q_w(x_k, y_k)\,\delta(x - x_k, y - y_k)$, where $Q_w$ is the volumetric discharge from the aquifer, $\delta$ the Dirac delta function, and $N$ the number of wells. If the product $\mathbf{q} \cdot \nabla\rho$ is assumed negligible, which is generally a reasonable assumption, particularly for homogeneous fluids, Darcy's law can be combined with (5.10) to yield

$$S\frac{\partial \bar{h}}{\partial t} - \nabla_{xy} \cdot (\mathbf{T} \cdot \nabla_{xy}\bar{h}) + Q(x, y) + q_z\Big|_{z2} - q_z\Big|_{z1} = 0 \qquad (5.11)$$

where $S \equiv lS_s$ is termed the storage coefficient, and $\mathbf{T} \equiv l\mathbf{K}$ the aquifer transmissivity. The last two terms in (5.11) represent the vertical flux from the confining beds which, as will be shown later, can be approximated using an analytical expression.

*The Approximating Equations*

To simulate two-dimensional horizontal groundwater flow, the linear parabolic partial differential equation (5.11) must be solved in conjunction with appropriate boundary and initial conditions. We begin by applying Galerkin's method as outlined in Chapters 3 and 4. The approximating integral equations are obtained from the orthogonality condition

$$\langle L\bar{h}, \phi_i\rangle = 0 \qquad (i = 1, 2, \ldots, M) \qquad (5.12)$$

where

$$L\bar{h} \equiv \nabla_{xy} \cdot (\mathbf{T} \cdot \nabla_{xy}\bar{h}) - q_z\Big|_{z2} + q_z\Big|_{z1} - Q(x, y) - S\,\partial\bar{h}/\partial t$$

Expansion of (5.12) and application of Green's theorem to the spatial derivative yields

$$\langle -\mathbf{T} \cdot \nabla_{xy}\bar{h}, \nabla_{xy}\phi_i\rangle + \int_{\Gamma} \mathbf{n} \cdot (\mathbf{T} \cdot \nabla_{xy}\bar{h})\varphi_i\,ds$$
$$+ \langle Q_T, \phi_i\rangle - \langle S\,\partial\bar{h}/\partial t, \phi_i\rangle = 0 \qquad (i = 1, 2, \ldots, M) \qquad (5.13)$$

where

$$Q_T \equiv q_z\Big|_{z1} - q_z\Big|_{z2} - Q(x, y).$$

It is interesting to note that through application of Green's theorem, we have eliminated the gradient of transmissivity from the equation and thereby simplified the computational scheme.

Assume a trial function of the form

$$\bar{h}(x, y, t) \simeq \sum_{j=1}^{M} h_j(t)\phi_j(x, y) \qquad (5.14)$$

where $h_j$ is the undetermined, time-dependent coefficient which has been shown to be the approximation to $\bar{h}(x_j, y_j, t)$ because of the way the basis functions are defined. Substitution of (5.14) into (5.13) gives

$$\sum_{j=1}^{M} h_j \langle \mathbf{T} \cdot \nabla_{xy}\phi_j, \nabla_{xy}\phi_i \rangle + \sum_{j=1}^{M} \langle S\phi_j, \phi_i \rangle (dh_j/dt) - \langle Q_T, \phi_i \rangle$$

$$- \int_\Gamma \mathbf{n} \cdot (\mathbf{T} \cdot \nabla_{xy}\bar{h})\phi_i \, ds = 0 \qquad (i = 1, 2, \ldots, M) \tag{5.15}$$

The last term in (5.15) is a flux boundary condition which can be written as

$$\int_\Gamma \mathbf{n} \cdot (\mathbf{T} \cdot \nabla_{xy}\bar{h})\phi_i \, ds = \int_\Gamma q_n \phi_i \, ds$$

where $q_n$ is the flux per unit length along $\Gamma$. When a flux or Neumann boundary is encountered, either $q_n$ is specified or, as in the case of a large diameter well, the total flux crossing $\Gamma$ is given. In either case, this integral is easily evaluated using the principles discussed in Section 4.11.2. In the case of a no-flow boundary, the flux is zero and this term vanishes. When a constant head or Dirichlet boundary is encountered no equation is generated and consequently, it is unnecessary to evaluate this term.

Equation (5.15) can be written using matrix notation as

$$[A]\{H\} + [B]\{dH/dt\} = \{F\} \tag{5.16}$$

where typical elements of matrices $[A]$, $[B]$, and $\{F\}$ are

$$a_{i,j} = \int_A \left( T_{xx} \frac{\partial \phi_j}{\partial x}\frac{\partial \phi_i}{\partial x} + T_{xy} \frac{\partial \phi_j}{\partial y}\frac{\partial \phi_i}{\partial x} + T_{yx} \frac{\partial \phi_j}{\partial x}\frac{\partial \phi_i}{\partial y} + T_{yy} \frac{\partial \phi_j}{\partial y}\frac{\partial \phi_i}{\partial y} \right) dA \tag{5.17a}$$

$$b_{i,j} = \int_A S\phi_j\phi_i \, dA \tag{5.17b}$$

$$f_i = \int_A Q_T\phi_i \, dA + \int_\Gamma q_n\phi_i \, ds \tag{5.17c}$$

Let us consider the evaluation of the integrals of (5.17a). Generally, the coordinate axes are arranged to be colinear with the principal components of the transmissivity tensor so that the cross derivatives vanish. This obviously is not a numerical restriction but is nevertheless a reasonable ploy inasmuch as the off-diagonal elements of the tensor are seldom, if ever, known in field problems. Even with this simplification, we are still faced with the problem of how to treat transmissivity when it is a function of space.

When direct integration methods are used, transmissivity and storage are generally assumed constant over each element. Because integration is performed over an element, these parameters are moved from under the integral. If one elects to use numerical integration, however, it is convenient to use a scheme called functional coefficients [3]. In this approach, the known parameter is approximated using basis functions to interpolate nodal values over the element. The transmissivity, for example, is replaced by the series

$$T_{xx}(x, y) = \sum_{k=1}^{M} T_{xx,k}\phi_k(x, y)$$

A typical term in (5.17a) would now read

$$\int_A T_{xx} \frac{\partial \phi_j}{\partial x}\frac{\partial \phi_i}{\partial x}\, dA = \int_A \sum_{k=1}^{M} T_{xx,k}\phi_k \frac{\partial \phi_j}{\partial x}\frac{\partial \phi_i}{\partial x}\, dA$$

For numerical integration, the functions $\phi_k$ as well as $\partial\phi_j/\partial x$ and $\partial\phi_i/\partial x$ are evaluated at the Gauss points. It should be kept in mind that, although this formulation is easily incorporated in a computer code, it does increase the order of the polynomial to be integrated and accordingly the number of Gauss points required for an exact integration. Thus two methods for approximating aquifer parameters may be used: one provides an abrupt change between elements and the other a gradational change. While this choice is of minor interest in horizontal flow problems where abrupt changes are seldom encountered, it may be advantageous to use a constant parameter over an element in cross-sectional or three-dimensional problems. Here beds of contrasting characteristics are in direct contact and this line of demarcation can be effectively represented by an element boundary.

The well discharge component of $Q_T$ is easily handled using finite elements. Because of the properties of the Dirac delta function, the integral $\int Q\phi_i\, dA$ is equal to the well discharge at node $i$ and is simply added directly to $f_i$.

The leakage terms $q_z|_{z_1}$ and $q_z|_{z_2}$ can be approximated analytically. Consider an aquitard of thickness $l^*$ with hydraulic conductivity $K^*$ and specific storage $S_s^*$ which lies above the confined aquifer with initial piezometric head $h_{i,0}$ and below an aquifer with constant head $h_{i,w}$. If a step head change $\Delta h$ is applied to the confined aquifer, the transient vertical flow in the aquitard is described by

$$\frac{\partial^2 h^*}{\partial z^2} - \frac{S_s^*}{K^*}\frac{\partial h^*}{\partial t} = 0$$

subject to the initial and boundary conditions

$$h^*(z, 0) = (h_{i,w} - h_{i,0})z/l^* + h_{i,0}$$
$$h^*(0, t) = \Delta h + h_{i,0}$$
$$h^*(l^*, t) = h_{i,w}$$

where we have set $z_2 = 0$ for convenience. The solution to this system of equations for the head distribution in the aquitard is

$$h^* = h_{i,0} + \Delta h + (h_{i,w} - h_{i,0} - \Delta h)\frac{z}{l^*} - \sum_{n=1}^{\infty}\left(\frac{2\Delta h}{n\pi}\right)\exp(-n^2\pi^2\alpha t)\sin\left(\frac{n\pi z}{l^*}\right)$$

where $\alpha = K^*/(S_s^* l^{*2})$. The total flux term at the boundary corresponding to $z_2$ is obtained by evaluating the spatial derivative of $h^*$ at $z = 0$ and noting that $q_z|_{z_2} = -K\,\partial h^*/\partial z|_0$ to obtain

$$q_z|_{z_2} = -\frac{K^*}{l^*}(h_{i,w} - h_{i,0} - \Delta h) + \frac{K^*\,\Delta h}{l^*}\sum_{n=1}^{\infty} 2\exp(-n^2\pi^2\alpha t)$$

In an actual situation, the change in head at $z = 0$, denoted $\Delta h$, will be a continuous function of time rather than an instananeous step change. However, if we consider the leakage problem to correspond roughly to a case where a step change in head is applied at one-half the elapsed time, then

$$q_z|_{z_2,t} \simeq -\frac{K^*}{l^*}(h_{i,w} - h_{i,0} - \Delta h_t) + \frac{K^*\,\Delta h_t}{l^*}\sum_{n=1}^{\infty} 2\exp(-n^2\pi^2\alpha t/2) \quad (5.18)$$

A similar expression can be obtained for $q_z|_{z_1}$. Amazingly enough, the approximation is reasonably good when compared to an analytical solution obtained by Hantush using his modified leaky aquifer theory [4, 5]. Because the expression for $q_z|_{z_2}$ contains the unknown head, the coefficient matrix $A$ will contain additional terms involving the information in Eq. (5.18).

Although the matrices $[A]$, $[B]$, and $\{F\}$ of Eq. (5.16) can now be evaluated, a set of ordinary differential equations must still be solved. One technique for obtaining the solution involves approximating the time derivative using an implicit finite difference scheme. The final equation then becomes (the case of $\varepsilon = 1$ in the formulation of Section 2.3)

$$[A]\{H\}_{t+\Delta t} + \frac{1}{\Delta t}[B]\{H_{t+\Delta t} - H_t\} = \{F\}_{t+\Delta t}$$

which can be rearranged to yield

$$\left([A] + \frac{1}{\Delta t}[B]\right)\{H\}_{t+\Delta t} = \frac{1}{\Delta t}[B]\{H\}_t + \{F\}_{t+\Delta t} \quad (5.19)$$

Because $[A]$ and $[B]$ are symmetric in this problem, matrix-solving techniques, such as Cholesky's method, which exploit this property of symmetry, should be used.

*The Water Table Aquifer*

When an aquifer is not overlain by a confining layer, the governing equation should be modified to account for the time dependence of the transmissivity. From its definition, we see that transmissivity is proportional to the saturated thickness of the aquifer. Because the saturated thickness in the areal two-dimensional water table case is linearly related to the head, one must consider a nonlinear analog of (5.11) given by

$$S_y(\partial\bar{h}/\partial t) - \nabla_{xy} \cdot (\mathbf{K}l(\bar{h}) \cdot \nabla_{xy}\bar{h}) + Q(x, y) - q_z \bigg|_{z1} = 0 \tag{5.20}$$

where $S_y$ is the specific yield and accounts for water removed from storage due to dewatering.

Although (5.20) is nonlinear, the transient solution is not highly sensitive to changes in saturated thickness. Accordingly, a numerical scheme can be devised wherein the saturated thickness is obtained by extrapolating head values from earlier known time levels to the current, unknown level. Experience shows that an extrapolation based on the latest two calculated heads provides satisfactory results. Notice, however, that this nonlinear system of equations is solved only approximately when this quasilinearization procedure is used.

A second approach that leads to a solution of this nonlinear system is based on the concept of iteration as outlined in Chapter 1. In this approach, the solution at a particular time level is obtained several times, with the coefficient matrix updated for each solution to reflect the most recent estimate of head and thereby saturated thickness. This procedure can be described for an implicit formulation in time by the following equation which is the analog of (5.19) except that $[A]$ is updated at each iteration to reflect a changing saturated thickness $l$:

$$\left([A]_{t+\Delta t} + \frac{1}{\Delta t}[B]\right)^n \{H\}_{t+\Delta t}^{n+1} = \frac{1}{\Delta t}[B]\{H\}_t + \{F\}_{t+\Delta t}^n \tag{5.21}$$

where the superscript $n$ indicates the iteration at time level $t + \Delta t$ and the subscript on $[A]$ denotes the time level at which the coefficients are evaluated. An error criterion is selected, such as

$$\{e\} = \{H\}_{t+\Delta t}^{n+1} - \{H\}_{t+\Delta t}^n$$

and Eq. (5.21) is recalculated for increasing $n$ until the absolute value of all the elements of $\{e\}$ lie within a specified tolerance. There are several techniques for extrapolating values from earlier iterations to obtain $[A]$ at the current iteration. A weighted average of the two most recent iterations is often effective:

$$\{H\}_{t+\Delta t}^{n+2} = (1 - \varepsilon)\{H\}_{t+\Delta t}^{n+1} + \varepsilon\{H\}_{t+\Delta t}^{n}$$

The weighting parameter $\varepsilon$ can be modified to achieve the most rapidly convergent stable solution ($\varepsilon = 0.7$ provides acceptable convergence for many problems).

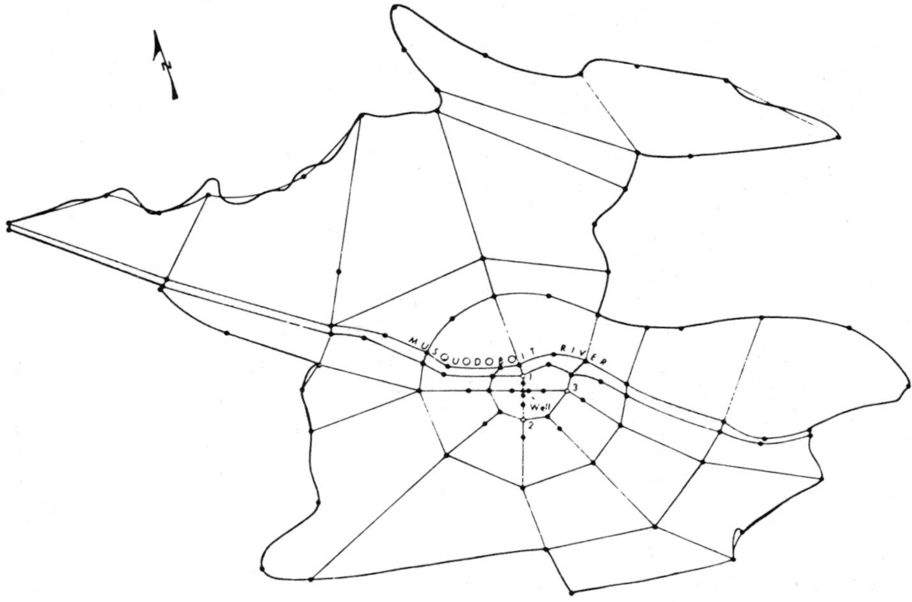


**Fig. 5.2.** Finite element configuration for aquifer analysis at Musquodoboit Harbor, Nova Scotia [36]: (▲) Pumping well; (○) observation wells; (●) nodes.

As an example of the application of finite element techniques in the analysis of areal groundwater flow, consider the element configuration of Fig. 5.2. This net was used to simulate an alluvial aquifer confined on all sides by relatively impermeable formations [6]. The aquifer is divided into two segments by a river that is simulated using a leaky confining bed approach wherein the stream bed acts as a confining layer extending over the local area beneath the river. Notice that in this problem the mixed element formulation is used to advantage by allowing for a higher concentration of nodes near the pumping well where large drawdowns are anticipated.

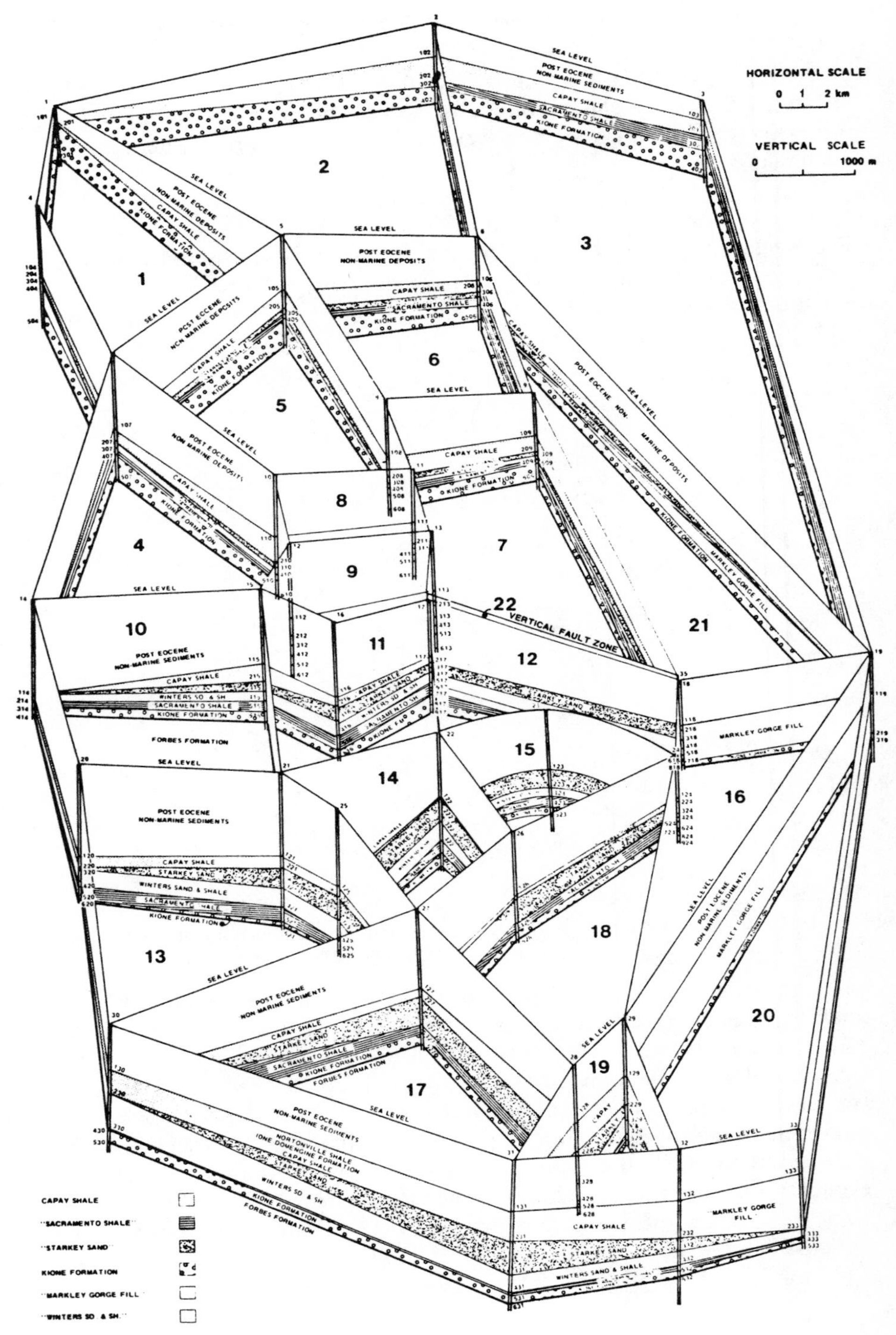
HORIZONTAL SCALE
0 1 2 km
VERTICAL SCALE
0 1000 m
SEA LEVEL
POST EOCENE NON-MARINE SEDIMENTS
CAPAY SHALE
SACRAMENTO SHALE
KIONE FORMATION
VERTICAL FAULT ZONE
MARKLEY GORGE FILL
FORBES FORMATION
STARKEY SAND
WINTERS SAND & SHALE
NORTONVILLE SHALE
IONE DOMENGINE FORMATION
CAPAY SHALE
"SACRAMENTO SHALE"
"STARKEY SAND"
KIONE FORMATION
"MARKLEY GORGE FILL"
"WINTERS SD. & SH."

### 5.2.3 Three-Dimensional Flow

The simulation of three-dimensional flow in a confined aquifer with no free boundary condition is a straightforward extension of the two-dimensional case. The principal impediment to three-dimensional models is not the complexity of the formulation nor the additional computer programming effort, but rather the considerable increase in computer storage and computational effort required to simulate a field problem. A few examples of three dimensional simulations appear in the literature [7–9]. The effective use of three-dimensional mixed isoparametric elements is clearly depicted in Fig. 5.3. The Sutter Basin in California is subdivided into 123 elements with 232 nodes. To assist in minimizing errors due to incorrect input data, several interesting features are incorporated in this model including a novel nodal numbering convention. Note that only the upper layer of 22 surface elements is illustrated in Fig. 5.3.

### 5.2.4 Radial Flow to a Well

One of the earliest applications of the finite element method in porous flow involved radial flow to a well. This problem is interesting primarily because it requires the formulation of finite elements in radial coordinates [10, 11]. The governing equation is

$$\nabla_{rz} \cdot (\mathbf{K} \cdot \nabla_{rz} h) = S_s \frac{\partial h}{\partial t}$$

where

$$\nabla_{rz}^2 h \equiv \frac{\partial^2 h}{\partial r^2} + \frac{1}{r^2}\frac{\partial^2 h}{\partial \theta^2} + \frac{1}{r}\frac{\partial h}{\partial r} + \frac{\partial^2 h}{\partial z^2}$$

Introducing a trial function defined in $(r, z, \theta)$ coordinates, we have

$$h \simeq \hat{h} = \sum_{j=1}^{M} h_j(t)\phi_j(r, \theta, z)$$

---

**Fig. 5.3.** Subdivision of the Sutter Basin area into subdomains for three-dimensional finite element formulation [8].

and imposing the orthogonality conditions of Galerkin's method yields

$$\sum_{j=1}^{M} h_j\langle \mathbf{K} \cdot \nabla_{rz}\phi_j , \nabla_{rz}\phi_i\rangle + \sum_{j=1}^{M} \langle S_s\phi_j , \phi_i\rangle \frac{dh_j}{dt}$$

$$- \int_{\Gamma} \mathbf{n} \cdot (\mathbf{K} \cdot \nabla_{rz} h)\phi_i \, ds = 0 \qquad (i = 1, 2, \ldots, M) \qquad (5.22a)$$

where $\langle \phi_j, \phi_i\rangle \equiv \int_z \int_r \int_\theta \phi_j \phi_i \, d\theta r \, dr \, dz$.

Let us now consider an axisymmetric triangular element such as illustrated in Fig. 5.4. If the flow does not depend on $\theta$, the finite element problem may be formulated in the $r$, $z$ plane, such that $\phi_j = \phi_j(r, z)$ and is a

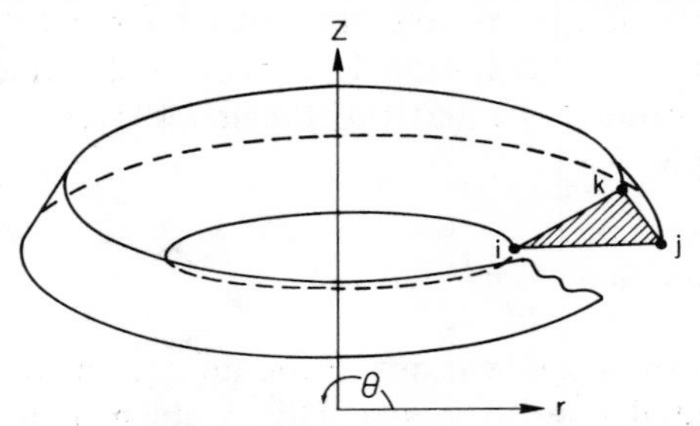


Fig. 5.4. Axisymmetric element with triangular cross section (after Neuman and Witherspoon [11]).

linear function over each triangle. Moreover, these functions are identical in form to those formulated for linear triangles in the Cartesian coordinate system:

$$\phi_n = (a_n + b_n r + c_n z)/2\Delta$$

where $a_i = r_j z_k - r_k z_j$, $b_i = z_j - z_k$, $c_i = r_k - r_j$, and

$$2\,\Delta = \begin{vmatrix} 1 & r_i & z_i \\ 1 & r_j & z_j \\ 1 & r_k & z_k \end{vmatrix} = \text{twice the area of triangle } ijk$$

and the nodal indices $i$, $j$, $k$ are cyclically permuted in the counterclockwise direction around the triangle.

Examination of Eq. (5.22a) reveals that the inner products involve both the basis functions and their derivatives. First consider integrals of the form

$$I = \int_z \int_r \frac{\partial \phi_j}{\partial r}\frac{\partial \phi_i}{\partial r} r \, dr \, dz \qquad (5.22b)$$

If an average value of $r$ over an element is defined by

$$\bar{r} = \int r\, dr \Big/ \int dr = (r_i + r_j + r_k)/3 \tag{5.22c}$$

the integral (5.22b) becomes

$$I = \bar{r} \int_z \int_r \frac{\partial \phi_j}{\partial r} \frac{\partial \phi_i}{\partial r}\, dr\, dz$$

This is easily integrated using the formulas provided in Chapter 4. However, a different strategy should be developed to evaluate integrals of the form

$$I = \int_z \int_r \phi_j \phi_i r\, dr\, dz$$

because the integrand is a cubic polynomial in $r$.

The radial coordinate formulation is generally used to describe flow to a well in a problem involving radial symmetry. To incorporate the boundary condition at the well bore requires evaluation of the surface integral in Eq. (5.22a). This term represents the fluid entering the system through any external surface of the array of axisymmetric elements (fluxes across internal surfaces balance and the net internal flux for the system is zero). Taking into account the radial symmetry, we can write

$$-\int_\Gamma K_{rr}\, \partial h/\partial r \Big|_{r=r_w} \phi_i\, ds = \int_z q_r \phi_i\, dz$$

where the "well element" has the specific form indicated in Fig. 5.5. Although the vertical distribution of $q_r$ along the well bore is not known, the sum of fluxes entering the well must equal the total discharge $Q$. The distribution of the flux shown in Fig. 5.5 is consistent with the earlier discussion of

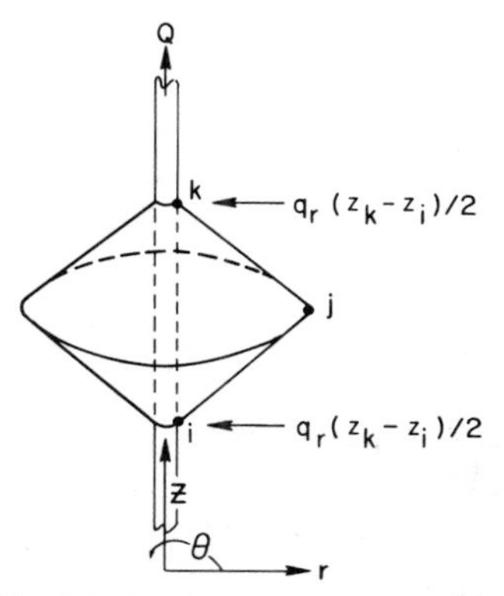


**Fig. 5.5.** Axisymmetric well element.

boundary flux terms in Chapter 4. As long as the distribution of $q_r$ can be assumed uniform along the well bore, Eq. (5.22a) may be solved using an appropriate finite difference discretization of the time derivative. When $q_r = q_r(t)$, we can use an average value over each time step.

In the case of a nonuniform flow field, the distribution of $q_r$ varies along the well bore and is time dependent. Thus there are $M + K$ unknowns where $K$ is the number of nodes along the well bore for which discharge values are required and node $K$ is nearest the surface. If the head in the well is assumed uniform for any given point in time, the number of unknowns is reduced to $M - K + 1$. Adopting a procedure suggested by Javandel and Witherspoon [12], one can initially guess at the discharge $\bar{Q}_K$ for the period $\Delta t$, where $Q_n$ is the discharge at node $n$. The equations can then be solved for the head and, subsequently, the discharge.

The calculated total discharge will be the sum of the values at each node

$$Q_1 = \sum_{k=1}^{K} \bar{Q}_k(\Delta t_1)$$

An improved estimate of $\bar{Q}_K$ for the next time step is obtained from the relationship

$$\bar{Q}_K(\Delta t_i) = (Q/Q_{i-1})\bar{Q}_K(\Delta t_{i-1})$$

where $Q$ is the constant prescribed discharge.

To adjust the computed drawdown for the inaccuracy due to any error in the original guess for $\bar{Q}_K$, the principle of superposition is invoked. Because the drawdown $s$ is proportional to discharge for any nodal point $j$,

$$s_j(t) = s_j(t_0) + Qf_j(t - t_0)$$

If the computed values $Q_i$ change from time step to time step, the drawdown is given by

$$s_j(t_1) = s_j(t_0) + Q_1\, f_j(t_1 - t_0)$$
$$s_j(t_2) = s_j(t_0) + Q_1\, f_j(t_2 - t_0) + (Q_2 - Q_1)f_j(t_2 - t_1)$$
$$\vdots$$
$$s_j(t_m) = s_j(t_0) + Q_1\, f_j(t_m - t_0) + \sum_{n=1}^{m-1} (Q_{n+1} - Q_n)f_j(t_m - t_n)$$

Because the drawdowns computed using $Q_1$ (which depends on the initial estimate $\bar{Q}_K$) and $Q$ would not necessarily be the same, the calculated drawdown is adjusted through the relationship

$$(Q/Q_1)\bar{s}_j(t_1) = (Q/Q_1)[Q_1\, f_j(t_1 - t_0)] = Qf_j(t_1 - t_0) = s_j(t_1)$$

where the initial drawdown is assigned the value zero. Using the corrected solutions $s_j(t_1)$, we can calculate new values of drawdown (or head) for $t_2 = t_1 + \Delta t$. Since we have used corrected values for drawdown, the solution at $t = t_2$ will reflect a discharge of $Q$ during $\Delta t_1$ and $Q_2$ during $\Delta t_2$. Thus we have for the second time step

$$\bar{s}_j(t_2) = Qf_j(t_2 - t_0) + (Q_2 - Q)f_j(t_2 - t_1)$$

If we allow $\Delta t_2 = \Delta t_1 = (t_1 - t_0)$, we obtain

$$\bar{s}_j(t_2) = s_j(t_2) + [(Q_2 - Q)/Q]s_j(t_1)$$

which can be rearranged to provide the corrected drawdown at $t = t_2$ of

$$s_j(t_2) = \bar{s}_j(t_2) - [(Q_2 - Q)/Q]s_j(t_1)$$

The general recurrence formula for adjusting the calculated values at time $t_m$ is

$$s_j(t_m) = \bar{s}_j(t_m) - [(Q_m - Q)/Q]s_j(t_n)$$

where $n < m$, and $t_n = t_m - t_{m-1}$.

*An Example Problem*

A finite element solution for radial flow to a well located in a nonhomogeneous medium was obtained by Neuman and Witherspoon [11]. The finite element net of Fig. 5.6 represents a radial cross section through a three-dimensional axisymmetric medium. The rectangles are divided by the computer code into triangles which are used in the approximation scheme. To obtain an accurate solution in the vicinity of the well without resorting to

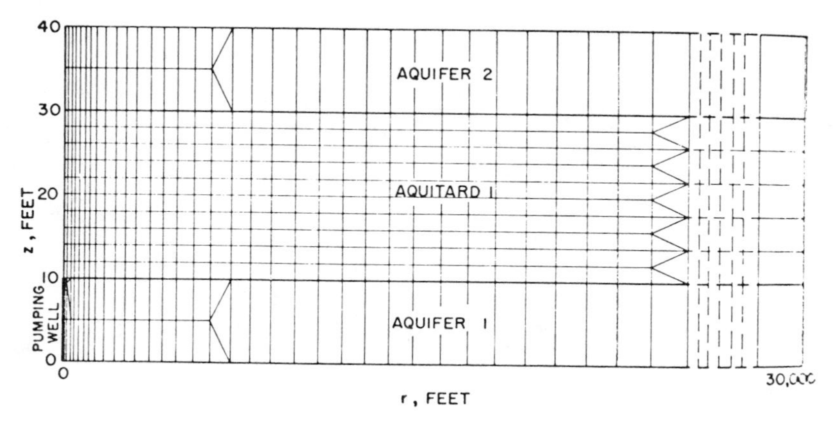


**Fig. 5.6.** Finite element net for axisymmetric two-aquifer system [11].

an unreasonable number of nodal points, the elements increase in size along the radial dimension from the $z$ axis outward. The complete net contains 542 nodal points and 501 elements.

## 5.3 Flow of Nonhomogeneous Fluids in Saturated Porous Media

### 5.3.1 *Two-Dimensional Horizontal Flow*

To simulate the flow of nonhomogeneous fluids in saturated porous media requires, in general, the simultaneous solution of the equations of mass transport and groundwater flow. For dilute solutions, the equations can be considered linear, but when the concentration of an ion in solution significantly affects the fluid density, we must consider a nonlinear system. We will first discuss the mass transport equation which is common to both cases and then proceed to the more difficult problem of density-dependent flow.

*Equations of Mass Transport in Porous Media*

It can be shown that the mass transport equations written for a continuum and integrated over a saturated porous medium result in the equation

$$\partial(\theta\rho_i)/\partial t + \nabla \cdot (\rho_i \mathbf{q}_i) = 0 \tag{5.23a}$$

where $\rho_i$ is the mass concentration of species $i$, and $\mathbf{q}_i$ the mass phase average velocity of species $i$. Let us consider an areal two-dimensional problem in which the fluid density is independent of concentration. Once again, an equation must be formulated in terms of vertically integrated parameters. Consideration of each term of (5.23a) yields

$$\int_{z_1}^{z_2} \frac{\partial}{\partial t}(\theta\rho_i)\, dz = \frac{\partial}{\partial t}\int_{z_1}^{z_2} (\theta\rho_i)\, dz - \theta\rho_i\bigg|_{z_2}\frac{\partial z_2}{\partial t} + \theta\rho_i\bigg|_{z_1}\frac{\partial z_1}{\partial t}$$

$$\int_{z_1}^{z_2} \frac{\partial}{\partial x}(\rho_i q_{ix})\, dz = \frac{\partial}{\partial x}\int_{z_1}^{z_2} \rho_i q_{ix}\, dz - \rho_i q_{ix}\bigg|_{z_2}\frac{\partial z_2}{\partial x} + \rho_i q_{ix}\bigg|_{z_1}\frac{\partial z_1}{\partial x}$$

$$\int_{z_1}^{z_2} \frac{\partial}{\partial y}(\rho_i q_{iy})\, dz = \frac{\partial}{\partial y}\int_{z_1}^{z_2} \rho_i q_{iy}\, dz - \rho_i q_{iy}\bigg|_{z_2}\frac{\partial z_2}{\partial y} + \rho_i q_{iy}\bigg|_{z_1}\frac{\partial z_1}{\partial y}$$

$$\int_{z_1}^{z_2} \frac{\partial}{dz}(\rho_i q_{iz})\, dz = \rho_i q_{iz}\bigg|_{z_2} - \rho_i q_{iz}\bigg|_{z_1}$$

Thus the integrated form of (5.23a) is

$$\frac{\partial}{\partial t}(\overline{\theta\rho}_i l) + \frac{\partial}{\partial x}(\bar{\rho}_i \bar{q}_{ix} l) + \frac{\partial}{\partial y}(\bar{\rho}_i \bar{q}_{iy} l)$$

$$+ \left[\left\{\theta\rho_i\bigg|_{z_1}\frac{\partial z_1}{\partial t} - \rho_i q_{iz}\bigg|_{z_1}\right\} - \left\{\theta\rho_i\bigg|_{z_2}\frac{\partial z_2}{\partial t} - \rho_i q_{iz}\bigg|_{z_2}\right\}\right] = 0 \quad (5.23\text{b})$$

where, for simplicity of notation, the spatial derivatives of $z$ have been assumed to be zero and the overbarred quantities are vertically integrated averages. The two terms in braces are rather interesting. One can interpret the time variation of $z$ to represent the moving grains at the boundary of the aquifer and $\rho_i q_{iz}|_{z_1}$ to represent the mass flux at the boundary. The sum of the two terms, therefore, is the net mass flux contributed to the system at each boundary relative to the grains. It is interesting to compare this situation to the earlier flow equation wherein, although not specifically indicated, the flux at the boundary was specified relative to the grains by virtue of the fact that $\mathbf{q}$ is thusly defined.

To obtain the form of the transport equation that is in general use, we introduce the constitutive relationship

$$\bar{\rho}_i \bar{\mathbf{q}}_i = \bar{\rho}_i \bar{\mathbf{q}} - \bar{\mathbf{D}} \cdot \nabla_{xy} \bar{\rho}_i \quad (5.24)$$

where $\mathbf{q}$ is the mass phase average velocity of the fluid and $\bar{\mathbf{D}}$ the dispersion tensor. Combination of (5.24) and (5.23b) yields

$$\partial(\overline{\theta\rho}_i l)/\partial t + \nabla_{xy} \cdot (\bar{\mathbf{q}}\bar{\rho}_i l) - \nabla_{xy} \cdot (\bar{\mathbf{D}} \cdot \nabla_{xy} \bar{\rho}_i) + \rho_i q_z^*\bigg|_{z_2} - \rho_i q_z^*\bigg|_{z_1} = 0$$

where $q_z^*$ is the mass flux from the confining layer relative to the moving boundary.

Although the role played by the dispersion coefficient $\bar{\mathbf{D}}$ in the transport equation is similar to that played by transmissivity in the flow equation, the formulation of the dispersion tensor is considerably more complex. There does not appear to be a unique form of $\bar{\mathbf{D}}$ that will accurately represent all field situations. The development generally adopted is that given by Scheidegger [13] for a porous medium, isotropic with respect to dispersivity:

$$\begin{aligned} \bar{D}_{xx} &= \bar{D}_{\mathrm{L}}(\bar{q}_x \bar{q}_x/\bar{q}^2) + \bar{D}_{\mathrm{T}}(\bar{q}_y \bar{q}_y/\bar{q}^2) + \bar{D}_{\mathrm{d}}\bar{\tau} \\ \bar{D}_{yy} &= \bar{D}_{\mathrm{L}}(\bar{q}_y \bar{q}_y/\bar{q}^2) + \bar{D}_{\mathrm{T}}(\bar{q}_x \bar{q}_x/\bar{q}^2) + \bar{D}_{\mathrm{d}}\bar{\tau} \\ \bar{D}_{yx} &= \bar{D}_{xy} = (\bar{D}_{\mathrm{L}} - \bar{D}_{\mathrm{T}})\bar{q}_x \bar{q}_y/\bar{q}^2 \end{aligned} \quad (5.25)$$

where $\bar{D}_L = \bar{\varepsilon}_1 |\bar{\mathbf{q}}|$, $\bar{D}_T = \bar{\varepsilon}_2 |\bar{\mathbf{q}}|$, $\bar{D}_L$ is a longitudinal dispersion coefficient, $\bar{D}_T$ a transverse dispersion coefficient, $\bar{D}_d$ a molecular diffusion coefficient, $\bar{\tau}$ tortuosity, and $\bar{\varepsilon}_1$ and $\bar{\varepsilon}_2$ the dispersivity coefficients.

*The Approximating Equations*

As in the case of the flow equations, let us obtain a set of approximating algebraic equations using Galerkin's procedure. Applying the orthogonality conditions and assuming a trial function of the form†

$$\bar{\rho} \simeq \sum_{j=1}^{M} c_j(t)\phi_j(x, y)$$

one obtains

$$\langle \nabla_{xy} \cdot (\bar{\mathbf{q}}\bar{\rho}l), \phi_i \rangle - \langle \nabla_{xy} \cdot (\bar{\mathbf{D}}l \cdot \nabla_{xy}\bar{\rho}), \phi_i \rangle + \langle \partial(\bar{\theta}\bar{\rho}l)/\partial t, \phi_i \rangle$$
$$+ \left\langle \rho q_z^* \Big|_{z2}, \phi_i \right\rangle - \left\langle \rho q_z^* \Big|_{z1}, \phi_i \right\rangle = 0 \qquad (i = 1, 2, \ldots, M)$$

or, upon application of Green's theorem and introduction of the trial solution, this expression becomes

$$\sum_{j=1}^{M} c_j[\langle \nabla_{xy} \cdot (\bar{\mathbf{q}}l\phi_j), \phi_i \rangle + \langle l(\bar{\mathbf{D}} \cdot \nabla_{xy}\phi_j), \nabla_{xy}\phi_i \rangle]$$
$$- \int_\Gamma l\mathbf{n} \cdot (\bar{\mathbf{D}} \cdot \nabla_{xy}\rho)\phi_i \, ds + \sum_{j=1}^{M} (dc_j/dt)\langle \theta l \phi_j, \phi_i \rangle$$
$$+ \left\langle \rho q_z^* \Big|_{z2} - \rho q_z^* \Big|_{z1}, \phi_i \right\rangle = 0 \qquad (i = 1, 2, \ldots, M) \qquad (5.26a)$$

Before the solution of Eq. (5.26a) is considered, let us examine how this equation differs in its basic structure from the groundwater flow equation (5.15) considered earlier. The most fundamental difference is the nonsymmetric form of the convective term [the first in (5.26a)]. Because Galerkin's method has been applied, this lack of symmetry causes no difficulty whatever, but if a variational principle is used, the development becomes far less straightforward. Nevertheless, the nonsymmetry of the differential operator will be reflected in a nonsymmetric coefficient matrix which, unfortunately, increases the computational effort required to achieve a solution.

The surface integral appearing in (5.26a) represents a diffusive mass flux boundary condition. Generally, this term is replaced by a known value calculated by integrating the product of the prescribed flux and the $i$th basis

† Note that the subscript $i$ has been dropped from $\rho$ to avoid confusion with indices.

function over the boundary $\Gamma$. There are occasions, however, when a boundary condition involving the convective and diffusive mass flux must be represented. This can be readily achieved by applying Green's theorem to both the first- and second-order space derivatives to give

$$-\langle(\bar{\mathbf{q}}\bar{\rho}l), \nabla_{xy}\phi_i\rangle + \langle l\bar{\mathbf{D}}\cdot(\nabla_{xy}\bar{\rho}), \nabla_{xy}\phi_i\rangle + \langle\partial(\bar{\theta}\bar{\rho}l)/\partial t, \phi_i\rangle$$
$$+ \left\langle \rho q_z^*\Big|_{z2}, \phi_i\right\rangle - \left\langle \rho q_z^*\Big|_{z1}, \phi_i\right\rangle$$
$$+ \int_\Gamma [\bar{\rho}\bar{\mathbf{q}}l - \bar{\mathbf{D}}l\cdot(\nabla_{xy}\bar{\rho})]\cdot\mathbf{n}\phi_i\, ds = 0 \qquad (i = 1, 2, \ldots, M) \tag{5.26b}$$

The total mass flux across the boundary $\Gamma$ now appears in the line integral.

There is a second somewhat subtle but nevertheless important difference between the two approaches to representing the transport equation. The term $\nabla_{xy}\cdot(\bar{\mathbf{q}}\bar{\rho})$ is often expanded to yield $\bar{\mathbf{q}}\cdot\nabla_{xy}\bar{\rho} + \bar{\rho}\,\nabla_{xy}\cdot\bar{\mathbf{q}}$. The divergence of the velocity is often considered negligible except at a point sink where it takes on the value of the mass discharge at the node. However, when Green's theorem is applied to the first-order term, the divergence of the velocity (and consequently the mass discharge) does not appear explicitly in the integral equation.

Possibly the most critical aspect of simulating mass transport in the areal plane is the specification of the velocity distribution. In hypothetical situations, the velocity is usually specified as either a known function or a constant. The simulation of natural processes, however, requires evaluation of the velocity distribution as well as the concentration distribution. Heretofore only the head distribution has been calculated, so we must find a simple scheme for obtaining the velocity field from this information.

For situations characterized by a relatively uniform piezometric surface, there is indeed a technique using the common zero-order continuous basis functions which requires very little computational effort. Moreover, it provides a continuous velocity distribution everywhere within an element but not at the element boundaries. Darcy's law states that

$$\bar{\mathbf{q}} = -\bar{\mathbf{K}}\cdot\nabla_{xy}\bar{h} \tag{5.27}$$

and from the finite element approximation for $\bar{h}$ we have

$$\nabla_{xy}\bar{h}(x, y, t) \simeq \nabla_{xy}\left\{\sum_{j=1}^{M} h_j(t)\phi_j(x, y)\right\} = \sum_{j=1}^{M} h_j(t)\,\nabla_{xy}\phi_j(x, y) \tag{5.28}$$

Combination of (5.28) and (5.27) yields

$$\bar{\mathbf{q}} \simeq -\sum_{j=1}^{M} h_j(t)\bar{\mathbf{K}} \cdot \nabla_{xy}\phi_j(x, y) \tag{5.29}$$

Because $h_j(t)$ is known from the solution of (5.16), Eq. (5.29) can be easily evaluated and substituted into (5.26a). Here the advantage of numerical integration lies in the fact that the Gauss points are located within the element where velocity values are readily obtainable. Inherent in this approach, however, is the discontinuity in velocity at the element boundary which, unfortunately, leads to a violation of the conservation of mass in a local sense. When the velocity distribution is rapidly changing within the region, this lack of continuity leads to erratic concentration values which bear very little relation to the correct solution. It will be shown how this difficulty can be circumvented, but generally at the expense of increased computational effort.

As with the representation of transmissivity, there is a choice of techniques for specifying the dispersion coefficient. However, because of the dependence of this parameter on the mass average velocity, it is advantageous to use a functional representation.

The algebraic equations that approximate the transport equation can be written in matrix form as

$$[A]\{C\} + [B]\{dC/dt\} = \{F\} \tag{5.30}$$

where

$$a_{i,j} = \int_A \left( \left[ \sum_{k=1}^{M} \bar{q}_{xk}\phi_k l \frac{\partial \phi_j}{\partial x} + \sum_{k=1}^{M} \bar{q}_{yk}\phi_k l \frac{\partial \phi_j}{\partial y} \right] \phi_i + \left[ \sum_{k=1}^{M} l\bar{D}_{xxk}\phi_k \frac{\partial \phi_j}{\partial x}\frac{\partial \phi_i}{\partial x} \right.\right.$$
$$+ \sum_{k=1}^{M} l\bar{D}_{yxk}\phi_k \frac{\partial \phi_j}{\partial x}\frac{\partial \phi_i}{\partial y}$$
$$\left.\left. + \sum_{k=1}^{M} l\bar{D}_{xyk}\phi_k \frac{\partial \phi_j}{\partial y}\frac{\partial \phi_i}{\partial x} + \sum_{k=1}^{M} l\bar{D}_{yyk}\phi_k \frac{\partial \phi_j}{\partial y}\frac{\partial \phi_i}{\partial y} \right] \right) dA$$

$$b_{i,j} = \bar{\theta} l \int \phi_j \phi_i \, dA$$

$$f_i = -\int_\Gamma l\bar{\mathbf{q}}_{\mathrm{m}} \cdot \mathbf{n}\phi_i \, ds + \int_A \left( \rho q_z^* \Big|_{z1} - \rho q_z^* \Big|_{z2} \right) \phi_i \, dA$$

and $\bar{\mathbf{q}}_{\mathrm{m}}$ is a vertically integrated dispersive mass flux across $\Gamma$, the system boundary.

In obtaining these coefficients, $\nabla_{xy} \cdot \mathbf{q}$ has been assumed negligible compared to other terms in this equation and the porosity has been considered constant over an element. Although the aquifer thickness is herein treated to be a separate parameter, in practice an effective parameter, say dispersivity, is defined by multiplying the vertically averaged value of the parameter by the thickness.

When Green's theorem is applied to both the first- and second-order space derivatives as indicated in Eq. (5.26b), the elements of $[A]$ and $\{F\}$ in (5.30) become

$$a_{i,j} = \int_A \Bigg( - \left[ \sum_{k=1}^{M} \bar{q}_{xk} \phi_k l \frac{\partial \phi_i}{\partial x} + \sum_{k=1}^{M} \bar{q}_{yk} \phi_k l \frac{\partial \phi_i}{\partial y} \right] \phi_j$$

$$+ \left[ \sum_{k=1}^{M} l\bar{D}_{xxk} \phi_k \frac{\partial \phi_j}{\partial x} \frac{\partial \phi_i}{\partial x} + \sum_{k=1}^{M} l\bar{D}_{yxk} \phi_k \frac{\partial \phi_j}{\partial x} \frac{\partial \phi_i}{\partial y} \right.$$

$$\left. + \sum_{k=1}^{M} l\bar{D}_{xyk} \phi_k \frac{\partial \phi_j}{\partial y} \frac{\partial \phi_i}{\partial x} + \sum_{k=1}^{M} l\bar{D}_{yyk} \phi_k \frac{\partial \phi_j}{\partial y} \frac{\partial \phi_i}{\partial y} \right] \Bigg) \, dA$$

$$f_i = - \int_\Gamma [\bar{\rho} \mathbf{q} l - \bar{\mathbf{D}} l \cdot (\nabla_{xy} \bar{\rho})] \cdot \mathbf{n} \phi_i \, ds + \int_A \left( \rho q_z^* \Big|_{z_1} - \rho q_z^* \Big|_{z_2} \right) \varphi_i \, dA$$

It should be noted that in this form of the equation, the assumption that $\nabla_{xy} \cdot \mathbf{q}$ vanishes is unnecessary since the divergence of the velocity does not appear explicitly in (5.26b).

Because we have assumed that the fluid density is independent of the concentration equation, the solution to (5.30) depends on the solution to (5.16) but the converse does not hold. Accordingly, the two equations, in conjunction with Darcy's law are solved sequentially without iteration. Often, however, the pressure propagation described by the flow equation is very rapid relative to the changes in concentration described by the transport equation. Under these circumstances an optimal time step for one equation is generally not suitable for the other. This difficulty is circumvented by defining the time step for the head equation as a multiple of the time step for the concentration equation, for example,

$$\Delta t_\mathrm{p} = m \, \Delta t_\mathrm{c}$$

where $\Delta t_\mathrm{p}$ is the time step for head, $\Delta t_\mathrm{c}$ the time step for concentration, and the integer $m$ the number of times the transport equation is solved between solutions for groundwater flow.

Equation (5.30) once again represents a set of ordinary differential equations which can be solved by approximating the time derivative using finite difference techniques. Unfortunately, a solution to the transport equation is more difficult to obtain numerically than a solution to the flow equation. The problems are particularly severe when convective transport dominates over diffusive transport. Figure 5.7 shows typical solutions obtained using a variety of numerical schemes. In general, the solutions are characterized by two phenomena, overshoot and numerical dissipation.† Overshoot describes the erroneously high values of concentration encountered as one approaches a sharp front from the upstream side. The analogous behavior on the downstream side is called undershoot. Numerical dissipation characteristically smears the sharp front, thereby generating a solution indicative of a larger dispersion coefficient. In practice the two phenomena are closely related. When a scheme is developed to minimize numerical dissipation, overshoot is encountered, but when overshoot is controlled it is generally at the expense of increased numerical dissipation.

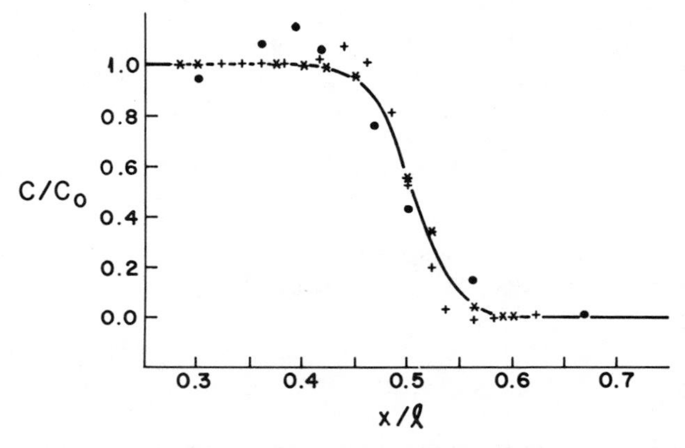


**Fig. 5.7.** Comparison of central difference (CDM) (●), Galerkin (∗), finite element (Rayleigh–Ritz) (+), and analytical (——) solutions of the one-dimensional transport equations [15].

*The Use of First-Order Continuous Functions*

Recently isoparametric elements based on first-order continuous basis functions have been introduced for the solution of the mass transport equation [14]. In test problems, this approach has provided accurate solutions with less computational effort than standard finite element approaches. Moreover, the problems associated with discontinuous velocity fields may be circumvented if the first derivative of hydraulic head (or fluid pressure) is continuous everywhere within the domain $D$, including interelement boundaries. The details of this numerical procedure are presented in Chapter 4. While this type of element has heretofore been used only in

† Note that numerical dissipation is also reported in the literature as numerical dispersion.

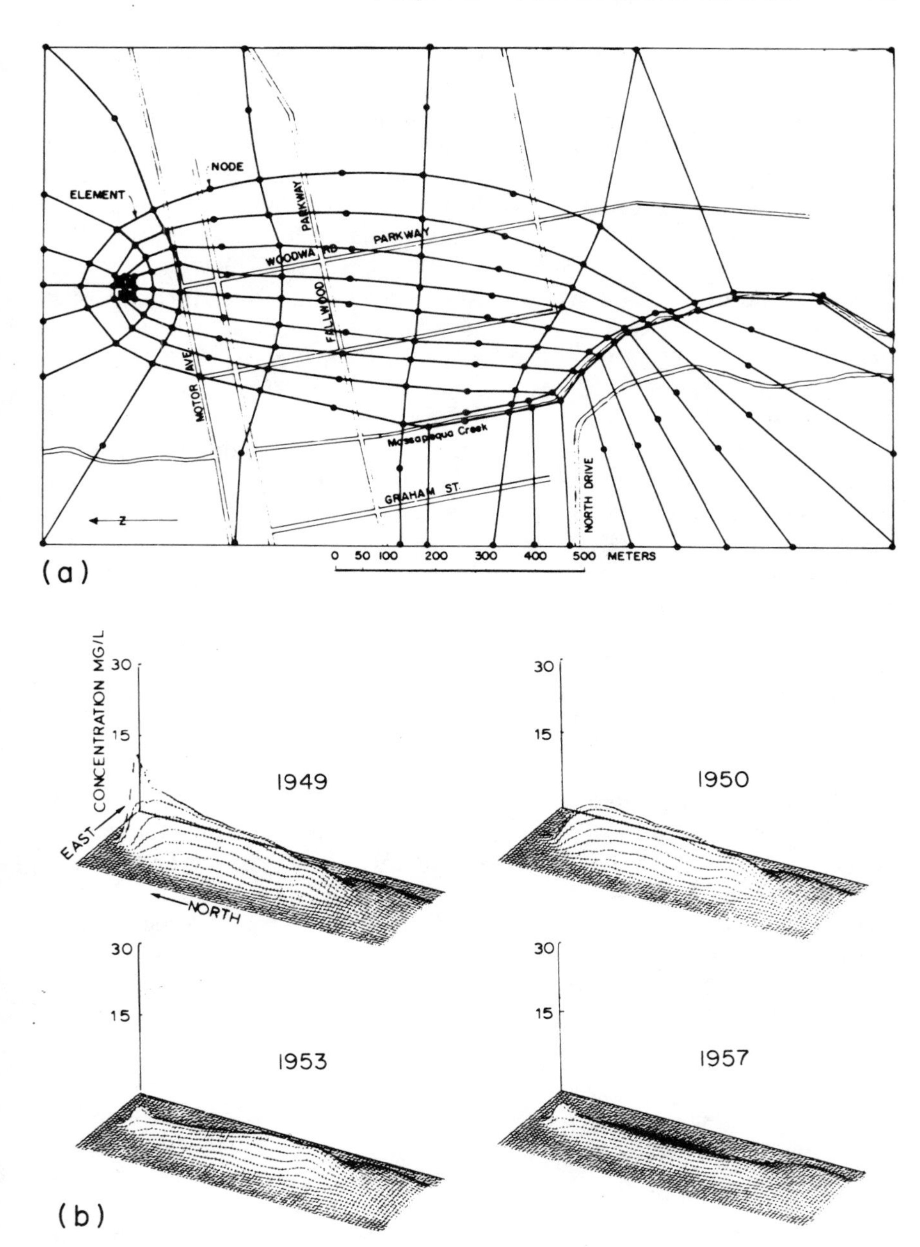


**Fig. 5.8.** (a) Finite element configuration used for groundwater flow and chromium transport equation. (b) Simulated areal distribution of contamination for the period 1949–1957 assuming that strength of contaminating source decreases by 75% after 1949.

model problems, it appears likely that this method will provide an effective means for the solution of field problems.

A typical finite element net for the solution of problems involving the areal, two-dimensional flow of nonhomogeneous fluids is illustrated in Fig. 5.8a [16]. Because this net is used for both the concentration transport equation and the flow equation, one has less freedom in selecting nodal and element locations than if two independent nets were being utilized. The problem depicted involves the movement of hexavalent chromium from a disposal pond near Motor Avenue, southward toward Massapequa Creek. Simulation of this system requires the solution of the steady state groundwater flow equation and the transient mass transport equation. The pond is treated as a constant concentration boundary and the creek as a leaky discharge area. Dirichlet nodes define the perimeter of the model. It is worth pointing out that when fluid leaves the system (here it discharges to the creek) the concentration of the discharging fluid is obviously the same as that in the aquifer, and consequently there is no sink term appearing in the transport equation (5.26a). However, when a source fluid enters the aquifer at a concentration different from the resident fluid, a source term is required. Several solutions for selected times after contamination began are presented in Fig. 5.8b.

### *5.3.2 Fourier Analysis of the Numerical Solution of the Mass Transport Equation*

While a qualitative description of the difficulties encountered in solving the transport equation has been presented, it is also useful to examine the phenomena quantitatively. One vehicle for such an analysis is an examination of the behavior of Fourier series components as they are propagated analytically and numerically [17]. A significant amount of information concerning the relative accuracy of various numerical schemes can be obtained by considering the one-dimensional transport equation given by

$$\frac{\partial c}{\partial t} + u\frac{\partial c}{\partial x} - D\frac{\partial^2 c}{\partial x^2} = 0 \tag{5.31}$$

where $u$, the velocity, and $D$, the dispersion coefficient, are assumed constant.

*Analytical Solution*

Let the general solution to (5.31) be represented by the Fourier series:

$$c = \sum_{n=-\infty}^{\infty} C_n \exp(\hat{\imath}\beta_n t + \hat{\imath}\sigma_n x) \tag{5.32}$$

where $-1 \leq (x - ut) \leq 1$, $\beta_n$ is the frequency of the $n$th component, $\sigma_n$ the spatial frequency or wave number, and $\hat{i} = (-1)^{1/2}$. Because Eq. (5.31) is linear, it is necessary to consider only one component of the summation represented by (5.32),

$$c \sim C_n \exp(\hat{i}\beta_n t + \hat{i}\sigma_n x) \tag{5.33}$$

To determine the analytical relationship between the wave number and the frequency, Eq. (5.33) is introduced into (5.31) to give

$$\beta_n + u\sigma_n - \hat{i}D\sigma_n^2 = 0$$

or, for the frequency $\beta_n$ as a function of $\sigma_n$,

$$\beta_n = \sigma_n(\hat{i}D\sigma_n - u) \tag{5.34}$$

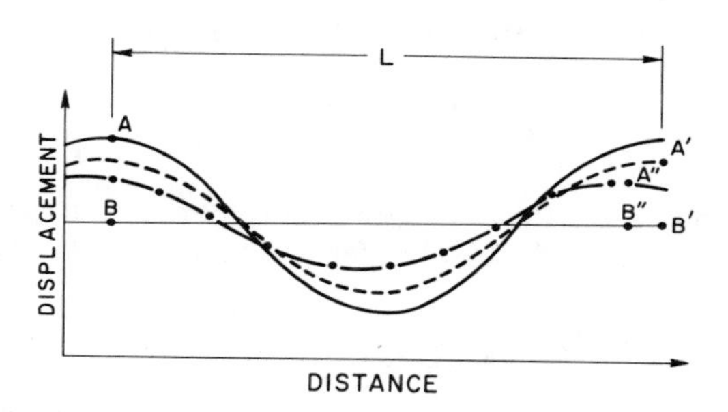


**Fig. 5.9.** Diagrammatic representation of typical waveform in Fourier analysis: $u$ is wave celerity, $A''B''/A'B'$ the amplitude ratio, and $2\pi[1 - (BB''/BB')]$ the phase lag (modified after Leendertse [17]). The curves shown are for the initial wave (——), the wave after time $L/u$ (- - -), and the computed wave after time $L/u$ (—·—).

Substitution of (5.34) into (5.33) yields the general solution to the governing equation

$$c \sim C_n \exp[\hat{i}\sigma_n(x - ut)]\exp[-D\sigma_n^2 t]$$

where the first exponential describes the translation and the second the amplitude modification of a Fourier component as a function of time. Accordingly, at an elapsed simulation time of $\Delta t$, the magnitude of the $n$th wave will decrease to $\exp(-D\sigma_n^2\,\Delta t)$ of its original amplitude and the wave will have advanced a distance $u\,\Delta t$. However, the amplitude and phase of this component, when propagated by a numerical scheme, do not necessarily coincide with these values as illustrated in Fig. 5.9. This discrepancy can be evaluated by comparing the Fourier analyses of the analytical and numerical solutions.

*Numerical Solution*

Although several numerical schemes for solving the transport equation will be compared, only the development for the linear finite element method will be considered in detail. The same general procedure is followed for each scheme although considerably more algebra is required in some cases than in others.

The approximating equations for the finite element method are generated using a Galerkin formulation as outlined in Chapter 3. In this procedure, the concentration is approximated by a finite series of the form

$$c \approx \hat{c} = \sum_{j=1}^{M} C_j(t)\phi_j(x) \tag{5.35}$$

where the $C_j$ are undetermined, time-dependent coefficients and the $\phi_j(x)$ are chapeau basis functions illustrated in Fig. 4.2 and defined for the node $j$ as

$$\phi_j(x) = \begin{cases} (x - x_{j-1})/(x_j - x_{j-1}) & (x_{j-1} \le x \le x_j) \\ (x_{j+1} - x)/(x_{j+1} - x_j) & (x_j \le x \le x_{j+1}) \end{cases} \tag{5.36}$$

Applying the Galerkin procedure, the approximating series (5.35) is substituted into the governing partial differential equation (5.31) to obtain

$$L\hat{c} \equiv \frac{\partial \hat{c}}{\partial t} + u\frac{\partial \hat{c}}{\partial x} - D\frac{\partial^2 \hat{c}}{\partial x^2} = \mathscr{R} \tag{5.37}$$

and the residual $\mathscr{R}$ is made orthogonal to each of the $M$ basis functions appearing in (5.35). Thus the following set of $M$ equations in $M$ coefficients results:

$$\int (L\hat{c})\phi_i(x)\,dx = 0 \qquad (i = 1, 2, 3, \ldots, M)$$

Combination of Eqs. (5.35)–(5.37), performance of the required integrations, and approximation of the time derivative of $C_j(t)$ using a finite difference representation, results in the following numerical analogue at an interior node:

$$\begin{aligned} &\frac{1}{6}\left[\frac{C_{i+1,k+1} - C_{i+1,k}}{\Delta t} + 4\frac{C_{i,k+1} - C_{i,k}}{\Delta t} + \frac{C_{i-1,k+1} - C_{i-1,k}}{\Delta t}\right] \\ &\quad + u\left[\varepsilon\frac{C_{i+1,k+1} - C_{i-1,k+1}}{2\,\Delta x} + (1-\varepsilon)\frac{C_{i+1,k} - C_{i-1,k}}{2\,\Delta x}\right] \\ &\quad - D\left[\varepsilon\frac{C_{i+1,k+1} - 2C_{i,k+1} + C_{i-1,k+1}}{(\Delta x)^2}\right. \\ &\quad \left. + (1-\varepsilon)\frac{C_{i+1,k} - 2C_{i,k} + C_{i-1,k}}{(\Delta x)^2}\right] = 0 \end{aligned} \tag{5.38}$$

where, as earlier, $\varepsilon$ is a weighting coefficient lying between 0 and 1, $x = i\,\Delta x$, $t = k\,\Delta t$, and $\Delta x$ and $\Delta t$ are assumed constant in space and time, respectively.

We can assume a solution to (5.38) similar in form to (5.32) such that

$$\hat{c} = \sum_{n=-\infty}^{\infty} C_n \exp(\hat{i}\beta'_n t + \hat{i}\sigma_n x) \tag{5.39}$$

where the frequency $\beta'_n$ obtained in the numerical solution is not necessarily equal to that obtained in the analytical solution. Because the undetermined coefficient $C_{i,k}$ in (5.38) represents the unknown function $\hat{c}$ at spatial node $i$ and the time level $k$, a Fourier series component of (5.39) is substituted in to give

$$\begin{aligned}
&\frac{1}{6\,\Delta t}\{(\lambda'_n - 1)[4 + 2\cos(\sigma_n\,\Delta x)]\} \\
&\quad + \frac{u}{2\,\Delta x}\{\varepsilon\lambda'_n[2\hat{i}\sin(\sigma_n\,\Delta x)] + (1-\varepsilon)[2\hat{i}\sin(\sigma_n\,\Delta x)]\} \\
&\quad - \frac{D}{(\Delta x)^2}\{\varepsilon\lambda'_n[2\cos(\sigma_n\,\Delta x) - 2] + (1-\varepsilon)[2\cos(\sigma_n\,\Delta x) - 2]\} = 0
\end{aligned}$$

where $\exp(\hat{i}\beta'_n\,\Delta t)$ has been denoted by $\lambda'$, which is an eigenvalue such that $C_{t+\Delta t} = \lambda' C_t$. This equation may be solved for the eigenvalue

$$\lambda'_n = \frac{\frac{1}{3}[2 + \cos(\sigma_n\,\Delta x)] - (1-\varepsilon)\{\mathscr{U}\hat{i}\sin(\sigma_n\,\Delta x) + 2\mathscr{D}[1 - \cos(\sigma_n\,\Delta x)]\}}{\frac{1}{3}[2 + \cos(\sigma_n\,\Delta x)] + \varepsilon\{\mathscr{U}\hat{i}\sin(\sigma_n\,\Delta x) + 2\mathscr{D}[1 - \cos(\sigma_n\,\Delta x)]\}} \tag{5.40}$$

where the dimensionless groups $u\,\Delta t/\Delta x$ and $D\,\Delta t/(\Delta x)^2$ have been replaced by $\mathscr{U}$ and $\mathscr{D}$, respectively.

Eigenvalue equations analogous to (5.40) appear in Table 5.1 for a finite difference approximation and for linear, quadratic, and Hermitian cubic finite element approximations to the spatial derivatives in the convective–dispersion equation. Furthermore, eigenvalue expressions are given for finite difference and linear finite element approximations to an equivalent two-equation form of the convective–dispersive equation:

$$\frac{\partial c}{\partial t} + \frac{\partial q}{\partial x} = 0 \qquad \text{and} \qquad q - uc + D\frac{\partial c}{\partial x} = 0$$

*Comparison of Analytical and Numerical Solutions*

Using the analytically derived frequency given by (5.34) we can write

$$\lambda_n = \exp(\hat{i}\beta_n\,\Delta t) = \exp(-\hat{i}\sigma_n u\,\Delta t - \sigma_n^2 D\,\Delta t) \tag{5.41a}$$

## Table 5.1

Numerical Schemes Used for Solution of the Convective–Dispersive Equation with Their Corresponding Eigenvalues

| | GOVERNING EQUATIONS | DISCRETIZED EQUATIONS | EIGENVALUES |
|---|---|---|---|
| SINGLE EQUATION FINITE ELEMENT | $\frac{\partial c}{\partial t} + u\frac{\partial c}{\partial x} - D\frac{\partial^2 c}{\partial x^2} = 0$ | $\frac{1}{6}\left[\frac{C_{i+1,k+1}-C_{i+1,k}}{\Delta t} + 4\frac{C_{i,k+1}-C_{i,k}}{\Delta t} + \frac{C_{i-1,k+1}-C_{i-1,k}}{\Delta t}\right]$ $+ u\left[\varepsilon\frac{C_{i+1,k+1}-C_{i-1,k+1}}{2\Delta x} + (1-\varepsilon)\frac{C_{i+1,k}-C_{i-1,k}}{2\Delta x}\right]$ $- D\left[\varepsilon\frac{C_{i+1,k+1}-2C_{i,k+1}+C_{i-1,k+1}}{(\Delta x)^2} + (1-\varepsilon)\frac{C_{i+1,k}-2C_{i,k}+C_{i-1,k}}{(\Delta x)^2}\right] = 0$ | $\dfrac{\frac{2+\cos(\sigma_n\Delta x)}{3} - (1-\varepsilon)\{\mathcal{U}\hat{i}\sin(\sigma_n\Delta x)+2\mathcal{D}[1-\cos(\sigma_n\Delta x)]\}}{\frac{2+\cos(\sigma_n\Delta x)}{3} + \varepsilon\{\mathcal{U}\hat{i}\sin(\sigma_n\Delta x) + 2\mathcal{D}[1-\cos(\sigma_n\Delta x)]\}}$ |
| SINGLE EQUATION FINITE DIFFERENCE | $\frac{\partial c}{\partial t} + u\frac{\partial c}{\partial x} - D\frac{\partial^2 c}{\partial x^2} = 0$ | $\frac{C_{i,k+1}-C_{i,k}}{\Delta t} + u\left[\varepsilon\frac{C_{i+1,k+1}-C_{i-1,k+1}}{2\Delta x} + (1-\varepsilon)\frac{C_{i+1,k}-C_{i-1,k}}{2\Delta x}\right]$ $- D\left[\varepsilon\frac{C_{i+1,k+1}-2C_{1,k+1}+C_{i-1,k+1}}{(\Delta x)^2} + (1-\varepsilon)\frac{C_{i+1,k}-2C_{i,k}+C_{i-1,k}}{(\Delta x)^2}\right] = 0$ | $\dfrac{1-(1-\varepsilon)\{\mathcal{U}\hat{i}\sin(\sigma_n\Delta x) + 2\mathcal{D}[1-\cos(\sigma_n\Delta x)]\}}{1+\varepsilon\{\mathcal{U}\hat{i}\sin(\sigma_n\Delta x) + 2\mathcal{D}[1-\cos(\sigma_n\Delta x)]\}}$ |
| TWO EQUATION FINITE ELEMENT | $\frac{\partial c}{\partial t} + \frac{\partial q}{\partial x} = 0$ <br> $q - uc + D\frac{\partial c}{\partial x} = 0$ | $\frac{1}{6}\left[\frac{C_{i+1,k+1}-C_{i+1,k}}{\Delta t} + 4\frac{C_{i,k+1}-C_{i,k}}{\Delta t} + \frac{C_{i-1,k+1}-C_{i-1,k}}{\Delta t}\right]$ $+ \varepsilon\frac{\mathcal{Q}_{i+1,k+1}-\mathcal{Q}_{i-1,k+1}}{2\Delta x} + (1-\varepsilon)\frac{\mathcal{Q}_{i+1,k}-\mathcal{Q}_{i-1,k}}{2\Delta x} = 0$ <br> $\frac{1}{6}\left[\mathcal{Q}_{i+1,k+1} + 4\mathcal{Q}_{i,k+1}+\mathcal{Q}_{i-1,k+1}\right] - \frac{1}{6}u\left[C_{i+1,k+1} + 4C_{i,k+1} + C_{i-1,k+1}\right]$ $+ D\left[\frac{C_{i+1,k+1}-C_{i-1,k+1}}{2\Delta x}\right]$ | $\dfrac{\frac{2+\cos(\sigma_n\Delta x)}{3} - (1-\varepsilon)\left[\mathcal{U}\hat{i}\sin(\sigma_n\Delta x) + \mathcal{D}\frac{3\sin^2(\sigma_n\Delta x)}{2+\cos(\sigma_n\Delta x)}\right]}{\frac{2+\cos(\sigma_n\Delta x)}{3} + \varepsilon\left[\mathcal{U}\hat{i}\sin(\sigma_n\Delta x) + \mathcal{D}\frac{3\sin^2(\sigma_n\Delta x)}{2+\cos(\sigma_n\Delta x)}\right]}$ |

| | | | |
|---|---|---|---|
| TWO EQUATION<br>FINITE DIFFERENCE | $\frac{\partial c}{\partial t} + \frac{\partial q}{\partial x} = 0$<br><br>$q - uc + D \frac{\partial c}{\partial x} = 0$ | $\frac{C_{i,k+1} - C_{i,k}}{\Delta t} + \varepsilon \frac{\mathcal{Q}_{i+1,k+1} - \mathcal{Q}_{i-1,k+1}}{2\Delta x} + (1-\varepsilon) \frac{\mathcal{Q}_{i+1,k} - \mathcal{Q}_{i-1,k}}{2\Delta x} = 0$<br><br>$\mathcal{Q}_{i,k+1} - u\, C_{i,k+1} + D \frac{C_{i+1,k+1} - C_{i-1,k+1}}{2\Delta x} = 0$ | $\frac{1 - (1-\varepsilon)\left[\mathcal{U}\hat{i} \sin(\sigma_n \Delta x) + \mathcal{D} \sin^2(\sigma_n \Delta x)\right]}{1 + \varepsilon\left[\mathcal{U}\hat{i} \sin(\sigma_n \Delta x) + \mathcal{D} \sin^2(\sigma_n \Delta x)\right]}$ |
| SINGLE EQUATION<br>QUADRATIC FINITE<br>ELEMENT | $\frac{\partial c}{\partial t} + u \frac{\partial c}{\partial x} - D \frac{\partial^2 c}{\partial x^2} = 0$ | FOR CORNER NODE i<br><br>$\frac{dC_i}{dt} = \frac{C_{i,k+1} - C_{i,k}}{\Delta t}$<br><br>$\frac{1}{10}\left[-\frac{dC_{i-2}}{dt} + 2\frac{dC_{i-1}}{dt} + 8\frac{dC_i}{dt} + 2\frac{dC_{i+1}}{dt} - \frac{dC_{i+2}}{dt}\right]$<br>$+ u\left[\varepsilon \frac{C_{i-2,k+1} - 4C_{i-1,k+1} + 4C_{i+1,k+1} - C_{i+2,k+1}}{4\Delta x}\right.$<br>$\left. + (1-\varepsilon) \frac{C_{i-2,k} - 4C_{i-1,k} + 4C_{i+1,k} - C_{i+2,k}}{4\Delta x}\right]$<br>$- D\left[\varepsilon \frac{-C_{i-2,k+1} + 8C_{i-1,k+1} - 14C_{i,k+1} + 8C_{i+1,k+1} - C_{i+2,k+1}}{4(\Delta x)^2}\right.$<br>$\left. + (1-\varepsilon) \frac{-C_{i-2,k} + 8C_{i-1,k} - 14C_{i,k} + 8C_{i+1,k} - C_{i+2,k}}{4(\Delta x)^2}\right] = 0$<br><br>FOR SIDE NODE i-1<br><br>$\frac{1}{10}\left[\frac{dC_{i-2}}{dt} + 8\frac{dC_{i-1}}{dt} + \frac{dC_i}{dt}\right] + u\left[\varepsilon \frac{C_{i,k+1} - C_{i-2,k+1}}{2\Delta x} + (1-\varepsilon) \frac{C_{i,k} - C_{i-2,k}}{2\Delta x}\right]$<br>$- D\left[\varepsilon \frac{C_{i,k+1} - 2C_{i-1,k+1} + C_{i-2,k+1}}{(\Delta x)^2} + (1-\varepsilon) \frac{C_{i,k} - 2C_{i-1,k} + C_{i-2,k}}{(\Delta x)^2}\right] = 0$ | $\Big\{ -\frac{5\varepsilon}{2} (1-\varepsilon)\, [\cos(2\sigma_n \Delta x) - 1]\, [\mathcal{U}^2 + 3\mathcal{D}^2]$<br>$+ (1-2\varepsilon)\, \mathcal{D}\, [\cos(2\sigma_n \Delta x) + 6.5]$<br>$+ \frac{1}{2} [\cos(2\sigma_n \Delta x) - 3] - \hat{i}\mathcal{U} \sin(2\sigma_n \Delta x)\, [(1-2\varepsilon) - 7.5\, \mathcal{D}\varepsilon\, (1-\varepsilon)]$<br>$\pm \Big( \mathcal{D}^2 [\cos(2\sigma_n \Delta x) + 6.5]^2 - \frac{5}{4} [1 - \cos(2\sigma_n \Delta x)] \cdot$<br>$[\mathcal{U}^2 + 3\mathcal{D}^2][3 - \cos(2\sigma_n \Delta x)] - \mathcal{U}^2 \sin^2(2\sigma_n \Delta x)$<br>$+ \hat{i}\mathcal{U} \sin(2\sigma_n \Delta x) \Big[ -2\mathcal{D}\big[\cos(2\sigma_n \Delta x) + 6.5\big]$<br>$+ \frac{15}{4} \mathcal{D}\big[\cos(2\sigma_n \Delta x) - 3\big]\Big]\Big)^{1/2} \Big\} \Big\{ \frac{5}{2} \varepsilon^2 [\cos(2\sigma_n \Delta x) - 1] \cdot$<br>$[\mathcal{U}^2 + 3\mathcal{D}^2] - 2\varepsilon\mathcal{D}[\cos(2\sigma_n \Delta x) + 6.5]$<br>$+ \frac{1}{2} [\cos(2\sigma_n \Delta x) - 3] + \hat{i}\mathcal{U} \sin(2\sigma_n \Delta x)[2\varepsilon - \frac{15}{2} \mathcal{D}\varepsilon^2] \Big\}^{-1}$ |

**Table 5.1**—*(continued)*

| | GOVERNING EQUATIONS | DISCRETIZED EQUATIONS | EIGENVALUES |
|---|---|---|---|
| HERMITIAN CUBICS<br>SINGLE EQUATION | $\frac{\partial c}{\partial t}+u\frac{\partial c}{\partial x}-D\frac{\partial^2 c}{\partial x^2}=0$ | $\frac{1}{70}\left[9\frac{dC_{i-1}}{dt}+52\frac{dC_i}{dt}+9\frac{dC_{i+1}}{dt}\right]+\frac{13\Delta x}{420}\left[\frac{dC'_{i-1}}{dt}-\frac{dC'_{i+1}}{dt}\right]$<br>$+u\left[\varepsilon\frac{C_{i+1,k+1}-C_{i-1,k+1}}{2\Delta x}+(1-\varepsilon)\frac{C_{i+1,k}-C_{i-1,k}}{2\Delta x}\right]$<br>$+\frac{u}{10}\Big[\varepsilon\,(-C'_{i+1,k+1}+2C'_{i,k+1}-C'_{i-1,k+1})$<br>$+(1-\varepsilon)\,(-C'_{i+1,k}+2C'_{i,k}-C'_{i-1,k})\Big]-\frac{6D}{5}\Big[\varepsilon\frac{C_{i+1,k+1}-2C_{i,k+1}+C_{i-1,k+1}}{(\Delta x)^2}$<br>$+(1-\varepsilon)\frac{C_{i+1,k}-2C_{i,k}+C_{i-1,k}}{(\Delta x)^2}\Big]+\frac{D}{5}\Big[\varepsilon\frac{C'_{i+1,k+1}-C'_{i-1,k+1}}{2\Delta x}$<br>$+(1-\varepsilon)\frac{C'_{i+1,k}-C'_{i-1,k}}{2\Delta x}\Big]=0$ | $1+\frac{B-2\varepsilon C\pm(B^2+4AC)^{1/2}}{2(-A-\varepsilon B+\varepsilon^2 C)}$<br>$A=[\frac{7}{6}\sin^2(\sigma_n\Delta x)-77+42\cos(\sigma_n\Delta x)]/49$<br>$B=[32U\hat{i}\sin(\sigma_n\Delta x)-2U\hat{i}\cos(\sigma_n\Delta x)\sin(\sigma_n\Delta x)$<br>$-282D+8D\cos^2(\sigma_n\Delta x)+64D\cos(\sigma_n\Delta x)]/7$<br>$C=[40\hat{i}\,DU\sin(\sigma_n\Delta x)-10\hat{i}\,DU\cos(\sigma_n\Delta x)\sin(\sigma_n\Delta x)$<br>$+120D^2-120D^2\cos(\sigma_n\Delta x)-30D^2\sin^2(\sigma_n\Delta x)$<br>$+12U^2-12U^2\cos(\sigma_n\Delta x)-U^2\sin^2(\sigma_n\Delta x)]$ |
| | | $\frac{13}{14}\left[\frac{\frac{dC_{i+1}}{dt}-\frac{dC_{i-1}}{dt}}{2\Delta x}\right]+\frac{1}{14}\left[-\frac{3}{2}\frac{dC'_{i+1}}{dt}+4\frac{dC'_i}{dt}-\frac{3}{2}\frac{dC'_{i-1}}{dt}\right]$<br>$+\frac{3u}{2}\left[\varepsilon\frac{C_{i+1,k+1}-2C_{i,k+1}+C_{i-1,k+1}}{(\Delta x)^2}+(1-\varepsilon)\frac{C_{i+1,k}-2C_{i,k}+C_{i-1,k}}{(\Delta x)^2}\right]$<br>$-\frac{u}{2}\left[\varepsilon\frac{C'_{i+1,k+1}-C'_{i-1,k+1}}{2\Delta x}+(1-\varepsilon)\frac{C'_{i+1,k}-C'_{i-1,k}}{2\Delta x}\right]$<br>$-\frac{D}{2}\Big\{\varepsilon\frac{C'_{i+1,k+1}-2\left[4C'_{i,k+1}-3\frac{(C_{i+1,k+1}-C_{i-1,k+1})}{2\Delta x}\right]+C'_{i-1,k+1}}{(\Delta x)^2}$<br>$+(1-\varepsilon)\frac{C'_{i+1,k}-2\left[4C'_{i,k}-3\frac{(C_{i+1,k}-C_{i-1,k})}{2\Delta x}\right]+C'_{i-1,k}}{(\Delta x)^2}\Big\}=0$ | |

as the analytical eigenvalue. Because

$$\exp(-\hat{i}\sigma_n u \, \Delta t) = \cos(\sigma_n u \, \Delta t) - \hat{i} \sin(\sigma_n u \, \Delta t)$$

and has a magnitude of unity, the ratio of the magnitude of the analytical wave after one time step to its magnitude at the beginning of the step will be

$$|\lambda_n| = \exp(-\sigma_n^2 D \, \Delta t) \tag{5.41b}$$

By a similar argument, it can be shown that the phase angle of the Fourier component after time $\Delta t$ is given by the relationship

$$\exp(\hat{i}\theta_n) = e^{\hat{i}\beta_n \Delta t}/|e^{\hat{i}\beta_n \Delta t}| = \lambda_n/|\lambda_n| \equiv \Lambda_n$$

where $\theta_n$ is the phase angle. The phase angle may be expressed in terms of $\Lambda_n$ through the relationship

$$e^{\hat{i}\theta_n} = \cos \theta_n + \hat{i} \sin \theta_n = \Lambda_n$$

which can be rewritten to yield

$$\theta_n = \cos^{-1}[\mathrm{Re}(\Lambda_n)] = \sin^{-1}[\mathrm{Im}(\Lambda_n)]$$

where Re and Im refer to the real and imaginary parts of a complex number, respectively. Another expression for $\theta_n$ may be obtained from the general expression for a complex number

$$r_n e^{\hat{i}\theta_n} = e^{\hat{i}\beta_n \Delta t} = \exp(-\hat{i}\sigma_n \, \Delta t u) \exp(-\sigma_n^2 D \, \Delta t)$$

Equation of the real and imaginary parts of this expression yields $\theta_n = -\sigma_n \Delta t u$. In summary, two expressions for the phase angle and the amplitude have been generated, one in terms of frequency and wave number and the other in terms of the eigenvalue $\lambda_n$. To demonstrate the ability of numerical schemes to propagate the Fourier components accurately, the eigenvalues of the numerical scheme will now be utilized to generate a comparison with the analytical solution.

Let the eigenvalue and phase angle of the $n$th Fourier component obtained from the numerical scheme be denoted by $\lambda_n'$ and $\theta_n'$, respectively. Now consider the computed and analytical components after a time such that the analytical wave has propagated one wavelength. The number of time steps required, $N_n$, is obtained from

$$N_n = L_n/u \, \Delta t = (L_n/\Delta x)(\Delta x/u \, \Delta t)$$

where $L_n$ is the wavelength of the $n$th component. Thus the ratio of the computed to actual amplitude after one wavelength is given by

$$\frac{\text{computed amplitude}}{\text{actual amplitude}} = \left[\frac{|\lambda'_n|}{\exp(-\sigma_n^2 D\,\Delta t)}\right]^{N_n}$$

$$= \left[\frac{|\lambda'_n|}{\exp(-4\pi^2\mathscr{D}(\Delta x/L_n)^2)}\right]^{L_n/\Delta x \mathscr{U}} \tag{5.41c}$$

where $\mathscr{U} = u\,\Delta t/\Delta x$ and $\mathscr{D} = D\,\Delta t/\Delta x^2$. The phase lag $\Theta_n$ after one complete wavelength is defined as

$$\Theta_n = (\theta'_n/\theta_n)2\pi - 2\pi = 2\pi[(\theta'_n/\sigma_n \mathscr{U}\,\Delta x) - 1] = \theta'_n N_n - 2\pi$$

In the following paragraphs, we will examine the results of a Fourier analysis performed on the finite difference and finite element approximations of the transport equation. Specifically, the stability, phase lag, and damping of numerically propagated waves as well as the numerical representation of the solution through Fourier series will be considered.

*Stability Analysis*

In the discussion of finite difference theory, we noted that some weighted average schemes are only conditionally stable. For stability, the maximum

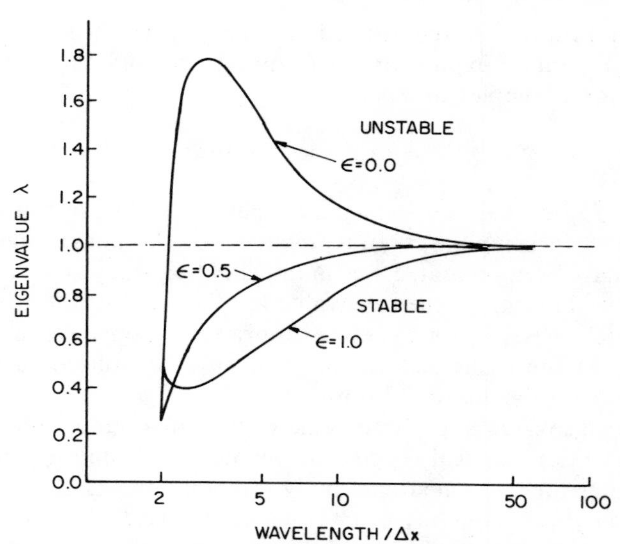


**Fig. 5.10.** Eigenvalues for finite element discretization of the transport equation with weighted finite difference approximations to the time derivative: $\varepsilon$ is the finite difference weighting coefficient, where $\mathscr{U} = 1.0$ and $\mathscr{D} = 0.1$.

eigenvalue of the amplification matrix (spectral radius) could not exceed unity. For small time steps or large values of $\Delta x$, this criterion is easily satisfied, but for an arbitrary choice of these parameters the stability is a function of the weighting coefficient $\varepsilon$ of Eq. (5.38). This is clearly evident in Fig. 5.10 where the eigenvalues for a finite element discretization are plotted for selected values of $\varepsilon$. "Explicit" ($\varepsilon = 0$), Crank–Nicholson ($\varepsilon = 0.5$), and implicit ($\varepsilon = 1.0$) formulations are presented. In this particular example, the explicit scheme is found to be unstable for the given values of $\mathscr{U}$ and $\mathscr{D}$ and any reasonable spatial mesh. In contrast, the remaining two schemes are stable for arbitrary mesh spacing.

*Phase Lag and Damping*

The eigenvalues for the linear finite element and finite difference methods are plotted in Fig. 5.11 for selected values of the weighting coefficient $\varepsilon$ with $\mathscr{D} = 0.069$ and $\mathscr{U} = 0.369$. As would be expected, the schemes are stable for all values of $\Delta x$, with $\varepsilon$ greater than 0.5. The phase lag for harmonics of varying wavelength when propagated through one wavelength $L_n$ are illustrated in Fig. 5.12. Characteristically, the harmonics of shorter

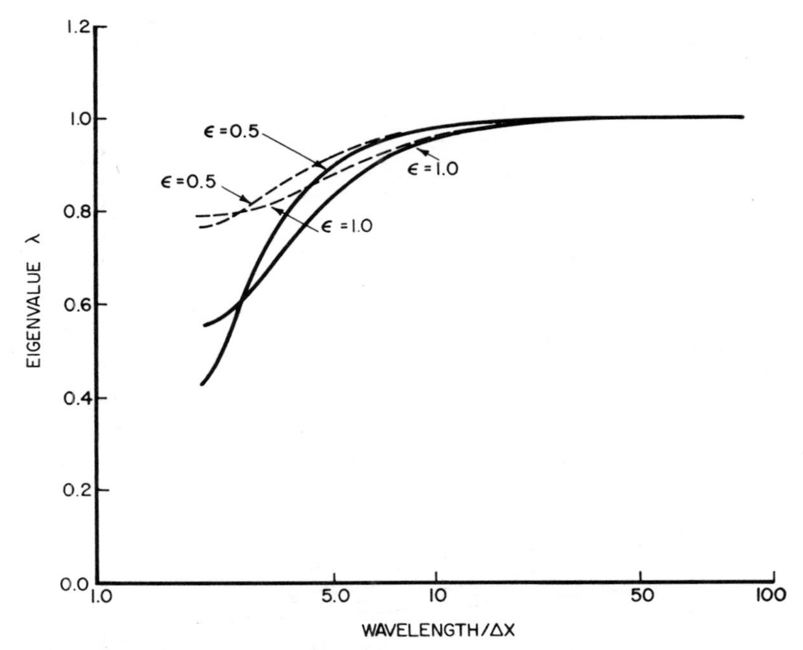


**Fig. 5.11.** Magnitudes of the eigenvalue $\lambda_n'$ for single equation finite element (——) and finite difference (- - -) methods with $\mathscr{D} = 0.069$ and $\mathscr{U} = 0.369$.

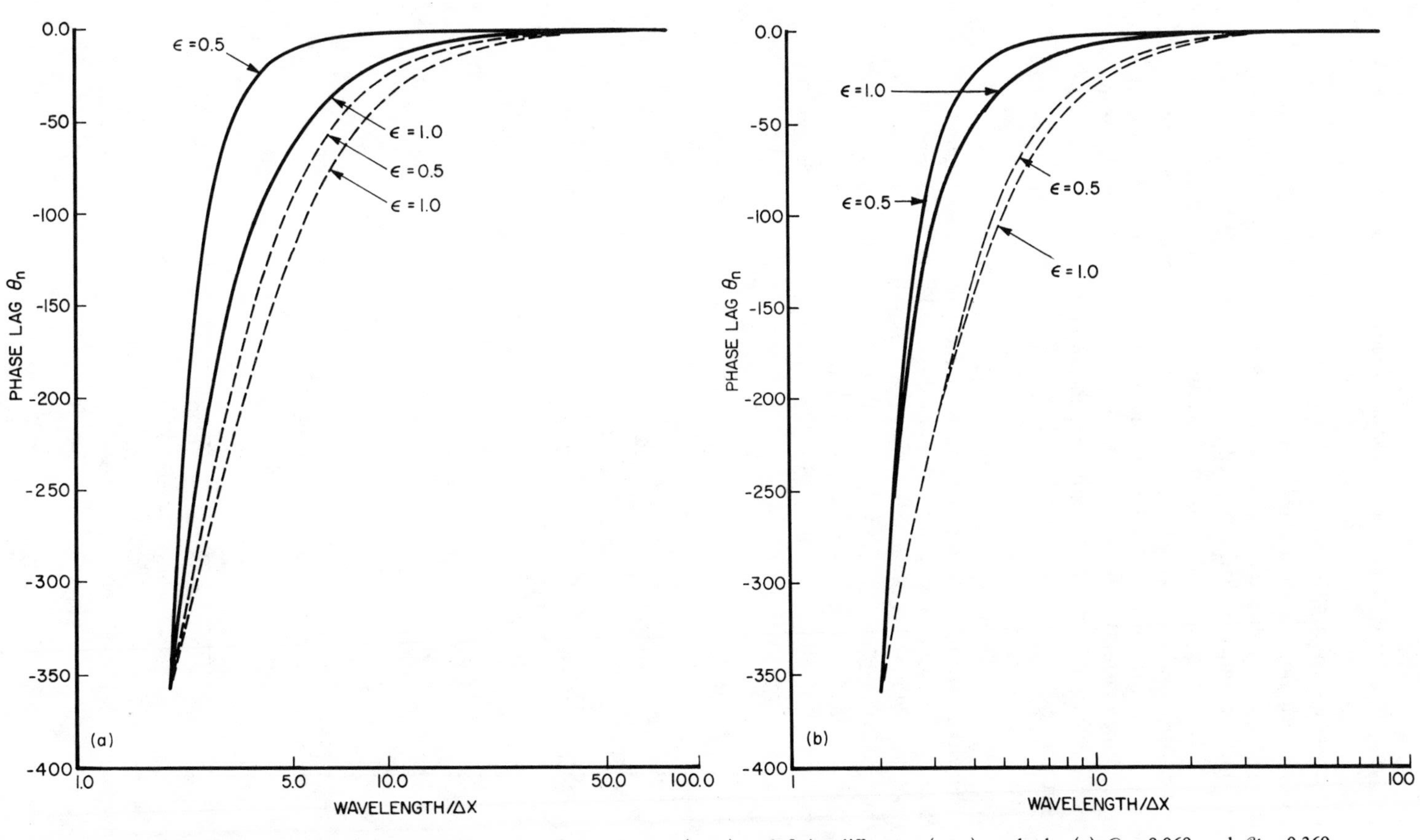


**Fig. 5.12.** Phase lag $\Theta_n$ for the single equation finite element (——) and finite difference (– – –) methods. (a) $\mathscr{D} = 0.069$ and $\mathscr{U} = 0.369$; (b) $\mathscr{D} = 0.0069$ and $\mathscr{U} = 0.369$.

wavelength are propagated poorly, while there is relatively little phase lag in the longer wavelengths. Although not as pronounced in this scheme as in others which will be considered, the centered in time difference approximation ($\varepsilon = 0.5$) exhibits less phase lag than the backward difference scheme

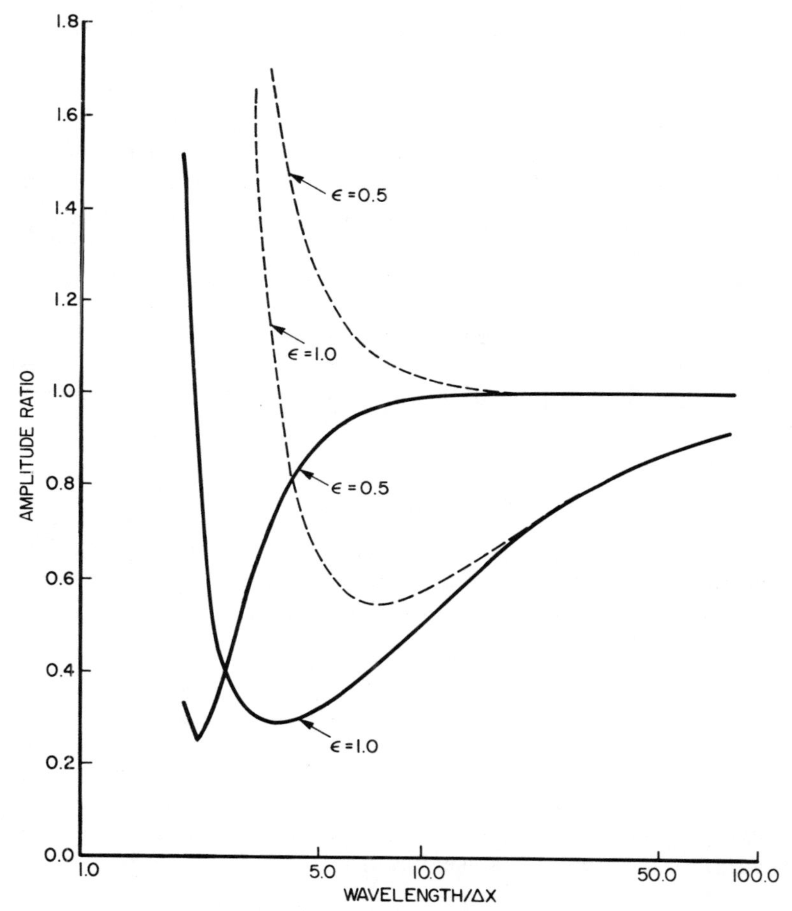


**Fig. 5.13.** Amplitude ratio for the single equation finite element (——) and finite difference (- - -) methods, where $\mathscr{D} = 0.069$ and $\mathscr{U} = 0.369$.

($\varepsilon = 1.0$). It is important to take into account the change in amplitude of the shorter wavelength harmonics in the evaluation of the significance of the phase lag. If wavelengths exhibiting unacceptable phase lag are damped out analytically as well as numerically, the solution will be less adversely affected than if these harmonics maintain or increase in amplitude. Figure 5.13

reveals that harmonics of shorter wavelength are more damped for larger values of $\varepsilon$. Thus phase lag (which is directly related to overshoot and undershoot) and numerical dissipation are both increased as $\varepsilon$ is increased from 0.5. Consequently, making a scheme more implicit causes a trade-off between propagating all the harmonics accurately and damping out the shorter wavelength components which exhibit unacceptable phase lag.

*Fourier Representation*

While the preceding analysis provides insight into the behavior of the harmonics appearing in the Fourier series, it does not illustrate the specific role played by phase lag and change in amplitude in the numerical solution of Eq. (5.31). This aspect of the problem can be examined by substituting the numerically determined eigenvalues in the Fourier series of Eq. (5.39). Because values of $\beta'_n$ are available from the earlier calculations, only the coefficients $C_n$ are required to evaluate the series (5.39). In the usual manner, $C_n$ are obtained from the initial conditions imposed on the system. An appropriate initial condition for the transport equation, and one that is convenient for the illustration of both overshoot and numerical dissipation, is given by the expression

$$c = 1 - H(z) \qquad \text{at } t = 0$$

where $z = x - ut$ $(-1 \leq z \leq 1)$ and $H(z)$ is the unit step function. The step or concentration front is thus located at $z = 0$ and it can be shown that the coefficients for the periodic Fourier series are given by

$$C_n = (\hat{\imath}/2n\pi)[1 - e^{\hat{\imath}n\pi}]$$

Evaluation of series (5.39) yields

$$C = \sum_{n=-\infty}^{\infty} [\hat{\imath}(1 - e^{\hat{\imath}n\pi})/2n\pi] \exp(\hat{\imath}\beta'_n t + \hat{\imath}n\pi x) \tag{5.41d}$$

Next the ability of selected numerical schemes to propagate accurately a concentration front will be considered using Eq. (5.41d).

*Comparison of Numerical Schemes*

A comparison of the finite element methods is at best, only indicative of behavior because of the many variations of the fundamental schemes. As a basis for comparison, the second-order correct, three-point schemes, outlined in Table 5.1 have been considered. Solutions obtained using both schemes in conjunction with the Fourier series of (5.41d) are indicated in Fig. 5.14 for $\mathscr{U} = 0.369$ and $\mathscr{D} = 0.069$ and 0.0069.

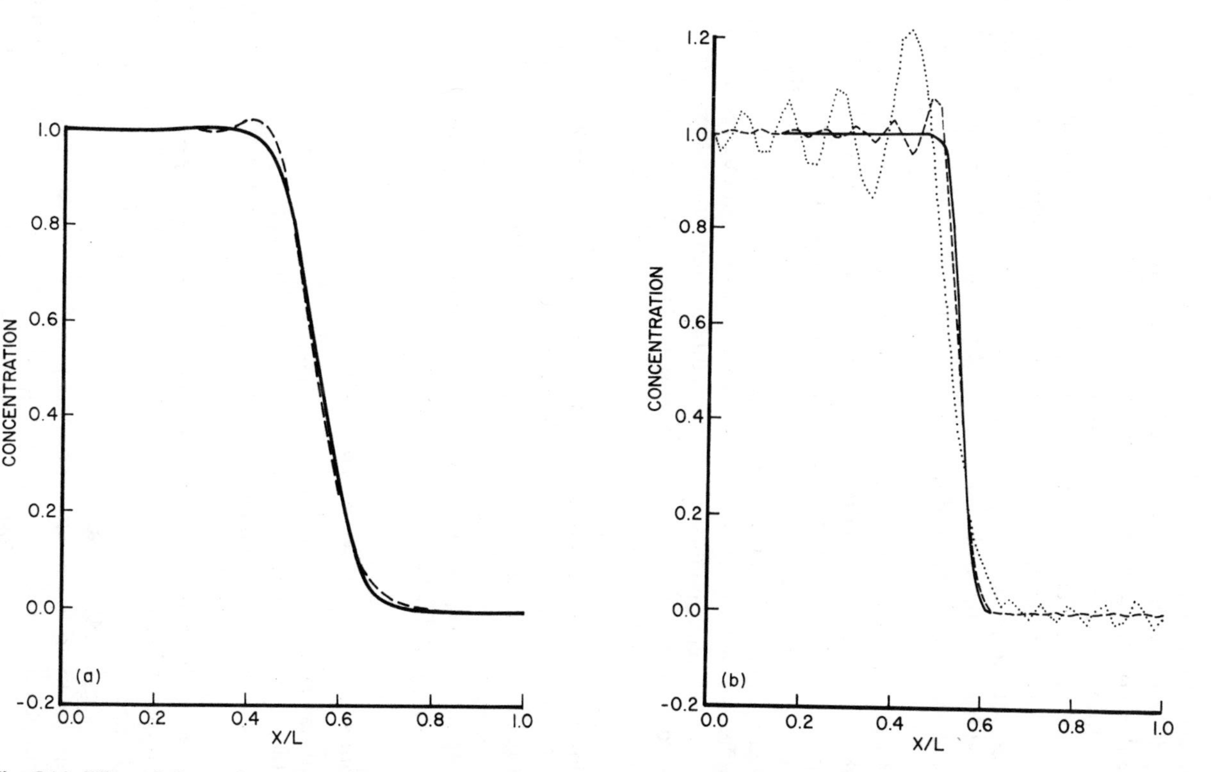


**Fig. 5.14.** Effect of $\mathcal{D}$ on the propagation of a concentration front by the single equation finite element and finite difference methods. (a) Finite element and analytic (——) and finite difference (- - -) curves for $\mathcal{D} = 0.069$, $\mathcal{U} = 0.369$ and $\varepsilon = 0.5$. (b) Finite element (- - -), analytic (——), and finite difference (···) curves for $\mathcal{D} = 0.0069$, $\mathcal{U} = 0.369$, and $\varepsilon = 0.5$.

*The Influence of the Weighting Coefficient $\varepsilon$*

It is well known that overshoot can be circumvented at the cost of numerical smearing by employing a first-order correct approximation for the time derivative. The impact of changing $\varepsilon$, the weighting factor, through the range from 0.5 to 1.0 is illustrated in Figs. 5.15a and 5.15b for the finite element and finite difference schemes, respectively. It is apparent that overshoot is a sensitive function of $\varepsilon$, particularly in the finite element method. Moreover, for the parameters considered here, overshoot can be effectively eliminated from the finite element solution by using $\varepsilon$ of 0.7 while generating about the same numerical dissipation as the centered finite difference scheme.

The markedly different response of the two numerical schemes is consistent with the earlier discussion of the propagation characteristics of each. While phase lag is a relatively sensitive function of $\varepsilon$ in the finite element method, this is not true for finite differences. Furthermore, the phase lag generated using $\varepsilon$ of 1.0 in the finite element schemes is still superior to all finite difference schemes. It is interesting to note that although increasing $\varepsilon$ in the finite difference method results in additional numerical damping of the higher-order harmonics, it is still far less effective than the finite element method in reducing overshoot.

Although overshoot may be virtually eliminated using appropriate weighting factors, the resulting numerical dissipation can cause misleading results. It was observed (see Fig. 5.16), for example, that when $\varepsilon$ of 1.0 was used in the finite element scheme, the solution obtained for $\mathscr{D}$ of 0.0069 was, in fact, coincident with the correct analytical solution for a $\mathscr{D}$ of 0.069. In addition, the numerical dissipation was of sufficient magnitude that the solutions for $\mathscr{D}$ of 0.00069 and $\mathscr{D}$ of 0.0069 were essentially identical.

*Evaluation of Several Numerical Schemes*

Analyses similar to those presented earlier for the finite element and finite difference schemes have been performed for several other numerical techniques. Some of these have appeared in the literature, while others apparently have not. While it would be impractical to present a complete analysis of each technique, solutions obtained using a variety of methods will be presented to illustrate their relative accuracy. All solutions presented were obtained using $\Delta x$ of 1/24, $\mathscr{D}$ of 0.0069, $\mathscr{U}$ of 0.369, and $\varepsilon$ of 0.5 and appear in Figs. 5.14b, 5.17a, and 5.17b.

While many of the schemes presented here have been discussed earlier, those involving upstream weighting in the finite difference scheme and dispersion correction in both schemes warrant additional comment. The upstream weighting scheme refers to the procedure whereby the convective

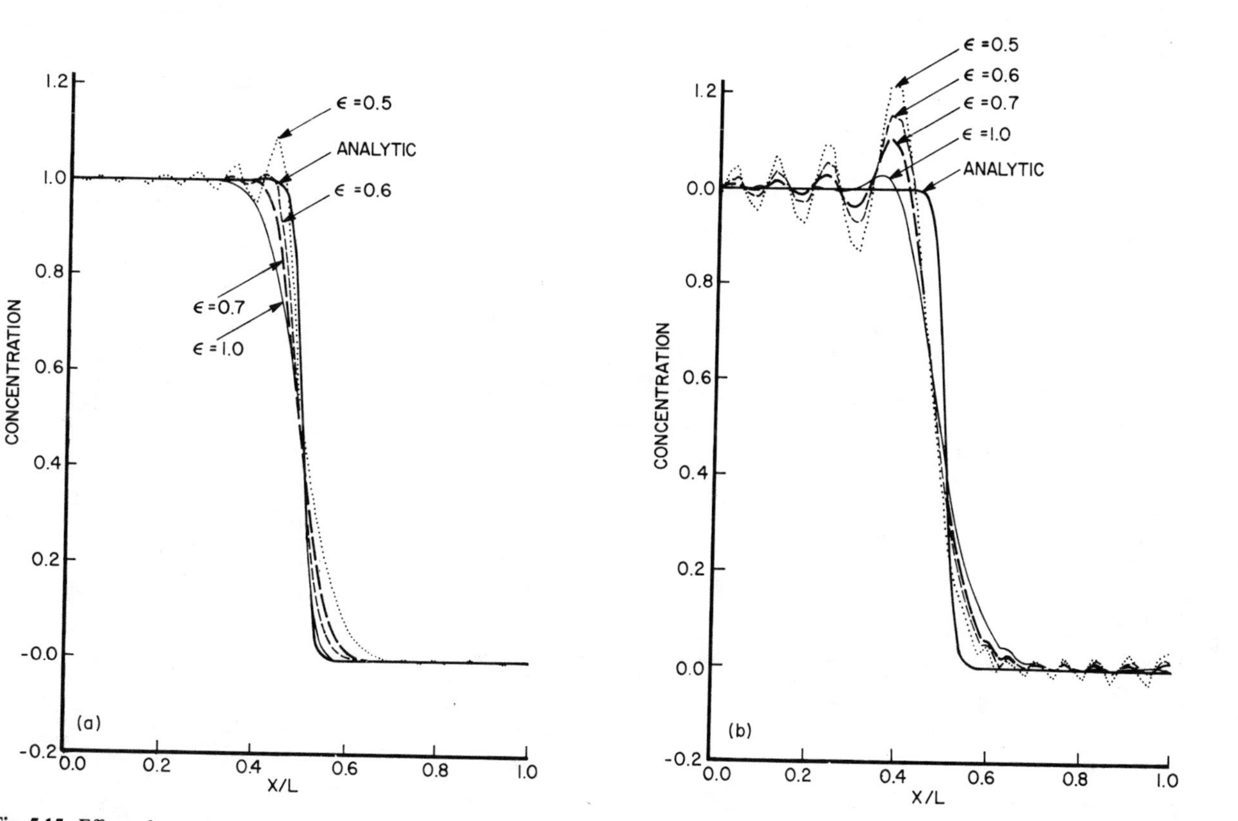


**Fig. 5.15.** Effect of $\varepsilon$ on the propagation of a concentration front by the single equation (a) finite element and (b) finite difference methods for $\mathscr{D} = 0.0069$ and $\mathscr{U} = 0.369$.

term is approximated using only nodes located at $i$ and $i - 1$ whereas the second derivative approximation uses $i - 1$, $i$, and $i + 1$. This is a first-order correct spatial difference approximation but is, nevertheless, sometimes used because it minimizes overshoot. The dispersion correction is a term added to the numerical expression to account for numerical dispersion attributable to

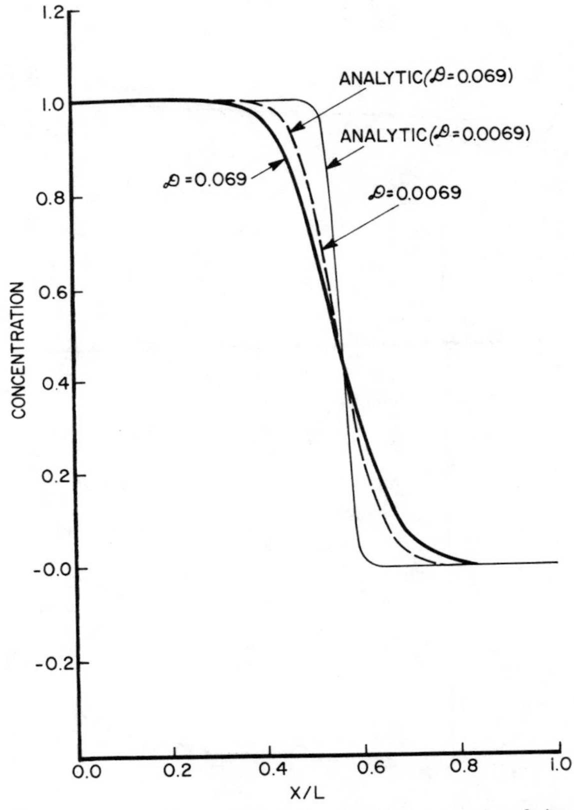


**Fig. 5.16.** Concentration profiles obtained by the single equation finite element method with $\varepsilon = 1.0$ and $\mathscr{U} = 0.369$ for various values of $\mathscr{D}$.

truncation of the Taylor series expansion. This technique has been described in detail by Chaudhari [18] and will not be developed here.

There are several general conclusions that can be drawn from an examination of the solutions to Eq. (5.31) presented in Figs. 5.14b, 5.17a, and 5.17b.

(a) The finite difference method considered is consistently inferior to finite element methods. Solutions obtained using small coefficients of dispersion exhibit unacceptable overshoot and excessive numerical smearing of the concentration front.

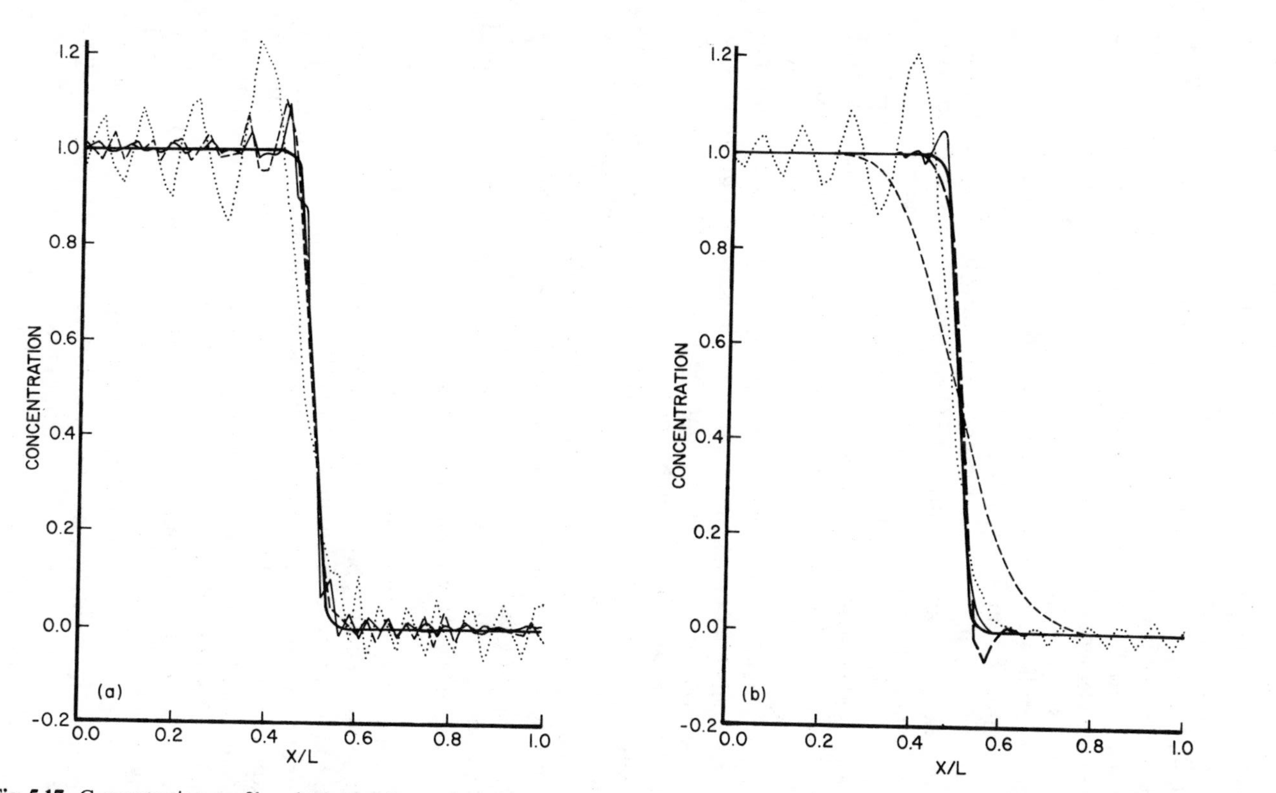


**Fig. 5.17.** Concentration profiles obtained using various finite element and finite difference computational schemes for $\mathscr{D} = 0.0069$, $\mathscr{U} = 0.369$, and $\varepsilon = 0.5$. (a) Two-equation finite difference (···); two-equation finite element (– – –); quadratic finite element (——). (b) Finite difference with dispersion correction (···) and upstream weighted convection (– – –); finite element with dispersion correction (— —); Hermite polynomial (——).

(b) The two-equation finite element and finite difference methods defined in Table 5.1 generated solutions generally inferior to the more familiar one-equation schemes.

(c) Quadratic finite elements generate a numerical solution that exhibits an oscillatory behavior because the solution is fourth-order accurate at the corner nodes but only second-order correct at midelement nodes [19].

(d) Finite element schemes based on Hermite polynomials provide solutions with a decreased amount of numerical smearing and overshoot. It should be noted, however, that this scheme generates two algebraic equations at each nodal point in the one-dimensional formulation.

(e) The correction for dispersion does not markedly affect the accuracy of the finite difference scheme but does substantially improve the ability of the finite element method to propagate solutions with small dispersion coefficients. An examination of the phase lag curves shows that the finite difference phase lag is relatively unaffected by the correction factor whereas the finite element curves indicate that shorter wavelengths are more accurately propagated. The introduction of undershoot in the "corrected" finite element scheme is consistent with an observed phase acceleration for low wave numbers (i.e., a phase acceleration of 25° is actually obtained) [20]. It should be pointed out, however, that the introduction of a correction factor results in a conditionally stable method.

(f) Upstream weighting of the convective term generates unacceptable solutions for convection-dependent transport. The amplification curves show that this is due to excessive damping over a wide range of harmonics.

(g) While overshoot may be minimized through an appropriate selection of the weighting factor $\varepsilon$, it is at the expense of increased smearing of the concentration front. When small dispersion coefficients are encountered, erroneous and misleading results may be generated.

It is apparent that the finite element scheme provides consistently more accurate solutions in that both numerical dispersion and numerical dissipation are less apparent in the finite element curves.

To understand why the finite element scheme is superior, it is necessary to reexamine Figs. 5.11–5.17. The eigenvalue plot, Fig. 5.11, reveals that the shorter wavelengths are more damped in the finite element numerical solution than in the finite difference scheme. Moreover, examination of Figs. 5.14–5.17 indicates that in comparison to the analytical solution these same wavelengths are damped due to the finite element numerical scheme. In contrast, the finite difference method amplifies the shorter wavelengths in comparison to the analytical solution for the second-order correct scheme ($\varepsilon = 0.5$). Examination of Fig. 5.12 indicates that although the finite element method generates a smaller phase lag for all wavelengths, neither scheme propagates the higher-order harmonics accurately. Thus one would expect to obtain a more accurate numerical solution in problems where these wavelengths are unimportant in the description of the concentration front. An example of such a problem is the one given in Fig. 5.14a. Although the poorly propagated small wavelengths are unimportant in the correct analytical solution, the finite difference scheme amplifies them and shifts them out of phase causing the overshoot problem and the slightly erroneous front. On the other hand, the finite element method damps these wavelengths so that they remain unimportant.

When $\mathscr{D}$ is decreased by an order of magnitude, as in Fig. 5.14b, the small wavelengths become more important to the correct description of the front. The poor propagation of these waves by both methods is the cause of the oscillations near the front. Once again, the superior propagation ability of the finite element method does yield a better solution than the finite difference method.

### 5.3.3 *Two-Dimensional Flow in the Vertical Plane (The Density-Dependent Case)*

Heretofore we have assumed that the fluid density is independent of the concentration of solutes. While this assumption holds for the majority of aquifer contamination problems, it is not universally true. One notable exception is the intrusion of salt water into coastal aquifers. Whereas this phenomenon has been generally treated as a problem involving the flow of immiscible fluids [2, p. 518], salt and fresh water are, in fact, miscible and to simulate their behavior accurately requires the solution of the mass transport equation.

*Governing Equations*

To obtain the specific discharge vector from the groundwater flow equations, the more general form of Darcy's law applying to nonhomogeneous fluids must be used:

$$\mathbf{q} = -(\mathbf{k}/\mu) \cdot (\nabla_{xy} p + \rho g \, \nabla_{xy} Z) \tag{5.41e}$$

where $Z$ is the elevation above some reference datum, and the $xy$ plane is vertical. Conservation of the saltwater solution requires that

$$\partial(\theta\rho)/\partial t + \nabla_{xy} \cdot (\rho \mathbf{q}) = 0 \tag{5.42}$$

It is possible to express the time derivative in (5.42) as a function of concentration and pressure. However, in simulating density-dependent flow, the movement of the solute is so slow relative to the pressure propagation that the pressure response can be considered instantaneous. Moreover, the time rate of change of density can be considered negligible compared to other terms in this equation. These assumptions lead to the flow equation

$$\rho \nabla_{xy} \cdot \mathbf{q} + \mathbf{q} \cdot \nabla_{xy} \rho = 0 \tag{5.43}$$

The appropriate transport equation for the solution is obtained by substituting the constitutive relationship

$$\rho_s \mathbf{q}_s = \rho_s \mathbf{q} - \mathbf{D} \cdot \nabla_{xy} \rho_s \tag{5.44}$$

into the species continuity equation

$$\partial(\theta \rho_s)/\partial t + \nabla_{xy} \cdot (\rho_s \mathbf{q}_s) = 0$$

Combination of this equation and (5.44) yields

$$\partial(\theta \rho_s)/\partial t + \rho_s \nabla_{xy} \cdot \mathbf{q} + \mathbf{q} \cdot \nabla_{xy} \rho_s - \nabla_{xy} \cdot (\mathbf{D} \cdot \nabla_{xy} \rho_s) = 0 \tag{5.45}$$

Examination of (5.41e), (5.43), and (5.45) reveals that there are four unknowns $p$, $\mathbf{q}$, $\rho_s$, and $\rho$† but only three equations. The additional condition required to obtain a solution to these equations is given by an empirical relationship developed for salt solutions which relates the salt concentration to the fluid density [21]

$$\rho = \rho_0 + (1 - E)\rho_s \tag{5.46}$$

where $\rho_0$ is the density of fresh water and $E$ is a constant and has a value of 0.3 for concentrations as high as sea water.

*The Approximating Equations*

We have already indicated the importance of a continuous velocity distribution when a solution to the transport equation is sought in a region of highly variable flow.‡ There are essentially two ways to obtain a continuous velocity field:

(1) use basis functions which have a continuous first derivative across element boundaries, or

(2) formulate a four-equation scheme which treats specific discharge as an undetermined coefficient [23]

Because an isoparametric element possessing first-order continuity is not commonly utilized, and because the flexibility of the isoparametric formula-

† $\theta$ has been assumed constant.

‡ Cheng [22] has obtained a steady flow solution to the variable density fluid problem using a discontinuous velocity formulation.

tion is of considerable importance, we will demonstrate the four-equation scheme.

The velocity field depends on concentration through the density relationship (5.46) and the concentration depends on the velocity through the convective transport term. Thus the system is nonlinear and it is necessary to solve the system of equations using an iterative scheme. Each iteration will consist of two steps: step 1 will require the solution of three simultaneous equations for flow, and step 2 will use the resulting specific discharge components to solve the mass transport equation.

*Step 1* Define the three operators

$$\begin{aligned} L_1(\mathbf{q}, \rho) &\equiv \rho \nabla_{xy} \cdot \mathbf{q} + \mathbf{q} \cdot \nabla_{xy}\rho = 0 \\ L_2(q_x, p) &\equiv q_x + \frac{k_{xx}}{\mu}\left(\frac{\partial p}{\partial x}\right) = 0 \\ L_3(q_y, p) &\equiv q_y + \frac{k_{yy}}{\mu}\left(\frac{\partial p}{\partial y} - \rho g\right) = 0 \end{aligned} \tag{5.47}$$

where the principal directions of $\mathbf{k}$ have been assumed to be in the vertical and horizontal directions, and the $y$ axis is parallel to the gravitational acceleration vector. Assume three trial solutions of the form

$$\begin{aligned} q_x \simeq \hat{q}_x &= \sum_{j=1}^{M} q_{xj}\phi_j(x, y) \\ q_y \simeq \hat{q}_y &= \sum_{j=1}^{M} q_{yj}\phi_j(x, y) \\ p \simeq \hat{p} &= \sum_{j=1}^{M} p_j\phi_j(x, y) \end{aligned} \tag{5.48}$$

For convenience, we have used the same bases for each trial function and they satisfy the essential boundary conditions. It may, however, prove more efficient to use lower-degree polynomials for the pressure than the specific discharge.

Applying the orthogonality conditions of the Galerkin method, we obtain

$$\langle L_1(\hat{\mathbf{q}}, \rho), \phi_i \rangle = 0, \qquad \langle L_2(\hat{q}_x, \hat{p}), \phi_i \rangle = 0, \qquad \langle L_3(\hat{q}_y, \hat{p}), \phi_i \rangle = 0$$
$$(i = 1, 2, \ldots, M) \tag{5.49}$$

Expansion of (5.49) and introduction of the trial functions (5.48) yields

$$[A]\{Q\} = \{F\} \tag{5.50}$$

where typical elements of $[A]$, $\{Q\}$, and $\{F\}$ are

$$[a_{i,j}] = \begin{bmatrix} \left\langle \sum_{k=1}^{M} \left( \rho_k \phi_k \frac{\partial \phi_j}{\partial x} + \rho_k \frac{\partial \phi_k}{\partial x} \phi_j \right), \phi_i \right\rangle & \left\langle \sum_{k=1}^{M} \left( \rho_k \phi_k \frac{\partial \phi_j}{\partial y} + \rho_k \frac{\partial \phi_k}{\partial y} \phi_j \right), \phi_i \right\rangle & 0 \\ \langle \phi_j, \phi_i \rangle & 0 & \left\langle \sum_{k=1}^{M} \left( \frac{k_{xx}}{\mu} \right)_k \phi_k \frac{\partial \phi_j}{\partial x}, \phi_i \right\rangle \\ 0 & \langle \phi_j, \phi_i \rangle & \left\langle \sum_{k=1}^{M} \left( \frac{k_{yy}}{\mu} \right)_k \phi_k \frac{\partial \phi_j}{\partial y}, \phi_i \right\rangle \end{bmatrix}$$

$$\{q_i\} = \begin{Bmatrix} q_{xi} \\ q_{yi} \\ p_i \end{Bmatrix}, \qquad \{f_i\} = \begin{Bmatrix} 0 \\ 0 \\ \left\langle \dfrac{k_{yy} g}{\mu} \sum_{k=1}^{M} \rho_k \phi_k, \phi_i \right\rangle \end{Bmatrix}$$

In formulating the elements of $[A]$, functional coefficients have been used whenever possible. Specifically, they are used to represent the known parameters $\rho$, $(k_{xx}/\mu)$, and $(k_{yy}/\mu)$.† When a cross-sectional model is characterized by abrupt changes in permeability, it may be advantageous to specify $(\mathbf{k}/\mu)$ as constant over an element and locate an element boundary at the interface between contrasting beds.

Because no contour integrals appear in the approximating equations, the vector $\{F\}$ will contain only information on Dirichlet boundaries, sources, and sinks, and the body-force term associated with the vertical component of specific discharge.

Careful examination of the matrix $[A]$ reveals two undesirable properties: it is not symmetric and contains zero elements along the diagonal. The latter property can be a nuisance inasmuch as many equation-solving schemes pivot on diagonal elements. Consequently, either the equations must be rearranged so that zeros do not occur on the diagonal or a form of Gaussian elimination that searches each row for the largest element prior to pivoting must be used.

Before leaving this set of equations, it should be pointed out that, due to the large size of this matrix $[3(M - P) \times 3(M - P)]$, $P$ being the number of Dirichlet nodes, it is sometimes necessary to utilize peripheral storage (specifically the disk) when using a noniterative equation solver. This is particularly important in solving three-dimensional problems of any reasonable size.

† Density is assumed known from the last iteration.

*Step 2* Having obtained specific discharge from step 1, the transport equation may now be solved. Let us define a fourth trial function

$$\rho_s \simeq \hat{\rho}_s = \sum_{j=1}^{M} c_j(t)\phi_j(x, y) \tag{5.51}$$

Applying Galerkin's procedure to Eq. (5.45), introducing trial function (5.51), and using functional coefficients, one obtains

$$[B]\{C\} + [D]\{dC/dt\} = \{F\} \tag{5.52}$$

where typical nonzero elements of $[B]$, $[D]$, and $\{F\}$ are

$$\begin{aligned} b_{i,j} = \sum_{k=1}^{M} \Bigg\{ & \left\langle \left[ q_{xk} \frac{\partial \phi_k}{\partial x} + q_{yk} \frac{\partial \phi_k}{\partial y} \right] \phi_j, \phi_i \right\rangle \\ & + \left\langle \left[ q_{xk} \phi_k \frac{\partial \phi_j}{\partial x} + q_{yk} \phi_k \frac{\partial \phi_j}{\partial y} \right], \phi_i \right\rangle \\ & + \left\langle D_{xxk} \phi_k \frac{\partial \phi_j}{\partial x}, \frac{\partial \phi_i}{\partial x} \right\rangle + \left\langle D_{yxk} \phi_k \frac{\partial \phi_j}{\partial x}, \frac{\partial \phi_i}{\partial y} \right\rangle \\ & + \left\langle D_{xyk} \phi_k \frac{\partial \phi_j}{\partial y}, \frac{\partial \phi_i}{\partial x} \right\rangle + \left\langle D_{yyk} \phi_k \frac{\partial \phi_j}{\partial y}, \frac{\partial \phi_i}{\partial y} \right\rangle \Bigg\} \end{aligned}$$

$$d_{i,j} = \theta\langle \phi_j, \phi_i \rangle, \qquad f_i = \int_\Gamma \mathbf{q}_m \cdot \mathbf{n}\phi_i \, ds$$

The unknown coefficients $c_j$ are obtained by solution of (5.52) using procedures outlined earlier for the analogous two-dimensional areal equation (5.30).

Although velocity and concentration values have been calculated they are only approximate in that the velocity field is not necessarily compatible with the new concentrations. Without advancing in time, the velocity distribution is recomputed using updated densities and the cycle is completed by once again solving for the concentration. One continues this repetitive or iterative procedure until the successive values of concentration are within a specified tolerance, whereupon computation proceeds to the next time step. This iterative scheme is essentially the same as outlined earlier for areal analysis of water-table problems. In this example, a weighted mean of recent iterative values has been found to be the best means of updating the velocity coefficient matrices

$$c^{*(n)} = (1 - \varepsilon)c^{(n)} + \varepsilon c^{*(n-1)}$$

where $c^{*(n)}$ is used to update the coefficients to calculate $c^{n+1}$. For the first iteration $c^{*(1)} = (1 - \varepsilon)c^{(1)} + \varepsilon c^{(0)}$. Convergence should be based on sequential values of $c^{*(n)}$, not $c^{(n)}$. As in the earlier example, $\varepsilon = 0.7$ provided rapid convergence for problems tested.

*An Example Problem*

Simulation of the transient position of the saltwater front in coastal aquifers is an important hydrological problem. A typical problem of this kind is illustrated in Fig. 5.18a and the finite element net used to approxi-

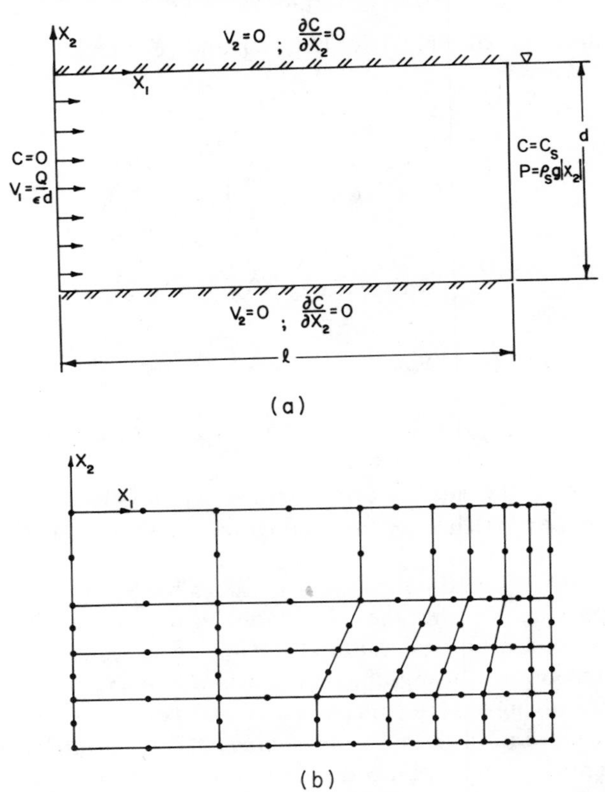


**Fig. 5.18.** (a) Diagrammatic representation of saltwater intrusion test problem. (b) Finite element net used to solve saltwater intrusion problem.

mate the system is shown in Fig. 5.18b. Although simulation of this system is relatively straightforward, one should be aware of two difficulties. The first factor to be considered is the continuity of the velocity. Attempts to use discontinuous velocities in this transient problem were totally unsuccessful. The resulting lack of continuity generated large errors in concentration and a convergent solution was unobtainable. A second difficulty was initially encountered at the outflow face $x_1 = l$. In preliminary runs, a Dirichlet condition was specified, namely $c = c_0$, where $c_0$ is the concentration of sea water. However, fluid exiting this face was forced by this condition to convect mass from the system at concentration $c_0$. In actual fact, the exiting fluid always has a concentration less than $c_0$, consequently, there is an erroneous loss of mass from the system not accounted for by the mathematics. To circumvent this problem, it was necessary to redefine an exit boundary as a Neumann condition of zero diffusive mass flux. A typical solution to this saltwater intrusion problem is presented in Fig. 5.19.

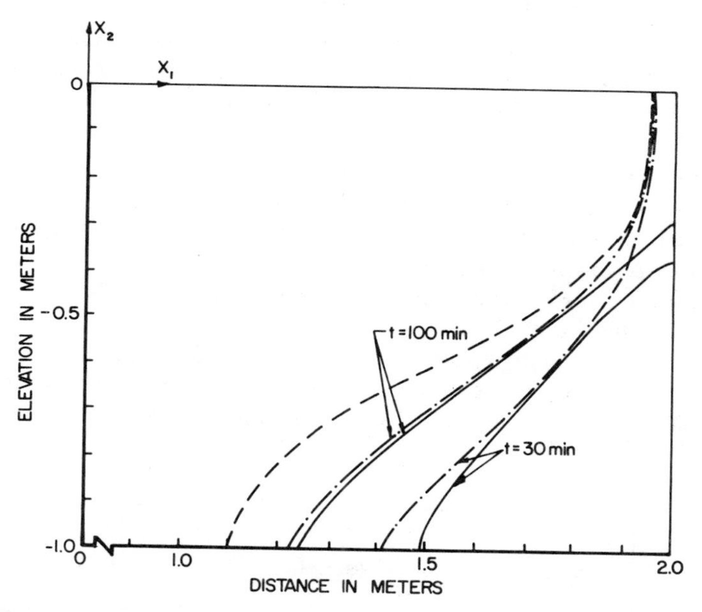


**Fig. 5.19.** Solution to saltwater intrusion problem with $Q = 6.6 \times 10^{-5}$ m²/sec, $D = 6.6 \times 10^{-6}$ m²/sec, $K = 1.0 \times 10^{-2}$ m/sec, $C_S = 35$ kg/m³, and $\varepsilon = 0.35$. (——) Present analysis; (—·—) Pinder and Cooper [4]; (- - -) Henry [45].

*A Field Problem*

While it is apparent that the technology presented above provides accurate solutions for a relatively simple sample problem, it remains to demonstrate the applicability of this approach to field problems. In this section, the Galerkin finite element method is used to calculate the transient position of the saltwater front in the Biscayne aquifer, Florida. The simulation utilizes field data obtained before and after a rainfall which sharply increased the height of the water table. The model is initially calibrated to provide the observed chloride distribution prevailing prior to the rainfall, which was then used as an initial condition for the transient problem.

*Governing Equations*

A cross section of the Biscayne aquifer is considered and the variables are referred to a Cartesian coordinate system. The distribution of the chloride ion in this two-dimensional space can be described by the following species-transport equation which is analogous to (5.23a).

$$\nabla \cdot (\mathbf{D} \cdot \nabla c) - \nabla \cdot (\mathbf{q}c) - \partial(c\theta)/\partial t = 0 \tag{5.53}$$

and the dispersion coefficient $\mathbf{D}$ is computed according to definition (5.25). Equations (5.41e) and (5.42) are once again solved simultaneously for pressure and velocity and (5.53) is evaluated separately.

*Simulation*

The technique which proved to be successful in predicting the transient position of the saltwater front in a theoretical case is now applied to an actual field problem. Data published by Kohout [24–26] and Kohout and Klein [27] for the Biscayne aquifer in the Cutler area, near Miami, Florida, are used for this study.

The Biscayne aquifer in this area is a nonartesian aquifer consisting of limestone and calcareous sandstone, which extends to an average depth of 100 ft below sea level (Fig. 5.20). Field data indicate that the saltwater front

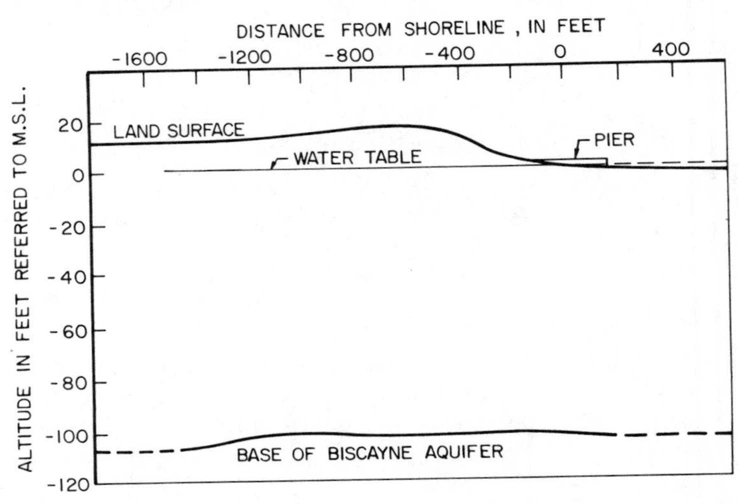


**Fig. 5.20.** Cross section of the Biscayne aquifer in the Cutler area (after Kohout and Klein [27]).

in the Biscayne aquifer is dynamically stable at a position as much as 8 miles seaward of that computed by the Ghyben–Herzberg principle (Fig. 5.21). It is hypothesized that this discrepancy is in part due to a cyclic flow of salt water induced by the dispersion of salts. In 1958, samples were collected and pressure heads were measured at a test site located in the Cutler area before and after a heavy rainfall which sharply increased the elevation of the water

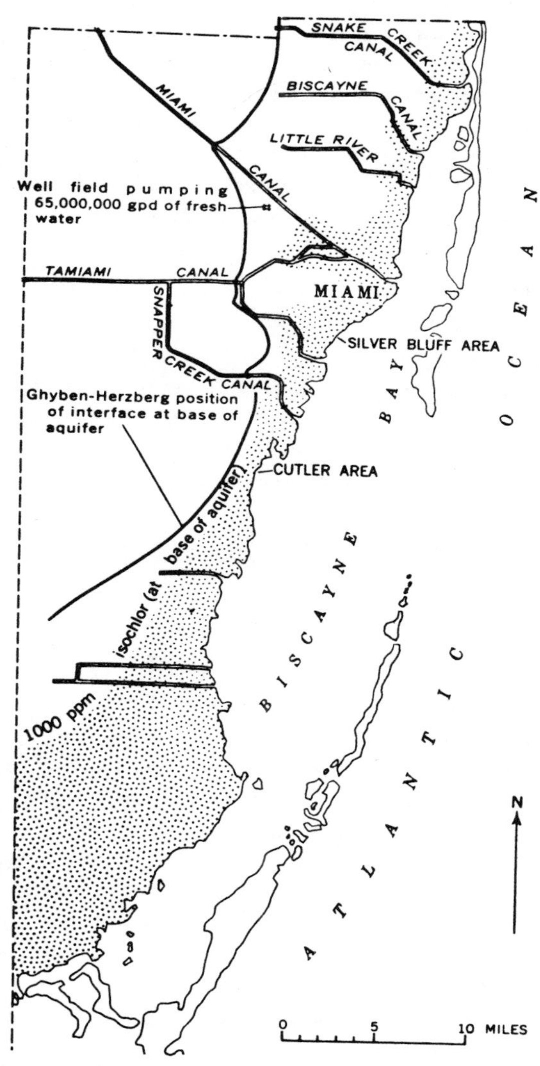


**Fig. 5.21.** Theoretical and actual position of the saltwater front in the Cutler area (after Kohout [46]).

table. Measurements showed that the distribution of chloride before the pulse of recharge could be considered as indicative of steady state conditions. This assumption is clearly justified by the fact that the concentration distribution before the rainfall [27] is virtually identical to the distribution observed several months after the end of recharge [24].

The first step in the analysis is the application of the numerical procedure presented in the preceding section to simulate the observed steady state conditions prevailing before the rainfall. The concentration distribution obtained is used thereafter as initial data for the transient problem. Were accurate pressure and concentration values available from field measurements for all points in the system, it would be unnecessary to simulate initial conditions; unfortunately, this was not the case. Moreover, numerical simulation of the steady state conditions observed in the field lends credence to the choice of time-independent hydrological parameters.

*Finite Element Model*

The finite element grid used to discretize the domain of interest is shown in Fig. 5.22. The mesh contains 96 quadrilateral elements with linear sides and 117 nodal points. It extends 1600 ft landward and 600 ft seaward off the

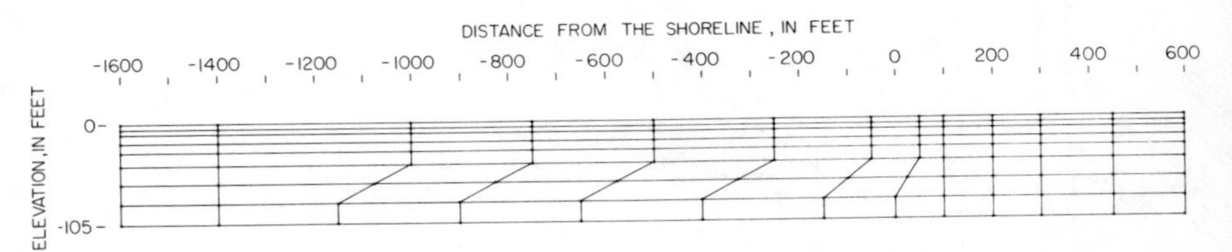


**Fig. 5.22.** Finite element model of the Biscayne aquifer.

shore which is located at the point $x_1 = 0$. Following Kohout and Klein [27], the aquifer is assumed to be homogeneous with a hydraulic conductivity of $1.35 \times 10^{-2}$ m/sec. The model runs indicate, however, that simulation of the observed chloride distribution requires a smaller value of this coefficient, and the hydraulic conductivity was taken equal to $0.45 \times 10^{-2}$ m/sec in the horizontal direction. Actual field measurements of the vertical hydraulic conductivity were not available, but it is believed that the ratio between vertical and horizontal components of this parameter is possibly greater than 100 in this part of the Floridian aquifer. In this study, the hydraulic conductivity was taken equal to $0.09 \times 10^{-4}$ m/sec in the vertical direction. Because experimental values of the dispersion coefficients were unavailable, the longitudinal and transverse dispersivities were found through a trial and error procedure to equal 6.66 and 0.666 m, respectively. The porosity of the Cutler area is approximately 0.25.

*Boundary Conditions*

The boundary conditions used in conjunction with the governing equations are shown schematically in Fig. 5.23. In solving the flow equations, a constant hydrostatic pressure distribution is assumed to prevail on the left and right vertical boundaries of the aquifer ($x_1 = -1600$ ft and $x_1 = 600$ ft). The normal velocity is set equal to zero at the bottom of the aquifer ($x_2 = -105$ ft). The values for pressure reported by Kohout [24] and Kohout and Klein [27] are plotted on Fig. 5.24 for the steady state distribution and for

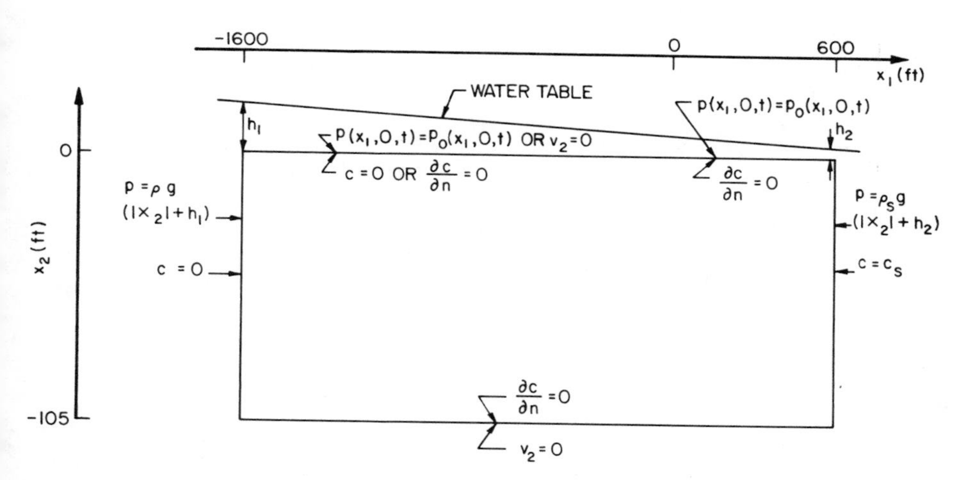


**Fig. 5.23.** Boundary conditions for flow and mass transport equations.

the week following the rainfall (May 29th). The pressure head data indicate that the aquifer is likely to contain a region of higher permeability in its central portion. The same assertion has been made by Kohout [25] but he did not evaluate the extent and the magnitude of this high permeability region. Consequently, in the present analysis, the boundary condition specified on the top of the aquifer is

$$\begin{aligned} &p(x_1, 0, t) \qquad \text{given for} \quad 0 \le x_1 < 600 \text{ ft} \\ &v_2 = 0 \qquad \text{for} \quad -1600 \text{ ft} < x_1 < 0 \end{aligned} \tag{5.54}$$

In solving for mass transport, it is assumed that the concentration is zero on the left vertical boundary of the aquifer ($x_1 = -1600$ ft) and that it has a constant value $c_s$ on the seaward boundary ($x_1 = 600$ ft), where $c_s$ is the seawater chloride concentration (19,000 ppm). On the horizontal boundaries of the aquifer ($x_2 = -105$ ft and $x_2 = 0$), the normal concentration gradient $\partial c/\partial n$ is assumed to vanish. Results of runs made using instead a mixed boundary condition at $x_2 = 0$, namely,

$$\begin{aligned} c &= 0 \qquad \text{for} \quad -1600 < x_1 < 0 \\ \partial c/\partial n &= 0 \qquad \text{for} \quad 0 < x_1 < 600 \text{ ft} \end{aligned} \tag{5.55}$$

indicate that both representations of boundary conditions along $x_2 = 0$ provide essentially the same results.

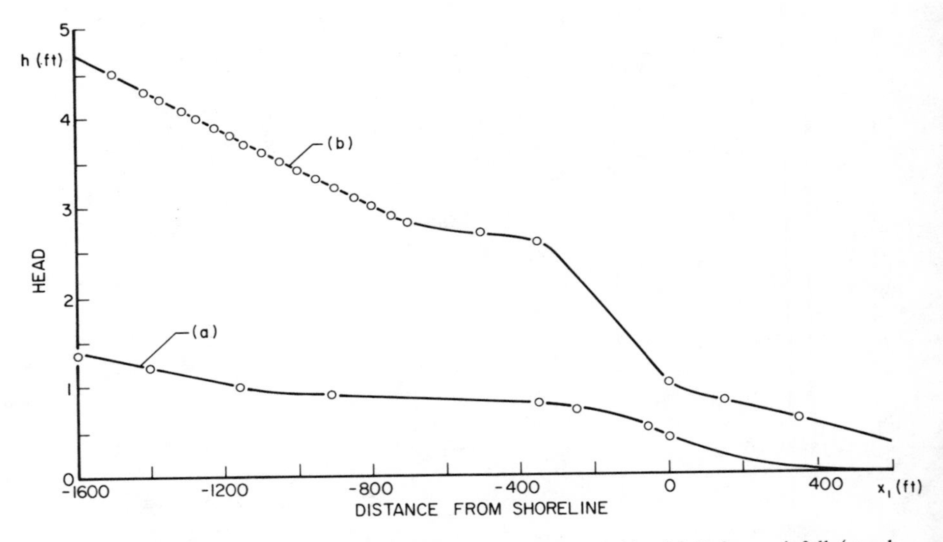


**Fig. 5.24.** Specified head distribution below the water table. (a) Before rainfall (steady state); (b) one week after rainfall. The open circles indicate observed values.

*Steady State Problem*

Results for the steady state problem are shown in Fig. 5.25 together with the field data reported by Kohout and Klein [27]. There is satisfactory agreement between the observed and experimental values. The principal differences concern the position of the 0.75 (14,200 ppm) and 0.95 (18,000 ppm) isochlors. The discrepancy may be due to the occurrence of a region of higher permeability mentioned earlier which is not taken into account in the numerical model, because its location and extent are unknown. Figure 5.25 also indicates that the observed isochlors have a some-

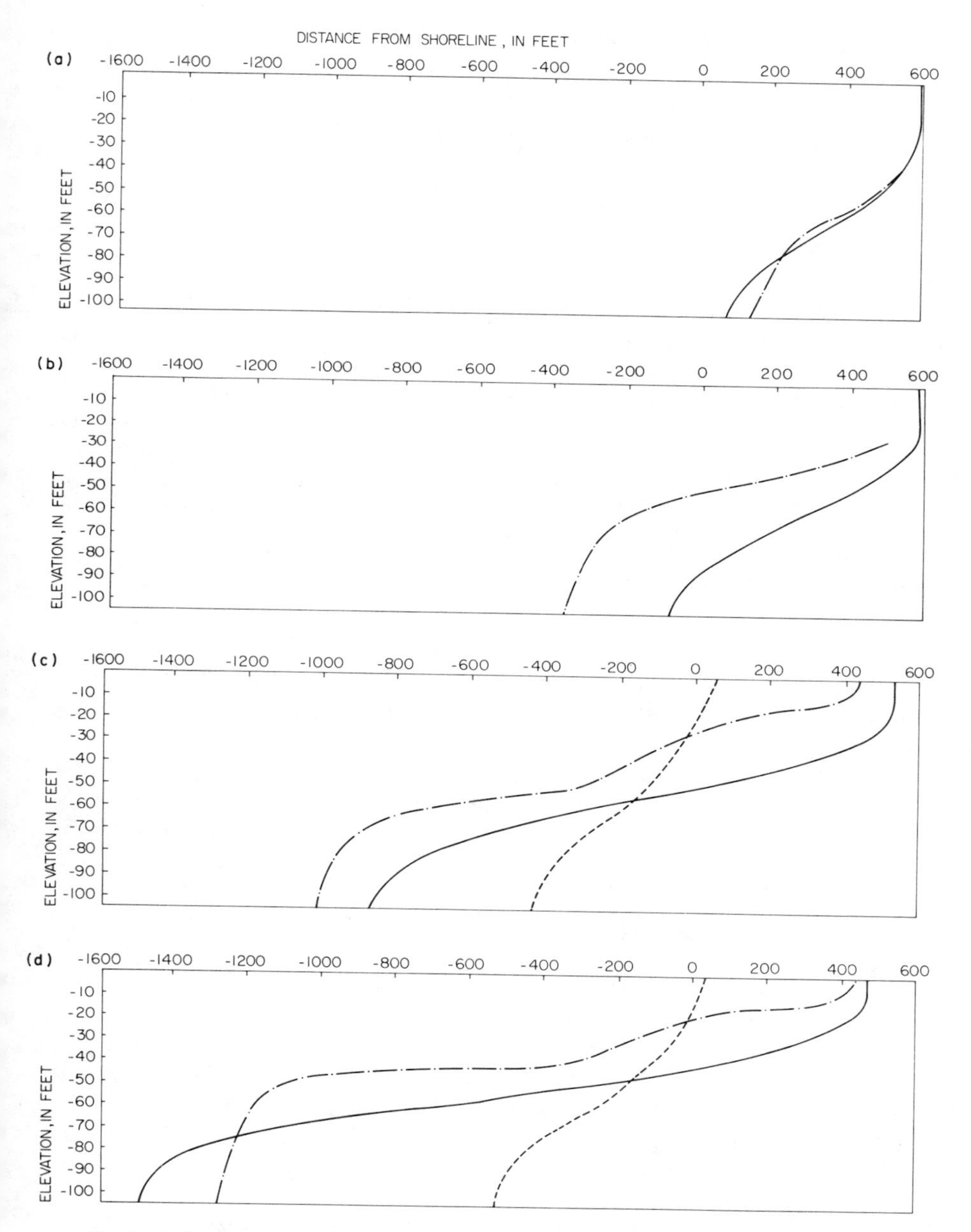


**Fig. 5.25.** Steady state concentration distribution, (a) 0.98 isochlor (18,500 ppm); (b) 0.95 isochlor (18,000 ppm); (c) 0.75 isochlor (14,200 ppm); (d) 0.5 isochlor (9500 ppm). (——) Present analysis, (—·—) after Kohout and Klein [27], (- - -) after Lee and Cheng [28].

what sharper front, which is probably due to a poor estimate of the ratio between vertical and horizontal permeabilities used in the simulation. The same steady state problem has been considered by Lee and Cheng [28] who present results corresponding to the following choice of parameters:

$$a = Q_f/(K_f \gamma d) = 0.04, \qquad P_e = Q_f/D = 100, \qquad \gamma = (\rho_s - \rho_f)/\rho_f = 0.025 \tag{5.56}$$

where $Q_f$ is the average discharge of freshwater per unit width, $K_f$ the hydraulic conductivity, $D$ a constant dispersion coefficient, $\rho_s$ and $\rho_f$ the salt and fresh water densities, respectively, and $d$ the characteristic thickness of the aquifer. The results presented by Lee and Cheng [28] are plotted in Fig. 5.25 where it is assumed that these authors have used the 19,000 ppm isochlor to normalize results for comparison with Kohout's experimental values [28]. It should be noted that Lee and Cheng take the field data prevailing several months after the rain as indicative of steady state conditions. As mentioned earlier, these values are essentially identical to the ones reported prior to the recharge, and used in this discussion.

*Transient Problem*

The steady state concentration distribution shown in Fig. 5.25 serves as an initial condition for the transient problem. The hydrographs presented by Kohout and Klein [27] indicate that the rainfall produces a large increase in freshwater head, followed by a rapid decline to a new level higher than the prerain level. The value of the pressure specified at the Dirichlet nodes is thus changed as the solution progresses through time from the initial conditions (occurrence of the rainfall) to the conditions prevailing a week after the rainfall (Fig. 5.24). Figure 5.26 represents the relative elevation of the potentiometric surface assumed between the time $t = 0$ and $t = 180$ hr. The stepwise representation is considered a reasonable representation of the observed head changes.

The results of the transient analysis obtained at $t = 180$ hr. are shown in Fig. 5.27 together with the field values reported by Kohout and Klein [27] for May 29th, 1958 ($t = 172$ hr). By comparing Fig. 5.27 to Fig. 5.25 [27] it can be seen that the heavy recharge produces a seaward movement of the saltwater front. It appears also that the displacement of the isochlors predicted with the numerical model is in satisfactory agreement with the experimental observations. As in the steady state case, the differences between the observed and simulated 0.75 (14,200 ppm) and 0.95 (18,000 ppm) isochlors are thought to be due to inhomogeneities occurring in the central part of the aquifer.

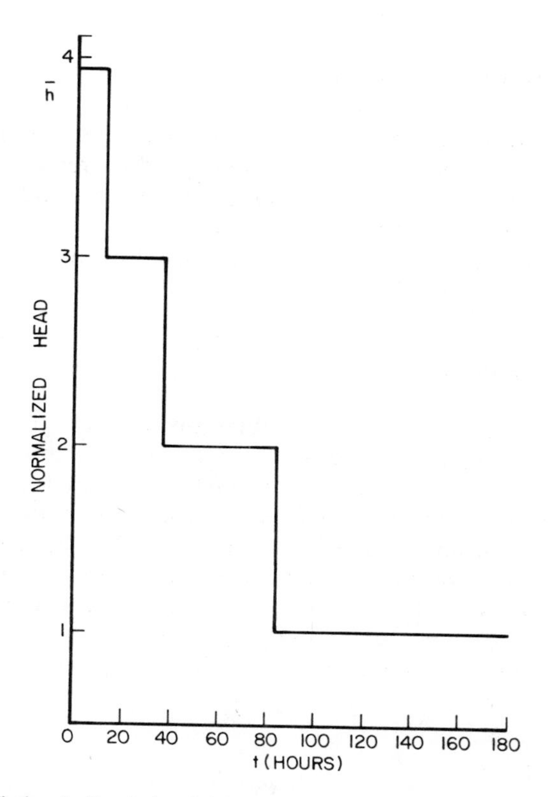


**Fig. 5.26.** Relative decline in head during the transient solution, where all values are normalized using the head at $t = 180$ hr as a reference.

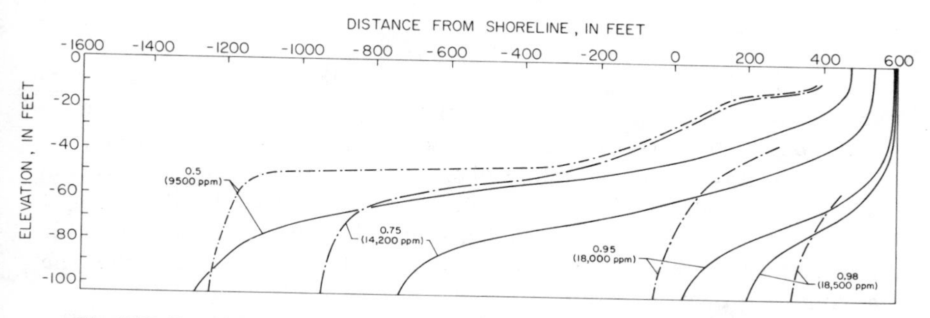


**Fig. 5.27.** Transient concentration distribution ($t = 180$ hours). (——) Present analysis, (—·—) after Kohout and Klein [27].

## 5.4 Saturated–Unsaturated Flows

Thus far, only single-phase flow through a porous medium has been considered. There are, however, many problems which require simulation of multiphase flow. Perhaps the most important of these, at least the one which has received the most attention, involves the flow of water through a porous medium containing continuous channels occupied by air.

### *5.4.1 Multiphase Equation Formulation*

There are two approaches to describing this physical system. The more accurate and unfortunately more complex approach requires the simultaneous solution of two partial differential equations, one written for the water phase and the other for the air. While this system of equations has been successfully solved using finite difference methods [29, 30], solutions using a finite element approach are lacking. There are, however, Galerkin-based solutions available for an analogous problem that is encountered in multiphase petroleum reservoir simulation [31].

The partial differential equations describing isothermal two-phase flow in porous media are [29]

$$\nabla \cdot \left( \frac{\rho_w k_{rw} \mathbf{k}}{\mu_w} \cdot (\nabla p_w + \rho_w g \, \nabla Z) \right) = \theta \frac{\partial (S_w \rho_w)}{\partial t} \tag{5.57a}$$

for the water phase and

$$\nabla \cdot \left( \frac{\rho_a k_{ra} \mathbf{k}}{\mu_a} \cdot (\nabla p_a + \rho_a g \, \nabla Z) \right) = \theta \frac{\partial (S_a \rho_a)}{\partial t} \tag{5.57b}$$

for the air phase, where $p_w$ and $p_a$ are water and air phase pressures, respectively; $Z$ the height above a reference plane; $S_w$ and $S_a$ water and air phase saturations, respectively; $\rho_w$ and $\rho_a$ water and air phase densities, respectively; $\mathbf{k}$ the absolute permeability of the porous medium; $k_{rw}$ and $k_{ra}$ the relative permeabilities of water and air phases, respectively; $\theta$ the porosity (assumed here to be time independent); and $\mu_w$ and $\mu_a$ the water and air phase viscosities, respectively. Examination of Eqs. (5.57) reveals that the two equations contain six unknowns $p_w$, $p_a$, $k_{rw}$, $k_{ra}$, $S_w$, and $S_a$; therefore, four auxiliary equations are required. The first is obtained from the stipulation that only air and water are present in the pores, such that

$$S_w + S_a = 1.0 \tag{5.58}$$

The second equation follows from the definition of capillary pressure which states that in a partially saturated porous medium, there is a difference between the pressure of water and air described by

$$p_c = p_a - p_w \tag{5.59}$$

where $p_c$ is the capillary pressure. Although this condition relates $p_a$ to $p_w$, it introduces a new dependent variable $p_c$ into the problem and therefore, three additional equations must still be formulated. These are empirically determined and relate the relative permeabilities and capillary pressure to saturation:

$$k_{rw} = f_1(S_w) \tag{5.60a}$$

$$k_{ra} = f_2(S_a) \tag{5.60b}$$

$$p_c = f_3(S_w) \tag{5.60c}$$

Typical saturation and relative permeability curves are presented in Figs. 5.28a, b. The final set of equations is highly nonlinear and considerable care must be exercised in generating numerical solutions.

There are several possible avenues of attack for solving (5.57). The approach to be presented herein is similar to that employed in solving the multiphase petroleum reservoir problem [31]. The first step is to express the saturations $S_w$ and $S_a$ of (5.57) in terms of the dependent variables $p_a$ and $p_w$ through application of the chain rule:

$$\frac{\partial S_w}{\partial t} = \frac{\partial S_w}{\partial p_c}\frac{\partial p_c}{\partial t} = \frac{\partial S_w}{\partial p_c}\left(\frac{\partial p_a}{\partial t} - \frac{\partial p_w}{\partial t}\right)$$

and

$$\frac{\partial S_a}{\partial t} = \frac{\partial S_a}{\partial p_c}\frac{\partial p_c}{\partial t} = \frac{\partial S_a}{\partial p_c}\left(\frac{\partial p_a}{\partial t} - \frac{\partial p_w}{\partial t}\right) = -\frac{\partial S_w}{\partial p_c}\left(\frac{\partial p_a}{\partial t} - \frac{\partial p_w}{\partial t}\right)$$

where the coefficient $\partial S_w/\partial p_c$ is obtained from (5.60c).

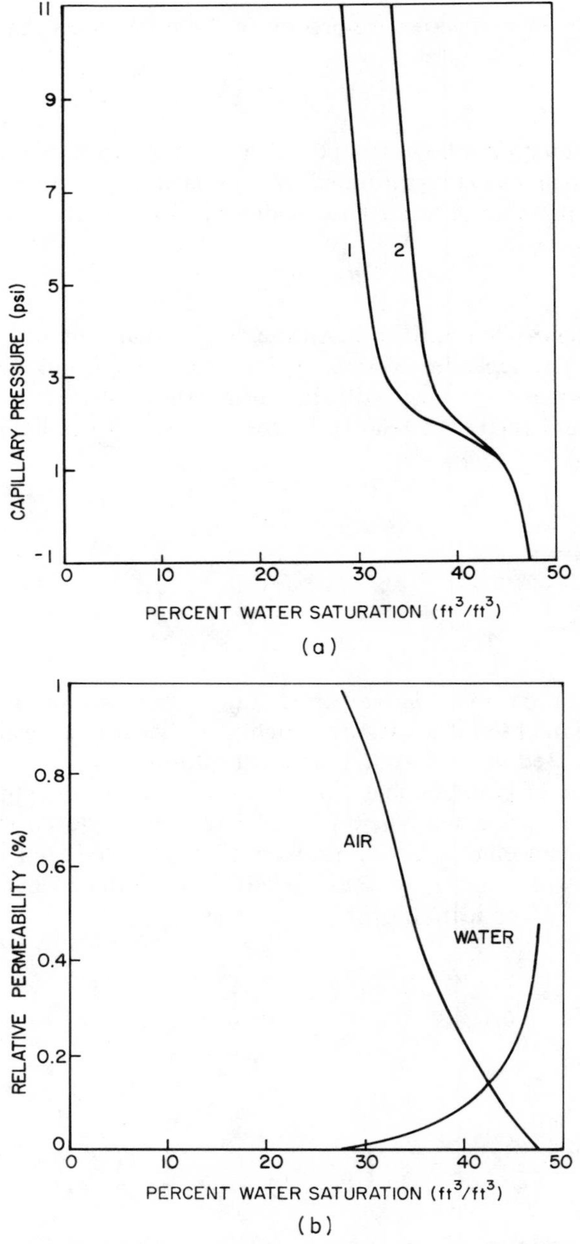


**Fig. 5.28.** (a) Typical saturation curve for air–water system [29]; (b) Typical relative permeability curve for air–water system [29].

### 5.4.2 *Application of Galerkin Method*

The formulation of the Galerkin approximation may be simplified by introduction of the following operators obtained from (5.57):

$$L_w(p_w, p_a) \equiv \nabla \cdot \left(\frac{\rho_w k_{rw} \mathbf{k}}{\mu_w} \cdot (\nabla p_w + \rho_w g \nabla Z)\right) - \theta \rho_w \frac{\partial S_w}{\partial p_c}\left(\frac{\partial p_a}{\partial t} - \frac{\partial p_w}{\partial t}\right) = 0 \tag{5.61a}$$

$$L_a(p_w, p_a) \equiv \nabla \cdot \left(\frac{\rho_a k_{ra} \mathbf{k}}{\mu_a} \cdot (\nabla p_a + \rho_a g \nabla Z)\right) + \theta \rho_a \frac{\partial S_w}{\partial p_c}\left(\frac{\partial p_a}{\partial t} - \frac{\partial p_w}{\partial t}\right) = 0 \tag{5.61b}$$

where the dependence of $\rho_w$ and $\rho_a$ on time has been considered negligible.

Let the two dependent variables $p_a$ and $p_w$ be approximated, respectively, by the finite series

$$p_w \simeq \hat{p}_w = \sum_{j=1}^{M} P_{w,j}(t)\phi_{w,j}(x, y) \tag{5.62a}$$

and†

$$p_a \simeq \hat{p}_a = \sum_{j=1}^{N} P_{a,j}(t)\phi_{a,j}(x, y) \tag{5.62b}$$

where $P_{w,j}$ and $P_{a,j}$ are undetermined coefficients and $\phi_{w,j}$ and $\phi_{a,j}$ are basis functions. In practice, it is generally possible to assume $\phi_{w,j}$ and $\phi_{a,j}$ to be the same set of functions, i.e., $\phi_{w,j} = \phi_{a,j} = \phi_j$ and $M = N$.

As in earlier examples, $P_a$ and $P_w$ must be selected such that $L_w(\hat{p}_w, \hat{p}_a)$ and $L_a(\hat{p}_w, \hat{p}_a)$ are orthogonal to the set of functions $\phi_i$, $i = 1, 2, \ldots, N$. This requirement is expressed mathematically by

$$\langle L_w(\hat{p}_w, \hat{p}_a), \phi_i \rangle = 0 \qquad (i = 1, 2, \ldots, N) \tag{5.63a}$$

$$\langle L_a(\hat{p}_w, \hat{p}_a), \phi_i \rangle = 0 \qquad (i = 1, 2, \ldots, N) \tag{5.63b}$$

Substitution of (5.61) and (5.62) into (5.63) and application of Green's

† Note that we have chosen to consider two space dimensions only for convenience; the approach is easily extended to three space dimensions.

theorem to those integrals containing second-order spatial derivatives yields (for two space dimensions)

$$\langle \mathbf{M}_w \cdot (\nabla_{xy}\hat{p}_w + \rho_w g \nabla_{xy} Z), \nabla_{xy}\phi_i\rangle + \left\langle \theta\rho_w \frac{\partial S_w}{\partial p_c}\left(\frac{\partial \hat{p}_a}{\partial t} - \frac{\partial \hat{p}_w}{\partial t}\right), \phi_i\right\rangle$$

$$-\int_\Gamma \mathbf{M}_w \cdot (\nabla_{xy}\hat{p}_w + \rho_w g \nabla_{xy} Z) \cdot \mathbf{n}\, \phi_i\, ds = 0 \qquad (i = 1, 2, \ldots, N) \quad (5.64a)$$

$$\langle \mathbf{M}_a \cdot (\nabla_{xy}\hat{p}_a + \rho_a g \nabla_{xy} Z), \nabla_{xy}\phi_i\rangle - \left\langle \theta\rho_a \frac{\partial S_w}{\partial p_c}\left(\frac{\partial \hat{p}_a}{\partial t} - \frac{\partial \hat{p}_w}{\partial t}\right), \phi_i\right\rangle$$

$$-\int_\Gamma \mathbf{M}_a \cdot (\nabla_{xy}\hat{p}_a + \rho_a g \nabla_{xy} Z) \cdot \mathbf{n}\, \phi_i\, ds = 0 \qquad (i = 1, 2, \ldots, N) \quad (5.64b)$$

where $\mathbf{M}_w \equiv \rho_w k_{rw} \mathbf{k}/\mu_w$ and $\mathbf{M}_a \equiv \rho_a k_{ra} \mathbf{k}/\mu_a$.

Combination of (5.62) and (5.64) yields the matrix equation

$$\begin{bmatrix} a_{1,1} & 0 & a_{1,2} & 0 & a_{1,3} & 0 & \cdots & a_{1,N} & 0 \\ 0 & b_{1,1} & 0 & b_{1,2} & 0 & b_{1,3} & \cdots & 0 & b_{1,N} \\ a_{2,1} & 0 & a_{2,2} & 0 & a_{2,3} & 0 & \cdots & a_{2,N} & 0 \\ \vdots & \vdots & \vdots & \vdots & \vdots & \vdots & \vdots & \vdots & \vdots \\ a_{N,1} & 0 & \cdot & \cdot & \cdot & & 0 & a_{N,N} & 0 \\ 0 & b_{N,1} & & \cdot & \cdot & \cdot & b_{N,N-1} & 0 & b_{N,N} \end{bmatrix} \begin{Bmatrix} P_{w1} \\ P_{a1} \\ P_{w2} \\ \vdots \\ P_{wN} \\ P_{aN} \end{Bmatrix}$$

$$- \begin{bmatrix} c_{1,1} & -c_{1,1} & c_{1,2} & -c_{1,2} & \cdots & c_{1,N} & -c_{1,N} \\ -d_{1,1} & d_{1,1} & -d_{1,2} & d_{1,2} & \cdots & -d_{1,N} & d_{1,N} \\ c_{2,1} & -c_{2,1} & c_{2,2} & -c_{2,2} & \cdots & c_{2,N} & -c_{2,N} \\ \vdots & \vdots & \vdots & \vdots & & \vdots & \vdots \\ c_{N,1} & -c_{N,1} & c_{N,2} & -c_{N,2} & \cdots & c_{N,N} & -c_{N,N} \\ -d_{N,1} & d_{N,1} & -d_{N,2} & d_{N,2} & \cdots & -d_{N,N} & d_{N,N} \end{bmatrix} \begin{Bmatrix} dp_{w1}/dt \\ dp_{a1}/dt \\ dp_{w2}/dt \\ \vdots \\ dp_{wN}/dt \\ dp_{aN}/dt \end{Bmatrix} = \begin{Bmatrix} f_{w1} \\ f_{a1} \\ f_{w2} \\ \vdots \\ f_{wN} \\ f_{aN} \end{Bmatrix}$$

$$[A]\{P\} + [B]\{dP/dt\} = \{F\} \quad (5.65)$$

The elements appearing in $[A]$, $[B]$, and $\{F\}$ are defined as

$$a_{i,j} \equiv \langle \mathbf{M}_w \cdot \nabla_{xy}\phi_j, \nabla_{xy}\phi_i\rangle, \qquad b_{i,j} \equiv \langle \mathbf{M}_a \cdot \nabla_{xy}\phi_j, \nabla_{xy}\phi_i\rangle$$

$$c_{i,j} \equiv \left\langle \theta\rho_w \frac{\partial S_w}{\partial p_c}\phi_j, \phi_i\right\rangle, \qquad d_{i,j} \equiv \left\langle \theta\rho_a \frac{\partial S_w}{\partial p_c}\phi_j, \phi_i\right\rangle$$

$$f_{wi} = \int_\Gamma \mathbf{M}_w \cdot (\nabla_{xy}\hat{p}_w + \rho_w g \nabla_{xy} Z) \cdot \mathbf{n}\, \phi_i\, ds - \langle \mathbf{M}_w \cdot \rho_w g \nabla_{xy} Z, \nabla_{xy}\phi_i\rangle$$

$$f_{ai} = \int_\Gamma \mathbf{M}_a \cdot (\nabla_{xy}\hat{p}_a + \rho_a g \nabla_{xy} Z) \cdot \mathbf{n}\, \phi_i\, ds - \langle \mathbf{M}_a \cdot \rho_a g \nabla_{xy} Z, \nabla_{xy}\phi_i\rangle$$

and the required integrations are, in general, performed once again numerically using Gauss points. Usually the contour integrals are known through boundary specifications or are deleted from the algebraic equations through the condensation process for Dirichlet nodes.

While it is possible to use a variety of schemes for solving Eq. (5.65), a relatively simple and satisfactory procedure uses the finite difference representation of the time derivative. In this approach, (5.65) is rewritten as

$$[A](\varepsilon\{P\}^{n+1} + [1-\varepsilon]\{P\}^{n}) + [B](1/\Delta t)(\{P\}^{n+1} - \{P\}^{n}) = \varepsilon\{F\}^{n+1} + (1-\varepsilon)\{F\}^{n} \tag{5.66}$$

where $[A]$ and $[B]$ are evaluated at $(n+\frac{1}{2})$ and unconditional stability generally requires $\frac{1}{2} \leq \varepsilon \leq 1$. When $\varepsilon$ is selected to be 1, the scheme is a fully implicit or backward difference approximation and when $\varepsilon = \frac{1}{2}$, the scheme becomes a central difference or Crank–Nicolson approximation. Equation (5.66) is rearranged in the computational algorithm so that unknown values appear on the left-hand side and known information on the right. The final expression thus becomes

$$(\varepsilon[A] + (1/\Delta t)[B])\{P\}^{n+1} = ([\varepsilon - 1][A] + (1/\Delta t)[B])\{P\}^{n} + \{F\} \tag{5.67}$$

The solution of (5.67) appears relatively straightforward, and would indeed be, were it not for the highly nonlinear behavior of the coefficients appearing in $[A]$ and $[B]$. There are several avenues of approach one can consider in solving this set of nonlinear algebraic equations. The most common approach involves an iterative procedure which can be generally written as

$$[C]^{m}\{P\}^{n+1,m+1} = \{D\}^{m} + \{F\}^{m} \tag{5.68}$$

where

$$[C]^{m} = \varepsilon[A]^{m} + (1/\Delta t)[B]^{m}$$
$$\{D\}^{m} = ([\varepsilon - 1][A]^{m} + (1/\Delta t)[B]^{m})\{P\}^{n}$$

Computationally, one first solves Eq. (5.68) for $\{P\}^{n+1,m+1}$ using initial conditions to evaluate the coefficients of $[C]^{m}$, $\{D\}^{m}$, and $\{F\}^{m}$. Next $\{P\}^{n+1,m+1}$ is used to update $[C]^{m+1}$, $[D]^{m+1}$, and possibly $\{F\}^{m+1}$. Note that $\{P\}^{n}$ remains unchanged during the iterative cycle. Equation (5.68) is solved once again for $\{P\}^{n+1,m+2}$ and this cycle is repeated until the difference between successive iterations is within a specified tolerance, i.e.,

$$|P_i^{n+1,m+1} - P_i^{n+1,m}| < E \qquad (i = 1, 2, \ldots, N)$$

When this condition is met, $\{P\}^{n}$ is updated to $\{P\}^{n+1}$ and the iterative cycle is initiated once again solving first for $\{P\}^{n+2,1}$.

An interesting alternative formulation has been found effective in the solution of the nonlinear equations describing surface water hydrodynamics.

Rewrite Eq. (5.67) in the form

$$\varepsilon[A]^{n+1}\{P\}^{n+1} + (1-\varepsilon)[A]^{n}\{P\}^{n} + (1/\Delta t)[B]^{n+\varepsilon}\{P\}^{n+1} - (1/\Delta t)[B]^{n+\varepsilon}\{P\}^{n}$$
$$= \varepsilon\{F\}^{n+1} + (1-\varepsilon)\{F\}^{n}$$

This approach, which appears somewhat less attractive than (5.66), clearly recognizes that the $[B]$ matrix is stationed at the same level in time as the time derivative and should be evaluated at that level. In contrast, the $[A]$ coefficient matrix is evaluated at the same time level as the associated $\{P\}$ matrix. From this point forward, the iterative aspects of the solution procedure are analogous to the scheme described by Eq. (5.68). It is worth noting that there are two approaches to calculating a coefficient matrix at the $n + \varepsilon$ time level. One can either compute the coefficients at the $n$ and $n + 1$ levels using the corresponding values of the dependent variables and interpolate to obtain the coefficients at $n + \varepsilon$ or, alternatively, use interpolation to obtain a value of the dependent variable at the $n + \varepsilon$ level and then compute the corresponding coefficient. Work to date suggests the latter scheme generates a more rapid rate of convergence.

## 5.5 Saturated–Unsaturated Flow of a Single Fluid

In this section we will simplify the problem and assume that the equation describing the dynamics of the air phase plays a minor role in defining water movement through the unsaturated zone. Utilizing this assumption, soil scientists have developed a single-equation formulation which has found application not only in soil science but also in hydrology. While the majority of numerical solutions using this second scheme are, once again, based on a finite difference formulation [32, 33], finite element solutions have also been obtained [8, 34, 35].

### *5.5.1 Single-Equation Formulation*

One approach to solving the unsaturated flow equations, while neglecting the air phase dynamics, is simply to consider only (5.64a) and disregard the pressure in the air phase. Provided the porous medium remains unsaturated, this is the most general formulation because it permits the fluid density to be a function of ion concentration or temperature as well as pressure. This formulation is generally abandoned, however, for a simpler scheme based on hydraulic head. The governing equations using this approach can be derived at several levels of sophistication [34, 35]. For simplicity, we will take (5.57a), modified to include variable porosity $\theta$, as a point of departure:

$$\nabla \cdot [(\rho_w k_{rw} \mathbf{k}/\mu_w) \cdot (\nabla p_w + \rho_w g \nabla Z)] = \partial(\theta S_w \rho_w)/\partial t \tag{5.57c}$$

If the fluid density $\rho_w$ is assumed to be a function of pressure only, Hubbert's potential may be defined by

$$\phi^* = gH = \int_{p_{w0}}^{p_w} dp'_w/\rho_w(p'_w) + g \int_{Z_0}^{Z} dz' \tag{5.69}$$

where $p_{w0}$ is the pressure at the reference elevation $Z_0$. Differentiation of (5.69) yields

$$\nabla H = (\nabla p_w/\rho_w g) + \nabla Z \tag{5.70}$$

and substitution of this equation in (5.57c) leads to

$$\nabla \cdot (\rho_w \mathbf{K} k_{rw} \cdot \nabla H) = \partial(\theta S_w \rho_w)/\partial t \tag{5.71}$$

where $\mathbf{K} \equiv \rho_w g \mathbf{k}/\mu_w$ and is called the saturated hydraulic conductivity. The right-hand side of (5.71) can be restated in terms of hydraulic head through the use of additional constitutive assumptions. Expansion of the time derivative in (5.71) yields

$$\frac{\partial(\theta S_w \rho_w)}{\partial t} = \theta \rho_w \frac{\partial S_w}{\partial t} + \theta S_w \frac{\partial \rho_w}{\partial t} + \rho_w S_w \frac{\partial \theta}{\partial t} \tag{5.72}$$

The empirical relationship adopted by Cooley [37] to treat the change in formation porosity is [34]

$$c_f = \frac{1}{\theta} \frac{\partial \theta}{\partial p_w} = \frac{1}{\rho_w g \theta} \frac{\partial \theta}{\partial H}$$

and the last term in (5.72) becomes

$$\rho_w S_w \frac{\partial \theta}{\partial t} = \rho_w S_w \frac{\partial \theta}{\partial H} \frac{\partial H}{\partial t} = \rho_w S_w c_f \rho_w g \theta \frac{\partial H}{\partial t} \tag{5.73}$$

To evaluate the temporal changes in fluid density the equation of state is evoked whereby

$$\partial \rho_w/\partial t = \rho_{w0} \beta \, \partial H/\partial t \tag{5.74}$$

and $\beta$ is a coefficient describing the fluid compressibility. Substitution of (5.74) and (5.73) into (5.72) yields

$$\frac{\partial(\theta S_w \rho_w)}{\partial t} = \theta \rho_w \frac{\partial S_w}{\partial t} + \theta S_w \rho_w \beta \frac{\partial H}{\partial t} + \rho_w S_w c_f \rho_w g \theta \frac{\partial H}{\partial t} \tag{5.75}$$

Combination of (5.75) and (5.71) and introduction of the assumption that the gradient of fluid density is negligible, reduces the flow equation to its final form

$$\nabla \cdot (\mathbf{K}k_{rw} \cdot \nabla H) = \theta \frac{\partial S_w}{\partial t} + S_w S_s \frac{\partial H}{\partial t} \tag{5.76}$$

where $S_s = \theta(\beta + \rho_w c_f g)$. Generally speaking, the fluid compressibility $\beta$ is several orders of magnitude smaller than the formation compressibility $c_f$. When this is indeed the case, $\theta S_w \, \partial \rho_w / \partial t$ can be considered negligible.

In practical problems, it is convenient to rewrite (5.76) in terms of pressure head $h$, elevation head $Z$, and volumetric moisture content $\Theta$ rather than in terms of $H$ and $S_w$. The volumetric moisture content $\Theta$ is defined as $\Theta = S_w \theta$ and the specific moisture capacity $C$ as $C = \partial\Theta/\partial h$. Since Eq. (5.70) can be rewritten as

$$\nabla H = \nabla h + \nabla Z$$

Eq. (5.76) becomes

$$\nabla \cdot [\mathbf{K}k_{rw} \cdot \nabla(h + Z)] = \theta \frac{\partial(\Theta/\theta)}{\partial t} + \left(\frac{\Theta}{\theta}\right) S_s \frac{\partial(h + Z)}{\partial t}$$

This expression is generally simplified further to yield

$$\nabla \cdot [\mathbf{K}k_{rw} \cdot (\nabla h + \mathbf{k})] = C \frac{\partial h}{\partial t} + \left(\frac{\Theta}{\theta}\right) S_s \frac{\partial h}{\partial t} \tag{5.77}$$

where $\mathbf{k}$ is the unit vector in the vertical direction. Herein we have assumed that one coordinate plane is horizontal and that porosity $\theta$ is time independent. This latter assumption is, unfortunately, somewhat inconsistent with the earlier hypothesis regarding formation compressibility. However, this form of the equation is the one that heretofore has been considered in finite element solutions.

### *5.5.2 Application of Galerkin's Method*

The formulation of the approximating algebraic equations using Galerkin's method for the single equation development is analogous to that presented earlier for the two-equation approach. The pressure head $h$ is approximated as

$$h \sim \hat{h} = \sum_{j=1}^{N} P_j(t)\phi_j(x, y) \tag{5.78}$$

where $P_j$ is a time-dependent, undetermined coefficient. Equation (5.77) may be reduced to the two-dimensional form

$$Lh \equiv \nabla_{xy} \cdot (\mathbf{K}_{rw} \cdot \nabla_{xy} h) + \nabla_{xy} \cdot (\mathbf{K}k_{rw} \cdot \mathbf{k}) - [C + (\Theta/\theta)S_s]\, \partial h/\partial t = 0 \qquad (5.79)$$

where we have assumed a cross-sectional formulation.

Galerkin's method requires that the coefficients $P_j$ be determined such that $L\hat{h}$ is orthogonal to each of the bases $\phi_j$; this requirement can be expressed as

$$\langle L\hat{h}, \phi_i \rangle = 0 \qquad (i = 1, 2, \ldots, N)$$

Substitution of (5.79) and (5.78) into this equation and application of Green's theorem to the resulting expression yields

$$\langle (\mathbf{K}'(\hat{h}) \cdot \nabla_{xy}\hat{h}, \nabla_{xy}\phi_i \rangle + \langle [C + (\Theta/\theta)S_s]\, \partial \hat{h}/\partial t, \phi_i \rangle$$
$$= \int_\Gamma \mathbf{n} \cdot \mathbf{K}'(\hat{h}) \cdot [\nabla_{xy}\hat{h} + \mathbf{k}]\, \phi_i\, ds \qquad (i = 1, 2, \ldots, N) \qquad (5.80)$$

where $\mathbf{K}' \equiv \mathbf{K}k_{rw}$ and $\mathbf{n}$ is an outwardly directed unit normal vector on $\Gamma$. Equation (5.80) may be expressed in matrix form as

$$[A]\{P\} + [B]\{dP/dt\} = \{F\} \qquad (5.81)$$

where

$$a_{i,j} \equiv \langle \mathbf{K}'(\hat{h}) \cdot \nabla_{xy}\phi_j, \nabla_{xy}\phi_i \rangle, \qquad b_{i,j} \equiv \langle [C + (\Theta/\theta)S_s]\phi_j, \phi_i \rangle$$

$$f_i \equiv \int_\Gamma \mathbf{n} \cdot \mathbf{K}'(\hat{h}) \cdot [\nabla_{xy}\hat{h} + \mathbf{k}]\phi_i\, ds$$

This expression is similar to the standard equation for groundwater flow established earlier, but is now highly nonlinear because of the dependence of both $\mathbf{K}'$ and $\Theta$ on the pressure head.

To evaluate (5.81), schemes for time discretization and iteration must be decided on. In general, (5.81) may be written to account for either time-centered or backward approximations in time as was done for the two-equation formulation

$$(\varepsilon[A] + (1/\Delta t)[B])\{P\}^{n+1}$$
$$= ([\varepsilon - 1][A] + (1/\Delta t)[B])\{P\}^n + \varepsilon\{F\}^{n+1} + (1 - \varepsilon)\{F\}^n \quad (5.82)$$

The coefficient matrices are functions of the dependent variables $P$ and should be evaluated at time $t = t^n + \varepsilon\, \Delta t^n$. In practice, a time-centered coefficient matrix (i.e., $\varepsilon = \frac{1}{2}$) has been used even with a backward difference scheme (i.e., $\varepsilon = 1$) to effectively dampen oscillations that often arise in solving highly nonlinear systems. Another possible scheme, which was introduced earlier in the two equation formulation, gives a matrix expression wherein the space and time matrices are evaluated at different time levels. This form of (5.82) is given by

$$(\varepsilon[A]^{n+1} + (1/\Delta t)[B]^{n+\varepsilon})\{P\}^{n+1} = ([\varepsilon - 1][A]^n + (1/\Delta t)[B]^{n+\varepsilon})\{P\}^n + \varepsilon\{F\}^{n+1} + (1 - \varepsilon)\{F\}^n$$

where $[B]^{n+\varepsilon}$ is defined as the matrix $[B]$ evaluated at time level $n + \varepsilon$.

It should be noted that when the specific storage $S_s$ is zero, the time derivative disappears in the saturated zone and the equation becomes elliptic in form. To avoid difficulties associated with the correct specification of the right-hand side of (5.82) a backward difference scheme was suggested by Neuman [34]. In this case, the time-dependent part of the right-hand side of (5.82) vanishes and a solution for $\{P\}^{n+1}$ can be easily obtained. Neuman points out, however, that an exception to this rule arises at nodes that pass from a state of saturation to a state of incomplete saturation during one time step. When the elements of $[B]$ are evaluated at the $n + \varepsilon$ time level through a linear extrapolation over time, the values computed for these transition nodes may differ from zero and the general form of (5.82) cannot be solved without knowing the values of $\{P\}^{n+1}$. However, when the specific storage is zero, the elements of $[B]$ are influenced only by the moisture content $\Theta$. Consequently, changes in $[B]$ must occur only when $P$ values are negative because $C = \partial\Theta/\partial h$ is zero for positive $h$ (and therefore $P$). Thus one is justified in using a zero value for $P$ at the transition node when computing $[B]$ [34].

### 5.5.3 *Boundary Conditions*

The Dirichlet- and Neumann-type boundary conditions for unsaturated flow are treated in the same manner as for saturated conditions. When a constant pressure head is specified at a node, the corresponding row and column of the coefficient matrix is deleted. Constant flux conditions are accounted for through the surface integral appearing in $\{F\}$. This integral is nonzero only along the domain boundary $\Gamma$ where it represents the component of flow normal to $\Gamma$. There is, however, one type of boundary specification which has not been encountered heretofore; this is the seepage face. The seepage face is the external boundary of the saturated zone where the fluid moves out of the system (see Fig. 5.29). In general, the location and extent of the seepage face are unknown. It is therefore necessary to develop an iterative approach wherein an approximate seepage face is specified, the governing equations are solved, and the seepage face is modified to reflect information arising from the solution. This cyclic procedure is repeated until some criterion specified by the analyst is met. As an example, consider the following scheme presented by Neuman [34].† Suppose that it is desired to predict the position of the seepage face at time $t^{n+1}$, given its location at $t^n$. During the first iteration, $P$ (and consequently $h$) is set to zero along the initial length of the seepage face and a Dirichlet-type boundary is assumed. At all nodes with $P < 0$, a Neumann-type boundary is specified with the inwardly directed normal flux $Q$ set to zero. The solution should provide negative values of $Q$ at nodes where $P$ is prescribed as zero (i.e., outward

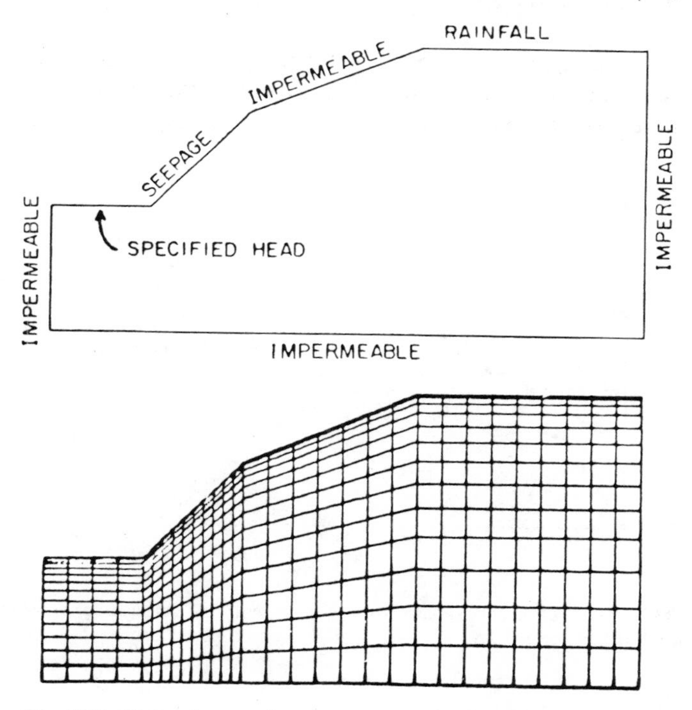


**Fig. 5.29.** Finite element discretization of an idealized water shed [42].

flow) and negative values of $P$ at nodes where $Q$ is prescribed to be zero. If, instead, a positive value of $Q$ is encountered at a node where $P = 0$, the value of $Q$ there is set equal to zero and, in the next iteration, this node is assumed to be on a Neumann type boundary. On the other hand, if a positive value of $P$ is encountered at a node where $Q = 0$, the corresponding value of $P$ is set to zero and the node is treated as a constant head node in the next iteration.

† Neuman performs spatial averaging for the time derivative which departs from finite element theory and closely resembles finite differences on an irregular net.

Neumann states that a necessary condition for convergence of this scheme is that the modification of the boundary condition proceed sequentially from node to node, starting at the saturated end of the seepage face. In addition, after having specified $Q = 0$ at any node during a given iteration, $Q$ at all the subsequent nodes along the seepage face must also be set to zero.

### 5.5.4 *Simulation of the Free Surface*

While it is generally desirable to solve the problem of groundwater flow with a free surface using the saturated–unsaturated flow equations described earlier, there may be occasions where it is advantageous to solve the sa-

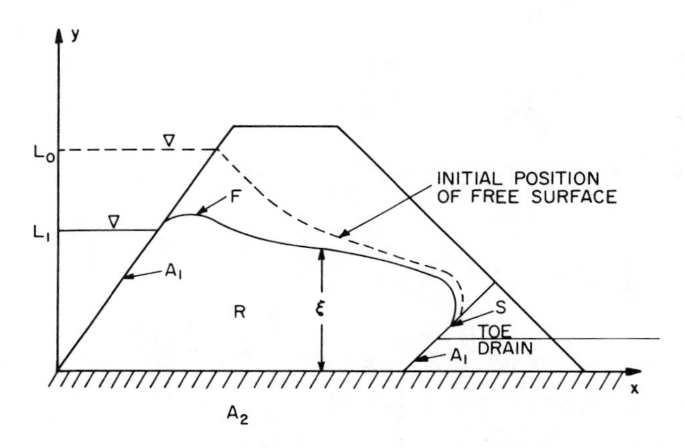


**Fig. 5.30.** Cross section of a dam with a transient free surface (after Neuman and Witherspoon [40]).

turated flow equations with the free surface specified as a moving boundary. This type of formulation has been outlined in finite element format in a series of papers by Neuman and Witherspoon [38–40]. This application of the finite element method is particularly interesting because a correct specification of the free surface in a transient simulation requires modification of the element configuration. The ability to modify mesh geometry easily is one of the attractive features of the finite element numerical approach.

The problem considered by Neuman and Witherspoon [40] is illustrated in Fig. 5.30. The water level in the reservoir drops from level $L_0$ to $L_1$ at a specified rate and the free surface responds by moving downward until a new equilibrium is established. The governing equations describing this system are

$$\begin{aligned}
\nabla_{xy} \cdot (\mathbf{K} \cdot \nabla_{xy} h) &= S_s \, \partial h/\partial t & & \text{(a)} \\
h(x, y, 0) &= h_0(x, y) & & \text{(b)} \\
\xi(x, 0) &= \xi_0(x) & & \text{(c)} \\
h(x, y, t) &= H_0(x, y, t) & \text{on } A_1 \quad & \text{(d)} \quad (5.83a) \\
\mathbf{K} \cdot \nabla_{xy} h \cdot \mathbf{n} &= -v(x, y, t) & \text{on } A_2 \quad & \text{(e)} \\
\xi(x, t) &= h(x, \xi, t) & \text{on } F \quad & \text{(f)} \\
\mathbf{K} \cdot \nabla_{xy} h \cdot \mathbf{n} &= (I\mathbf{j} - S_y(\partial \xi/\partial t)\mathbf{j}) \cdot \mathbf{j} & \text{on } F \quad & \text{(g)} \\
h(x, y, t) &= y & \text{on } S \quad & \text{(h)}
\end{aligned}$$

where $\xi$ is the elevation of the free surface above the horizontal datum plane from which head is measured, $I$ the net vertical specific rate of infiltration at free surface, $S_y$ the specific yield, $S_s$ the elastic specific storage, and $H_0$ the hydraulic head on the prescribed head boundary. Application of Galerkin's method of (5.83a), line (a), yields

$$\langle \nabla_{xy} \cdot (\mathbf{K} \cdot \nabla_{xy} h), \phi_i \rangle = \langle S_s \, \partial h/\partial t, \phi_i \rangle \qquad (i = 1, 2, \ldots, N)$$

Substitution of the approximating function

$$h \simeq \hat{h} = \sum_{j=1}^{N} H_j \phi_j(x, y)$$

and modification of the second-order term using Green's theorem yields

$$\begin{aligned}
\sum H_j \langle \mathbf{K} \cdot \nabla_{xy} \phi_j, \nabla_{xy} \phi_i \rangle &+ \sum (\partial H_j/\partial t) \langle S_s \phi_j, \phi_i \rangle \\
&- \int_\Gamma (\mathbf{K} \cdot \nabla_{xy} \hat{h}) \cdot \mathbf{n} \phi_i \, ds = 0 \qquad (5.83b)
\end{aligned}$$

Equation (5.83b) can be expressed in matrix form as

$$[A]\{H\} + [B]\{\partial H/\partial t\} = \{F\} \qquad (5.83c)$$

where typical elements of $[A]$, $[B]$, and $\{F\}$ are

$$a_{i,j} = \langle \mathbf{K} \cdot \nabla_{xy} \phi_j, \nabla_{xy} \phi_i \rangle, \quad b_{i,j} = \langle S_s \phi_j, \phi_i \rangle, \quad f_i = \int_\Gamma (\mathbf{K} \cdot \nabla_{xy} \hat{h}) \cdot \mathbf{n} \phi_i \, dS$$

The analogous equations developed by Neuman and Witherspoon [40] using a variational approach can be obtained by substituting 5.83a, lines (e) and (g), into the line integral definition of $f_i$ above. Because the nodal points move in response to the change in the geometry of the free surface, the partial derivative appearing in (5.83c) must be evaluated using the general definition of a total derivative

$$\frac{\partial h}{\partial t} = \frac{dh}{dt} - \frac{\partial h}{\partial x}\frac{dx}{dt} - \frac{\partial h}{\partial y}\frac{dy}{dt}$$

The expansion and contraction of the network, as reflected in the rate $dx/dt$, $dy/dt$, can be evaluated using geometric and constitutive relationships [40]. The total time derivative that eventually emerges can be effectively approximated using a Crank–Nicolson approximation.

The resulting set of algebraic equations are nonlinear and can be solved using a variety of iterative schemes. An efficient and stable iterative scheme can be found along with computed results in the work of Neuman and Witherspoon [40].

## 5.6 Saturated–Unsaturated Transport

Relatively little work has been reported on the simulation of mass and energy transport through the unsaturated zone using any numerical approach [41]. The recent work of Duguid and Reeves [42] appears to be the first attempt at solving this problem using a finite element scheme. Even in this case, however, the transport equation is considered independently of the unsaturated flow equation and consequently the more difficult coupled, nonlinear system is not represented.

The equation describing mass transport in a saturated–unsaturated porous medium can be written as

$$\frac{\partial}{\partial t}(c\Theta) + \nabla \cdot (\mathbf{q}c) - \nabla \cdot \{[\Theta \mathbf{D}(\Theta)/\theta] \cdot \nabla c\} = 0 \qquad (5.84)$$

where it is assumed that there are no chemical reactions, sources, or sinks, and that the term accounting for intergranular movement is small relative to other terms in the equation and can be neglected (see Bredehoeft and Pinder [43], for a detailed development of the porous medium mass transport equation). It is readily apparent that this expression is nearly identical to the expression developed for saturated flow, the only difference being the dependence of the dispersion coefficient on the moisture content. Unfortunately, relatively little is known about this functional relationship $\mathbf{D}(\Theta)$ and the expressions defining the dispersion coefficient for saturated flow (5.25) are used [42].

Thus the solution of (5.84) can be accomplished using the procedures developed earlier for single-phase transport recognizing, of course, that $\Theta$, unlike porosity, cannot be assumed relatively constant over time. It is important to note that an accurate solution to the transport equation for unsaturated flow once again requires a correct representation of velocity. Consequently, it may be necessary to either solve the three-equation formulation or employ a first-order continuous basis function finite element scheme to obtain a velocity distribution that satisfactorily conserves mass.

A natural extension of the above discussion leads to the problem of density-dependent flow. While this does not particularly complicate the solution of the transport equation, and computer codes are available that consider this situation, they do not appear to have been reported in the literature.

## References


1. M. K. Hubbert, The theory of groundwater motion, *J. Geol.* **48(8)**, Pt. 1, 785–944 (1940).
2. J. Bear, "Dynamics of Fluids in Porous Media." American Elsevier, New York, 1972.
3. G. F. Pinder, E. O. Frind, and S. S. Papadopulos, Functional coefficients in the analysis of groundwater flow, *Water Resources Res.* **9(1)**, 222–226 (1973).
4. J. D. Bredehoeft and G. F. Pinder, Digital analysis of areal flow in multiaquifer groundwater systems: A quasi three-dimensional model, *Water Resources Res.* **6(3)**, 883–888 (1970).
5. M. S. Hantush, Modification of the theory of leaky aquifers, *J. Geophys. Res.* **65(11)**, 3713–3725 (1960).
6. G. F. Pinder and J. D. Bredehoeft, Application of the digital computer for aquifer evaluation, *Water Resources Res.* **4(5)**, 1069–1093 (1968).
7. Y. H. Huang and S. L. Sonnenfeld, Analysis of unsteady flow toward an artesian well by three-dimensional finite elements, *Water Resources Res.* **10(3)**, 591–596 (1974).
8. S. K. Gupta, K. K. Tanji, and J. N. Luthin, A three-dimensional finite element ground water model, Water Resources Center Contrib. No. 152, Univ. California, Davis, California (1975).
9. M.-J. Verge, A Three-dimensional saturated-unsaturated groundwater flow model for practical application, PhD thesis, Dept. of Earth Sciences, University of Waterloo, Ontario, Canada (1975).
10. I. Javandel and P. A. Witherspoon, Application of the finite element method to transient flow in porous media, *Soc. Pet. Eng. J.* **8(3)**, 241–252 (1968).
11. S. P. Neuman and P. A. Witherspoon, Transient Flow of Ground Water to Wells in Multiple-Aquifer Systems, Rep. 69-1, Geotech. Eng., Univ. California, Berkeley, California (1969).
12. I. Javandel and P. A. Witherspoon, Analysis of Transient Fluid Flow in Multi-Layered Systems. Water Resources Center Contrib. No. 124, Univ. California, Berkeley, California (1968).
13. A. E. Scheidegger, General theory of dispersion in porous media, *J. Geophys. Res.* **66(10)**, 3273 (1961).
14. M. Van Genuchten, G. F. Pinder, and E. O. Frind, Simulation of two-dimensional contaminant transport with isoparametric Hermitian finite elements, *Water Resources Res.* (in press).

15. Y. M. Shum, Use of the finite element method in the solution of diffusion-convection equations, *Soc. Petrol. Eng. J.* **11(2)**, 139–144 (1971).
16. G. F. Pinder, A Galerkin-finite element simulation of groundwater contamination on Long Island, New York, *Water Resources Res.* **9(6)**, 1657–1669 (1973).
17. J. J. Leendertse, "A Water-Quality Simulation Model for Well-Mixed Estuaries and Coastal Seas," Vol. 1, Principles of Computation. Rand Corp. RM 6230-RC, 1970.
18. N. M. Chaudhari, An improved numerical technique for solving multi-dimensional miscible displacement equations, *Soc. Pet. Eng. J.* **11(3)**, 277–284 (1971).
19. W. G. Gray and G. F. Pinder, On the relationship between the finite element and finite difference methods, *Int. J. Numer. Methods Eng.* **10**, 893–923 (1976).
20. M. Van Genuchten and W. G. Gray, Analysis of some dispersion corrected numerical schemes for solution of the transport equation, *Int. J. Numer. Methods Eng.* (in press).
21. G. P. Baxter and C. C. Wallace, Changes in volume upon solution in water of the halogen salts of the alkali metals, *J. Amer. Chem. Soc.* **38**, 70–104 (1916).
22. R. T. Cheng, On the study of convective dispersion equation, *in* "Finite Element Methods in Flow Problems," pp. 29–48. Univ. Alabama Press, Montgomery, Alabama, 1974.
23. U. Meissner, A mixed finite element model for use in potential flow problems, *Int. J. Numer. Methods Eng.* **6**, 467–473 (1973).
24. F. A. Kohout, Cyclic flow of salt water in the Biscayne aquifer of southeastern Florida, *J. Geophys. Res.* **65**(7), 2133–2141 (1960).
25. F. A. Kohout, Flow pattern of fresh and salt water in the Biscayne aquifer of the Miami area, Florida, *Int. Assoc. Sci. Hydrol. Publ.* **52**, 440–448 (1960).
26. F. A. Kohout, Fluctuations of ground-water levels caused by dispersion of salts, *J. Geophys. Res.* **66**(8), 2429–2434 (1961).
27. F. A. Kohout and H. Klein, Effect of pulse recharge on the zone of diffusion in the Biscayne aquifer, *Int. Assoc. Sci. Hydrol. Publ.* **70**, 252–270 (1967).
28. C. H. Lee and R. T. Cheng, On seawater encroachment in coastal aquifers, *Water Resources Res.* **10**(5), 1039–1043 (1974).
29. D. Green, H. Dabiri, C. Weinaug, and R. Prill, Numerical modeling of unsaturated groundwater flow and comparison of the model to a field experiment. *Water Resources Res.* **6**, 862–874 (1970).
30. L. Van Phuc and H. J. Morel-Seytoux, Effect of soil air movement and compressibility on infiltration rates, *Soil Sci. Soc. Amer. Proc.* **36(2)**, 237–241 (1972).
31. C. L. McMichael and G. W. Thomas, Reservoir simulation by Galerkin's method, *Soc. Pet Eng. J.* **13**(3), 125–138 (1973).
32. J. Rubin, Theoretical analysis of two-dimensional transient flow of water in unsaturated and partly unsaturated soils, *Soil Sci. Soc. Amer. Proc.* **32**, 607–615 (1968).
33. R. A. Freeze, Three-dimensional, transient, saturated–unsaturated flow in a groundwater basin, *Water Resources Res.* **7(2)**, 347–366 (1971).
34. S. P. Neuman, Saturated–unsaturated seepage by finite elements, *J. Hydrol. Div. ASCE* **99**, HY12, 2233–2251 (1973).
35. M. Reeves and J. O. Duguid, Water Movement through Saturated–Unsaturated Porous Media: A Finite Element Galerkin Model. Oak Ridge Nat. Lab. Rep. 4927 (1975).
36. G. F. Pinder and E. O. Frind, Application of Galerkin's procedure to aquifer analysis, *Water Resources Res.* **8**(1), 108–120 (1972).
37. R. L. Cooley, A finite difference method for unsteady flow in variably saturated porous media: Application to a single pumping well, *Water Resources Res.* **7**(6), 1607–1625 (1971).
38. S. P. Neuman and P. A. Witherspoon, Variational principles for confined and unconfined flow of groundwater, *Water Resources Res.* **6(5)**, 1376–1382 (1970).
39. S. P. Neuman and P. A. Witherspoon, Finite element method of analyzing steady seepage with a free surface, *Water Resources Res.* **6**(3), 889–897 (1970).

40. S. P. Neuman and P. A. Witherspoon, Analysis of nonsteady flow with a free surface using the finite element method, *Water Resources Res.* **7**(3), 611–623 (1971).
41. E. Bresler, Simultaneous transport of solutes and water under transient unsaturated conditions, *Water Resources Res.* **9**(4), 975–986 (1973).
42. J. O. Duguid and M. Reeves, Material Transport through Porous Media: A Finite Element Galerkin Model. Oak Ridge Nat. Lab., Oak Ridge, Tennessee, ORNL-4928 (1976).
43. J. Bredehoeft and G. F. Pinder, Mass transport in flowing groundwater, *Water Resources Res.* **9**(1), 194–210 (1973).
44. G. F. Pinder and H. H. Cooper, Jr., A numerical technique for calculating the transient position of the saltwater front, *Water Resources Res.* **6**(3), 875–882 (1970).
45. H. R. Henry, Effects of dispersion of salt encroachment in coastal aquifers *in* "Sea Water in Coastal Aquifers" (H. H. Cooper, Jr., F. A. Kohout, H. R. Henry, and R. E. Glover, eds.). U.S. Geological Survey Water-Supply Paper 1613-C, C70-C84, (1964).
46. F. A. Kohout, The flow of fresh water and salt water in the Biscayne aquifer of the Miami area, Florida, *in* "Sea Water in Coastal Aquifers" (H. H. Cooper, Jr., F. A. Kohout, H. R. Henry, and R. E. Glover, eds.). U.S. Geological Survey Water-Supply Paper 1613-C, C12-C32 (1964).

# Chapter 6 | Lakes

## 6.1 Introduction

The finite element method has been applied to the modeling of lakes in which the density is constant. These models are useful in computing the steady state, wind-driven lateral flows and circulation patterns. Unsteady lacustrine flows and the "fall turnover" due to vertical density gradients have thus far not been modeled with finite elements. If the flow is density independent, a solution for the hydrodynamics may be obtained, and then the steady state and transient movement of thermal or material pollutants can be investigated. The equations used in these computations and the details of the application of the finite element method to the study of lakes will be discussed in the following sections.

## 6.2 The Equations Which Describe the Flow

The hydrodynamic equations used in lake circulation models are subsets of the equations described by Liggett and Hadjitheodorou [1]. The incompressible form of the momentum transport equation is

$$\frac{\partial \mathbf{v}}{\partial t} + \nabla \cdot (\mathbf{v}\mathbf{v}) + \mathbf{F}_C + \frac{1}{\rho}\nabla p - \mathbf{g} - \varepsilon\, \nabla^2 \mathbf{v} = 0 \tag{6.1}$$

where $\mathbf{v}$ is the velocity, $\mathbf{F}_C$ is the Coriolis force which accounts for the effects of the earth's rotation on the flow, $\mathbf{g}$ the gravity vector, $p$ the pressure, $\rho$ the fluid density, and $\varepsilon$ the eddy viscosity, assumed constant.

If the wind conditions are assumed to be time invariant, the velocity field will be steady and the time derivative in (6.1) is zero. When the inertial forces are much smaller than the Coriolis force, the second term in (6.1) may be neglected. The importance of this term can be examined by considering the Rossby number [2]

$$N_{\mathrm{Ro}} = |\mathbf{v}|/L(2\Omega \sin \theta) \tag{6.2}$$

which is the ratio of the inertial forces to the Coriolis force. In this definition $\Omega$ is the angular velocity of the earth, $\theta$ the latitude, and $L$ a characteristic length for the water body. Because this discussion of lakes will be concerned with only large-scale circulation, the characteristic length of the water body is chosen to reflect this concept.

Csanady [3], in his study of the North American Great Lakes, selected a characteristic length of 100 km. The value of $|\mathbf{v}|$ selected was 10 cm/sec and $2\Omega \sin \theta$ is roughly $10^{-4}$/sec at the latitude of the Great Lakes. Therefore, the Rossby number has a value $N_{\mathrm{Ro}} \cong 10^{-2}$, and the inertial terms are much less important than the Coriolis terms. Thus for steady flow, (6.1) takes the form

$$\mathbf{F}_{\mathrm{C}} - \mathbf{g} = -(1/\rho)\, \nabla p + \varepsilon\, \nabla^2 \mathbf{v} \tag{6.3}$$

The model will be further restricted by requiring the lake to be shallow. A shallow lake is one in which the depth is much less than a typical horizontal dimension and the Ekman [4] depth of frictional influence $\pi[\varepsilon/(\Omega \sin \theta)]^{1/2}$ is not much smaller than the depth of the lake. Thus the dominant frictional forces can be assumed to be the result of bottom friction; moreover horizontal diffusion of momentum can be neglected. The vector components of (6.3) in the $x$ (easterly) and $y$ (northerly) directions are, respectively,

$$-fv = -\frac{1}{\rho}\frac{\partial p}{\partial x} + \varepsilon \frac{\partial^2 u}{\partial z^2} \tag{6.4}$$

$$fu = -\frac{1}{\rho}\frac{\partial p}{\partial y} + \varepsilon \frac{\partial^2 v}{\partial z^2} \tag{6.5}$$

where $u$ is the easterly velocity, $v$ the northerly velocity, and $f$ the Coriolis parameter, $2\Omega \sin \theta$, assumed to be constant.

The requirement of a shallow lake allows us to assume that the vertical velocity and its derivatives in $x$ and $y$ are small compared to the horizontal velocities and can be neglected. Under these conditions the vertical pressure gradient is hydrostatic and the $z$ component of (6.3) becomes

$$g = -(1/\rho)(\partial p/\partial z) \tag{6.6}$$

where $z$ is positive upward.

The continuity equation is simply that for incompressible flow:

$$\frac{\partial u}{\partial x} + \frac{\partial v}{\partial y} + \frac{\partial w}{\partial z} = 0 \tag{6.7}$$

Equations (6.4)–(6.7) are the general forms of the point hydrodynamic equations used in the study of lake circulation. By subjecting them to various boundary condition constraints at the surface, they can be integrated to obtain the different forms of the equations actually used in lake models. These equations will be derived in the next few sections and their solution by the finite element method will be outlined.

## 6.3 Simplified Flow Model

### 6.3.1 Restrictions and Equations

This model is designed to solve the steady state equations of motion for a lake of constant depth but of arbitrary lateral topography. The development of the equations solved by this model will follow that of Cheng and Tung [5].

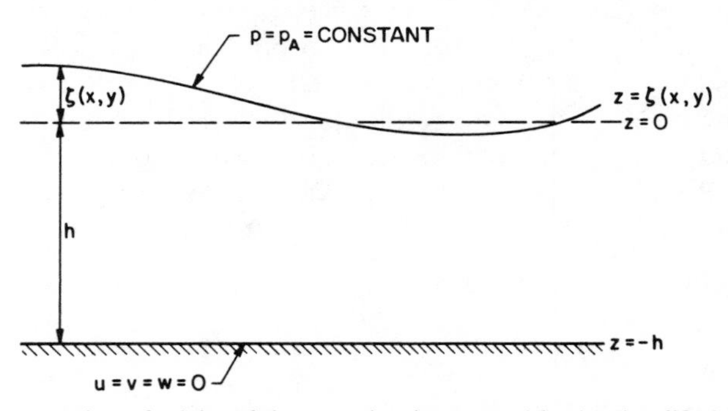


**Fig. 6.1.** Cross section of a lake of the type that is assumed in the simplified flow model.

The solution obtained is the stream function for the flow averaged over the depth. Figure 6.1 depicts the type of lake considered. No slip conditions are imposed at the bottom of the lake such that

$$u = v = w = 0 \quad \text{at} \quad z = -h \tag{6.8}$$

Let the equation for the surface of the lake be $F(x, y, z, t) = 0$. The fact that the surface of the lake is described by this equation through time can be expressed as

$$DF/Dt = 0 \tag{6.9}$$

where $D/Dt$ is the substantial derivative

$$\frac{D}{Dt} = \frac{\partial}{\partial t} + u\frac{\partial}{\partial x} + v\frac{\partial}{\partial y} + w\frac{\partial}{\partial z} \tag{6.10}$$

The equation for the surface is

$$F(x, y, z, t) = z - \zeta(x, y) \tag{6.11}$$

where $\zeta$, the surface elevation, is not a function of time because only the steady state case is being considered. Substitution of (6.11) into (6.9) yields the conditions

$$u_s \frac{\partial \zeta}{\partial x} + v_s \frac{\partial \zeta}{\partial y} - w_s = 0 \tag{6.12}$$

where $u_s$, $v_s$, and $w_s$ refer to velocities at the surface.

Now a mean velocity is defined such that

$$U = (1/h) \int_{-h}^{\zeta} u \, dz \tag{6.13}$$

and similarly

$$V = (1/h) \int_{-h}^{\zeta} v \, dz \tag{6.14}$$

Note that for $U$ and $V$ to be true averages, the integral would have to be multiplied by $1/(h + \zeta)$ rather than $1/h$. If $\zeta$ is assumed to be much smaller than $h$, the difference between $U$ and $V$ defined by (6.13) and (6.14) and the true mean can be neglected.

The continuity equation (6.7) can be integrated, subject to Eq. (6.12) and the constraints that $h$ is a constant and $w = 0$ at $z = -h$, to obtain

$$(\partial U/\partial x) + (\partial V/\partial y) = 0 \tag{6.15}$$

The pressure at the surface is assumed to be constant and atmospheric ($p_A$). Therefore Eq. (6.6) can be solved to obtain

$$p - p_A = \rho g(\zeta - z) \tag{6.16}$$

Substitution of this expression in Eqs. (6.4) and (6.5) yields

$$-fv = -g\frac{\partial \zeta}{\partial x} + \varepsilon\frac{\partial^2 u}{\partial z^2} \tag{6.17}$$

$$fu = -g\frac{\partial \zeta}{\partial y} + \varepsilon\frac{\partial^2 v}{\partial z^2} \tag{6.18}$$

Integration of these equations over the depth subject to definitions (6.13) and (6.14) yields

$$-fhV = -g(h+\zeta)\frac{\partial \zeta}{\partial x} + \varepsilon \frac{\partial u}{\partial z}\bigg|_{\zeta} - \varepsilon \frac{\partial u}{\partial z}\bigg|_{-h} \tag{6.19}$$

$$fhU = -g(h+\zeta)\frac{\partial \zeta}{\partial y} + \varepsilon \frac{\partial v}{\partial z}\bigg|_{\zeta} - \varepsilon \frac{\partial v}{\partial z}\bigg|_{-h} \tag{6.20}$$

where the last two terms in each of these equations account for the wind stress at $\zeta$ and the bottom stress at $-h$. At the free surface the wind stress is denoted by

$$W_x = \rho\varepsilon \frac{\partial u}{\partial z}\bigg|_{\zeta}, \qquad W_y = \rho\varepsilon \frac{\partial v}{\partial z}\bigg|_{\zeta} \tag{6.21}$$

where $W_x$ and $W_y$ will be specified functions. The bottom stresses are assumed to be linearly proportional to the mean velocity components, or

$$\rho\varepsilon \frac{\partial u}{\partial z}\bigg|_{-h} = \gamma U \qquad \text{and} \qquad \rho\varepsilon \frac{\partial v}{\partial z}\bigg|_{-h} = \gamma V \tag{6.22}$$

Substituting (6.21) and (6.22) back into (6.19) and (6.20) and recalling the assumption that $\zeta$ is much less than $h$ yields

$$-fhV = -gh\frac{\partial \zeta}{\partial x} + \frac{1}{\rho}W_x - \frac{\gamma}{\rho}U \tag{6.23}$$

$$fhU = -gh\frac{\partial \zeta}{\partial y} + \frac{1}{\rho}W_y - \frac{\gamma}{\rho}V \tag{6.24}$$

To eliminate the terms involving the pressure gradients, Eq. (6.23) is differentiated with respect to $y$, (6.24) is differentiated with respect to $x$, and the results are subtracted giving

$$\gamma\left(\frac{\partial U}{\partial y} - \frac{\partial V}{\partial x}\right) = \frac{\partial W_x}{\partial y} - \frac{\partial W_y}{\partial x} \tag{6.25}$$

The stream function $\psi$ is defined such that

$$U = \partial\psi/\partial y, \qquad V = -\partial\psi/\partial x \tag{6.26}$$

and the wind stress curl field, $(\partial W_x/\partial y) - (\partial W_y/\partial x)$, is denoted by $g(x, y)$ so that Eq. (6.25) becomes

$$\gamma\, \nabla^2\psi = g(x, y) \tag{6.27}$$

where $\nabla^2 = \partial^2/\partial x^2 + \partial^2/\partial y^2$, the two-dimensional Laplacian. We can define the dimensionless groups

$$\Psi = \psi\left(\frac{2h^2}{f\varepsilon}\right)^{1/2} \Big/ L^2, \qquad \Gamma = (\gamma L/T)(f\varepsilon/2h^2)^{1/2} \tag{6.28}$$

$$X = x/L, \qquad Y = y/L, \qquad G(X, Y) = \frac{g(x, y)}{(T/L)}$$

where $T$ is the characteristic wind stress and $L$ the characteristic lateral length. Equation (6.27) can be rewritten as

$$\Gamma\, \nabla^2\Psi = G(X, Y) \tag{6.29}$$

Estimates of the parameters in (6.28) appropriate for Lake Erie are given in Table 6.1.

**Table 6.1**

Typical parameter values for Lake Erie[a]

| Parameter | Value | Parameter | Value |
|---|---|---|---|
| $L$ | $3 \times 10^7$ cm | $f$ | $5 \times 10^{-5}$ rad/sec |
| $h$ | $2 \times 10^3$ cm | $\gamma$ | 0.0025 dyne-sec/cm$^3$ |
| $T$ | 1 dyne/cm | $\Gamma$ | $(1.0547)^{1/2} \cong 1.0$ |
| $\varepsilon$ | 30 cm$^2$/sec | | |

[a] After Cheng and Tung [5].

The boundary condition for Eq. (6.29) is the specification of $\Psi$ along a solid boundary. Because the normal velocity at a solid boundary is zero, the derivative of $\Psi$ with respect to the tangent to the boundary is zero and $\Psi$ must be a constant along the boundary. The value selected for this constant is arbitrary because the gradients of $\Psi$, which are the velocities, are unaffected by the magnitude of $\Psi$. For convenience we will select

$$\Psi = 0 \qquad \text{on } \lambda_0 \tag{6.30}$$

where $\lambda_0$ is the external boundary of the lake. When there are islands present in the lake, the selection of an appropriate value for $\Psi$ on the island boundaries is dependent on the value of $\Psi$ selected along $\lambda_0$. This point will be considered in detail in a later section.

### 6.3.2 Finite Element Formulation

Application of the method of weighted residuals to (6.29) yields

$$\iint_{\mathscr{A}} [-\nabla^2\Psi + (1/\Gamma)G(X, Y)]\phi_i \, dX \, dY = 0 \qquad (i = 1, \ldots, N) \quad (6.31)$$

where $\phi_i$ is the weighting function and $N$ is usually the number of nodes. Green's theorem is used to obtain

$$\iint_{\mathscr{A}} \left[\frac{\partial\Psi}{\partial X}\frac{\partial\phi_i}{\partial X} + \frac{\partial\Psi}{\partial Y}\frac{\partial\phi_i}{\partial Y} + \frac{1}{\Gamma}G(X, Y)\phi_i\right] dX \, dY - \int_{\lambda_0} \nabla\Psi\phi_i \cdot \mathbf{n} \, ds = 0 \qquad (i = 1, \ldots, N) \quad (6.32)$$

Galerkin's method requires that the weighting functions and the basis functions be identical so $\Psi$ is expanded such that

$$\Psi \cong \sum_{j=1}^{N} \Psi_j\phi_j \quad (6.33)$$

where $\Psi_j$ is the finite element solution for $\Psi$ at node $j$. Since $\Psi$ is specified at the boundary, there is no need to generate (6.32) when node $i$ is a boundary node. Therefore, the surface integral, which is equal to zero when $i$ is an interior node, will never have to be evaluated. Substitution of (6.33) into (6.32) results in

$$\iint_{\mathscr{A}} \left\{\sum_{j=1}^{N}\left[\Psi_j\frac{\partial\phi_j}{\partial X}\frac{\partial\phi_i}{\partial X} + \Psi_j\frac{\partial\phi_j}{\partial Y}\frac{\partial\phi_i}{\partial Y}\right] + \frac{1}{\Gamma}G(X, Y)\phi_i\right\} dX \, dY = 0 \quad (6.34)$$

$$(i = 1, \ldots, N \quad \text{and} \quad i \text{ is not a boundary node})$$

This equation can be written in matrix form as

$$[KI \;\vdots\; KB]\begin{Bmatrix}\Psi_1\\ \vdots\\ \Psi_I\\ \hdashline \vdots\\ \Psi_N\end{Bmatrix} = \begin{Bmatrix}E_1\\ \vdots\\ E_I\end{Bmatrix} \quad (6.35)$$

Where the $[KI \;\vdots\; KB]$ matrix has $I = N - B$ rows, where $B$ is the number of boundary nodes, and $N$ columns. If the first $I$ of the $\Psi$ values correspond to $\Psi$ values at interior nodes and are therefore unknown, then $[KI \;\vdots\; KB]$ and $\{\Psi\}$ are partitioned as outlined in Chapter 4 into sections dealing with known and unknown components. After substitution of the known values of $\Psi_{I+1}$ through $\Psi_N$ into (6.35), this equation can be rearranged to

$$[KI]\begin{Bmatrix}\Psi_1\\ \vdots\\ \Psi_I\end{Bmatrix} = \begin{Bmatrix}E_1\\ \vdots\\ E_I\end{Bmatrix} - [KB]\begin{Bmatrix}\Psi_{I+1}\\ \vdots\\ \Psi_N\end{Bmatrix} \tag{6.36}$$

In this problem, however, the $\Psi_{I+1}$ through $\Psi_N$ boundary nodes have a value of zero so the last product in (6.36) is zero. The equation that must be solved is

$$[KI]\begin{Bmatrix}\Psi_1\\ \vdots\\ \Psi_I\end{Bmatrix} = \begin{Bmatrix}E_1\\ \vdots\\ E_I\end{Bmatrix} \tag{6.37}$$

where

$$ki_{i,j} = \iint_{\mathscr{A}} \left[\frac{\partial \phi_j}{\partial X}\frac{\partial \phi_i}{\partial X} + \frac{\partial \phi_j}{\partial Y}\frac{\partial \phi_i}{\partial Y}\right] dX\, dY \tag{6.38}$$

and

$$e_i = -\iint_{\mathscr{A}} [\Gamma^{-1} G(X, Y)\phi_i]\, dX\, dY \tag{6.39}$$

It can be seen that the $[KI]$ matrix is symmetric and can therefore be solved by many efficient routines.

### 6.3.3 *Solution*

Cheng and Tung [5] solved Eq. (6.34) on the triangular grid of Lake Erie depicted in Fig. 6.2 using linear basis functions. This grid contains 516 elements and 308 nodal points, among which 214 are interior points. Con-

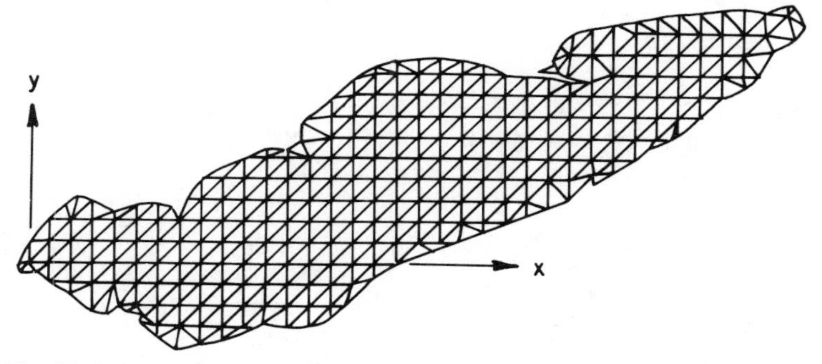


**Fig. 6.2.** Triangular finite element grid of Lake Erie (after Cheng and Tung [5]).

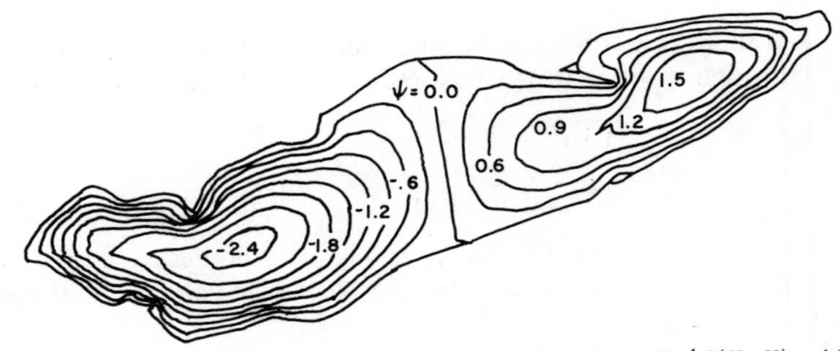


**Fig. 6.3.** Wind-driven mean circulation pattern in Lake Erie; $\Gamma^{-1}G(X, Y) = 1.0 - 2X$ (after Cheng and Tung [5]).

tour plots of the stream function obtained from two different forms of the forcing function $\Gamma^{-1}G(X, Y)$ are presented in Figs. 6.3 and 6.4. Although a comparison with field data is not made, the feasibility of using the finite element method and the ease with which it models an arbitrary geometry are demonstrated.

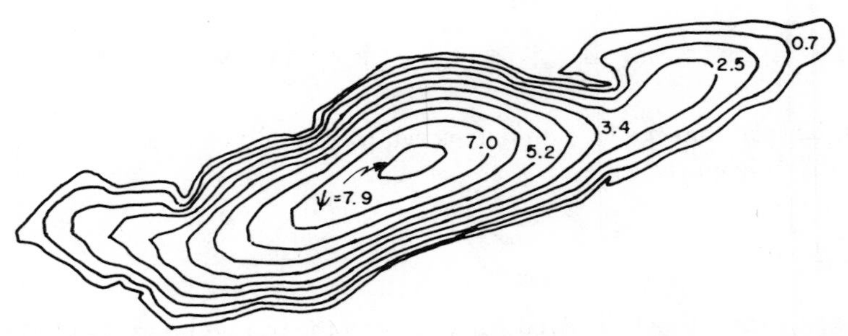


**Fig. 6.4.** Wind-driven mean circulation pattern in Lake Erie, $\Gamma^{-1}G(X, Y) = -1.0$ (after Cheng and Tung [5]).

### *6.3.4 Extension to Lakes with Islands*

The model can be extended to treat lakes that have islands in them [6]. There is no increased difficulty here due to the added irregularities in the geometry. The problem is that boundary conditions along the islands' shores are not known. Consider the grid in Fig. 6.5, which is used to model Lake Erie with some of its islands. Although we are free to specify $\Psi = 0$ on $\lambda_0$, the boundary values of $\Psi$ along the islands must be specified such that the gradient in $\Psi$ between boundaries is consistent with the flow in the lake. By imposing the constraint that the solution is free from singularities throughout the entire region enclosed by $\lambda_0$, the solution of the problem is uniquely determined.

Assume that there are $M$ interior boundaries in addition to the principal boundary $\lambda_0$. Since (6.29) is linear, its solution may be expressed as

$$\Psi = \Psi_{(0)}(X, Y) + \sum_{k=1}^{M} C_k \Psi_{(k)}(X, Y) \tag{6.40}$$

where $\Psi_{(0)}$ is a solution to (6.29) with boundary conditions

$$\Psi_{(0)} = 0 \qquad \text{on } \lambda_k \quad (k = 0, \ldots, M)$$

and the $\Psi_{(k)}$ are solutions to

$$\nabla^2 \Psi_{(k)} = 0 \qquad (k = 1, \ldots, M)$$

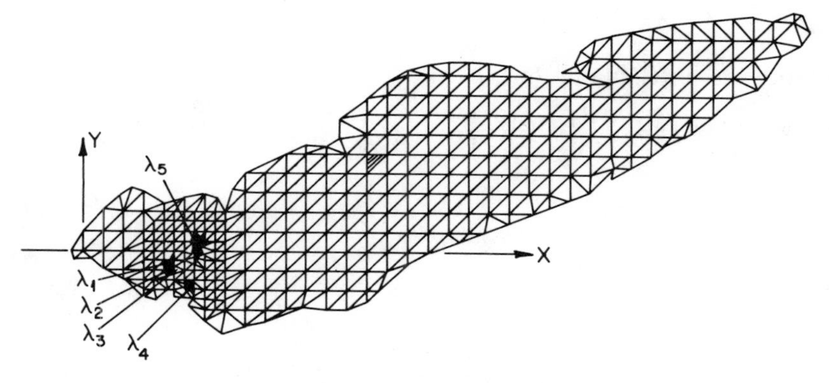


**Fig. 6.5.** Finite element grid for Lake Erie with islands included (after Cheng [6]). $\lambda_1$ = North Bass Island; $\lambda_2$ = Middle Bass Island; $\lambda_3$ = South Bass Island; $\lambda_4$ = Kelleys Island; $\lambda_5$ = Pelee Island.

with boundary conditions

$$\Psi_{(k)} = \delta_{k,l} \qquad \text{on } \lambda_l \quad (l = 0, \ldots, M)$$

where $\delta_{k,l}$ is the Kronecker delta. The solution for the $M + 1$ different $\Psi$s can be handled by the finite element method as $M + 1$ separate problems, or combined into one big problem. Since the $\nabla^2$ operator will have the same finite element representation for each $\Psi_{(k)}$, Eq. (6.36) will take the form

$$[KI] \begin{bmatrix} \Psi_{(0)1} & \Psi_{(1)1} & \cdots & \Psi_{(M)1} \\ \Psi_{(0)2} & \Psi_{(1)2} & \cdots & \Psi_{(M)2} \\ \vdots & \vdots & & \vdots \\ \Psi_{(0)I} & \Psi_{(1)I} & \cdots & \Psi_{(M)I} \end{bmatrix}$$

$$= \begin{bmatrix} E_1 & 0 & \cdots & 0 \\ E_2 & 0 & \cdots & 0 \\ \vdots & \vdots & & \vdots \\ E_I & 0 & \cdots & 0 \end{bmatrix} - [KB] \begin{bmatrix} \Psi_{(0)I+1} & \Psi_{(1)I+1} & \cdots & \Psi_{(M)I+1} \\ \vdots & \vdots & & \vdots \\ \Psi_{(0)N} & \Psi_{(1)N} & \cdots & \Psi_{(M)N} \end{bmatrix} \tag{6.41}$$

In this equation the last matrix product, which involves the boundary values of $\Psi_{(k)}$, must be carried along since some of these values are nonzero. Once Eq. (6.41) is solved, it is only necessary to obtain the values of $C_k$ in (6.40).

The solution given by (6.40) satisfies the governing differential equation (6.29) for any values of the $C$s. If (6.29) is integrated over the region covered by each island $\mathscr{A}_l$, $M$ equations are obtained:

$$\iint_{\mathscr{A}_l} \nabla^2\Psi \, d\mathscr{A} = \iint_{\mathscr{A}_l} (G/\Gamma) \, d\mathscr{A} \qquad (l = 1, \ldots, M) \tag{6.42}$$

The divergence theorem can be applied to obtain

$$\int_{\lambda_l} \left( \frac{\partial \Psi}{\partial X} dY - \frac{\partial \Psi}{\partial Y} dX \right) = \iint_{\mathscr{A}_l} \frac{G}{\Gamma} d\mathscr{A} \qquad (l = 1, \ldots, M) \tag{6.43}$$

Substitution of (6.40) into (6.43) yields

$$\sum_{k=1}^{M} \left\{ \int_{\lambda_l} \left( \frac{\partial \Psi_{(k)}}{\partial Y} dX - \frac{\partial \Psi_{(k)}}{\partial X} dY \right) C_k \right\} = - \int_{\lambda_l} \left( \frac{\partial \psi_{(0)}}{\partial Y} dX - \frac{\partial \Psi_{(0)}}{\partial X} dY \right) - \iint_{\mathscr{A}_l} \frac{G}{\Gamma}(X, Y) \, d\mathscr{A} \qquad (l = 1, \ldots, M) \tag{6.44}$$

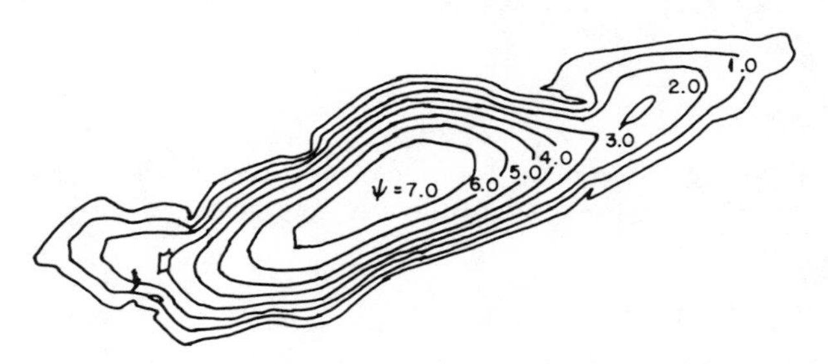


**Fig. 6.6.** Wind-driven circulation in Lake Erie with islands included corresponding to circulation in Fig. 6.4 without islands $\Gamma^{-1}G(X, Y) = -1.0$ (after Cheng [6]). $C_1 = 2.356$; $C_2 = 2.131$; $C_3 = 1.848$; $C_4 = 0.791$; $C_5 = 2.943$.

Since $\Psi_{(0)} \cdots \Psi_{(M)}$ are known, the $C_k$s can be determined from the $M$ equations (6.44) for a specified value of $\Gamma^{-1}G(X, Y)$. The primary difficulty with this procedure is that it requires a numerical differentiation of $\Psi_{(k)}$ for substitution into (6.44). This numerical differentiation can generate significant errors in the term that must be integrated in (6.44) and therefore introduce serious error into the values of $C_k$ obtained.

A plot of $\Psi$ obtained using this procedure on the triangular grid of Fig. 6.5 is presented in Fig. 6.6 for the case where $\Gamma^{-1}G(X, Y) = -Y$. Values of the $C_k$s are indicated in the figure caption.

## 6.4 Generalized Flow Model

### 6.4.1 *Restrictions and Equations*

This model also solves the steady state motion equations for a vertically averaged stream function in a lake with arbitrary lateral topography. It differs from the previous model in that the bathymetry of the lake may be arbitrary while the surface of the lake is level and coincident with $z = 0$ (Fig. 6.7). Furthermore the average stream function obtained can be used to obtain the lateral velocity profile as a function of depth as well as the averaged velocity.

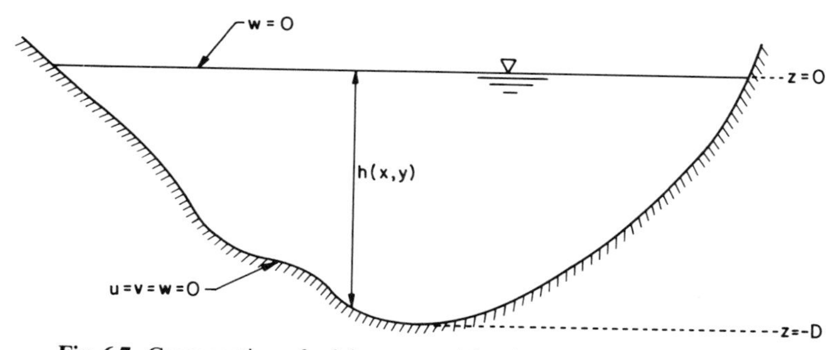


**Fig. 6.7.** Cross section of a lake assumed for the generalized flow model.

The boundary conditions imposed for the solution of the flow equations are

$$u = v = w = 0 \qquad \text{at} \quad \text{solid surfaces}$$

$$\rho\varepsilon\ \partial u/\partial z = W_x, \qquad \rho\varepsilon\ \partial v/\partial z = W_y \qquad \text{at} \quad z = 0 \tag{6.45}$$

$$w = 0 \qquad \text{at} \quad z = 0$$

Liggett and Hadjitheodorou [1] have detailed the rearrangement of (6.4)–(6.7) subject to (6.45). Because the analysis is long and involves a large number of complicated parameters, their approach will only be outlined here.

First the following dimensionless groups are defined:

$$X = x/L, \qquad Y = y/L, \qquad Z = z/D$$

$$U = fLu/gD, \qquad V = fLv/gD, \qquad W = fL^2w/gD^2, \qquad H = h/D,$$

$$P = p/\rho g D \tag{6.46}$$

where $L$ is a characteristic horizontal length, and $D$ the maximum depth of the lake. Substitution of these relations into (6.4)–(6.7) yields

$$-V = -\frac{\partial P}{\partial X} + E^2 \frac{\partial^2 U}{\partial Z^2} \tag{6.47}$$

$$U = -\frac{\partial P}{\partial y} + E^2 \frac{\partial^2 V}{\partial Z^2} \tag{6.48}$$

$$\frac{\partial P}{\partial Z} = -1 \tag{6.49}$$

$$\frac{\partial U}{\partial X} + \frac{\partial V}{\partial Y} + \frac{\partial W}{\partial Z} = 0 \tag{6.50}$$

and the boundary conditions are

$$\begin{gathered} U = V = W = 0 \quad \text{at} \quad \text{solid surfaces} \\ \partial U/\partial Z = \Delta, \qquad \partial V/\partial Z = \theta \qquad \text{at} \quad Z = 0 \\ W = 0 \quad \text{at} \quad Z = 0 \end{gathered} \tag{6.51}$$

where $E = (\varepsilon/f D^2)^{1/2}$ is the Ekman number, the ratio of the viscous forces to the Coriolis force, $\Delta = f LW_x/\varepsilon\rho g$, and $\theta = f LW_y/\varepsilon\rho g$. The analytic solutions to (6.47) and (6.48) are of the form

$$U = -\frac{\partial P}{\partial Y} + \frac{\partial P}{\partial Y} a_1(X, Y, Z) + \frac{\partial P}{\partial X} a_2(X, Y, Z) \tag{6.52}$$

$$V = \frac{\partial P}{\partial X} + \frac{\partial P}{\partial Y} a_3(X, Y, Z) + \frac{\partial P}{\partial X} a_4(X, Y, Z) \tag{6.53}$$

If average velocities are defined by

$$\bar{U} = (1/H) \int_{-h/D}^{0} U \, dZ \tag{6.54}$$

$$\bar{V} = (1/H) \int_{-h/D}^{0} V \, dZ \tag{6.55}$$

(6.52) and (6.53) may be integrated over the depth to obtain

$$\bar{U} = b_1 \frac{\partial P}{\partial X} + b_4 \frac{\partial P}{\partial Y} + b_2 \tag{6.56}$$

$$\bar{V} = -b_4 \frac{\partial P}{\partial X} + b_1 \frac{\partial P}{\partial Y} + b_3 \tag{6.57}$$

where $b_1$, $b_2$, $b_3$, and $b_4$ are functions of $X$ and $Y$. After introduction of the stream function defined as

$$\bar{U} = (1/H)(\partial\Psi/\partial Y) \tag{6.58}$$

$$\bar{V} = -(1/H)(\partial\Psi/\partial X) \tag{6.59}$$

into (6.56) and (6.57), the pressure derivatives may be solved for

$$\frac{\partial P}{\partial Y} = \frac{(b_4/H)\,\partial\Psi/\partial Y - (b_1/H)\,\partial\Psi/\partial X - (b_2 b_4 + b_1 b_3)}{b_1^2 + b_4^2} \tag{6.60}$$

$$\frac{\partial P}{\partial X} = \frac{(b_1/H)(\partial\Psi/\partial Y) + (b_4/H)(\partial\Psi/\partial X) - (b_1 b_2 - b_3 b_4)}{b_1^2 + b_4^2} \tag{6.61}$$

Elimination of the pressure gradients between (6.60) and (6.61) by cross differentiation yields the equation

$$\frac{\partial^2\Psi}{\partial X^2} + \frac{\partial^2\Psi}{\partial Y^2} + A(X, Y)\frac{\partial\Psi}{\partial X} + B(X, Y)\frac{\partial\Psi}{\partial Y} + C(X, Y) = 0 \tag{6.62}$$

with the boundary condition that $\Psi$ is a constant, arbitrarily chosen as zero, on the shore. The functions $A$, $B$, and $C$ depend on the physical parameters $E$, $\Delta$, and $\theta$ in the following way:

$$A = -\frac{H}{b_1}(b_1^2 + b_4^2)\left(\frac{\partial r}{\partial x} + \frac{\partial s}{\partial y}\right), \qquad B = -\frac{H}{b_1}(b_1^2 + b_4^2)\left(\frac{\partial r}{\partial y} + \frac{\partial s}{\partial x}\right)$$

$$C = -\frac{H}{b_1}(b_1^2 + b_4^2)\left[\frac{\partial}{\partial x}(q\theta) - \frac{\partial}{\partial y}(q\Delta) + \frac{\partial}{\partial y}(t\theta) + \frac{\partial}{\partial x}(t\Delta)\right]$$

$$b_1 = [\gamma(\beta + \kappa) + \varepsilon(\delta + \lambda)]/(\alpha H)$$

$$h_2 = [\sqrt{2}\,E\alpha + \beta - \delta]/(2\alpha\mu) - [(\beta\gamma + \delta\varepsilon)(\beta + \kappa) + (\beta\varepsilon - \delta\gamma)(\delta + \lambda)]/(\alpha H)$$

$$h_3 = (\beta + \delta)/(2\mu) + [(\beta\varepsilon - \delta\gamma)(\beta + \kappa) - (\beta\gamma + \delta\varepsilon)(\delta + \lambda)]/(\alpha H)$$

$$b_2 = h_2\theta + h_3\Delta, \qquad b_3 = h_3\theta - h_2\Delta$$

$$b_4 = [\varepsilon(\beta + \kappa) - \gamma(\delta + \lambda)]/(\alpha H) - 1$$

$$q = (b_1 h_3 + h_2 b_4)/(b_1^2 + b_4^2), \qquad r = (b_1/H)/(b_1^2 + b_4^2)$$

$$s = (b_4/H)/(b_1^2 + b_4^2), \qquad t = (h_3 b_4 - b_1 h_2)/(b_1^2 + b_4^2)$$

$$\alpha = 4(\cos^2\mu\cosh^2\mu + \sin^2\mu\sinh^2\mu)$$

$$\beta = \sqrt{2}\,Ee^{-\mu}(\sin\mu - \cos\mu)/2$$

$$\gamma = -2\sin\mu\sinh\mu, \qquad \delta = \sqrt{2}\,Ee^{-\mu}(\sin\mu + \cos\mu)/2$$

$$\varepsilon = 2\cos\mu\cosh\mu, \qquad \kappa = \sqrt{2}\,Ee^{\mu}(\sin\mu + \cos\mu)/2$$

$$\lambda = \sqrt{2}\,Ee^{\mu}(\sin\mu - \cos\mu)/2, \qquad \mu = \sqrt{2}\,h/2E$$

The coefficients $A$, $B$, and $C$ are functions of the lake bathymetry, and $C$ depends also on the surface shear stresses. It is interesting to note that by solving the equation for the stream function averaged through the depth, obtaining the lateral pressure gradients from (6.60) and (6.61), and then substituting these gradients into (6.52) and (6.53), it is possible to obtain values of $U$ and $V$ that are functions of $X$, $Y$, and $Z$ rather than only values averaged through $Z$.

The condition that $\Psi$ is constant on the lake boundary prescribes a zero vertically averaged velocity normal to the boundary. It does not, however, require that the point velocities normal to a vertical boundary be everywhere zero. Prohibiting vertical boundaries ensures that point normal velocities to the boundary are everywhere zero.

The definition of $\Psi$ in this model, (6.58) and (6.59), is quite different from that used in the simplified flow model in that $H$ here is a function of $X$ and $Y$. From this definition it is apparent that $H = 0$ represents a computational singularity. Therefore it is necessary to have a small nonzero depth all along the boundary. If the flow region is taken to be one bounded by a contour that is a small percent of the maximum depth, the flow exterior to this boundary and its effects on the interior region can be assumed to be negligibly small.

### *6.4.2 Finite Element Formulation*

Equation (6.62) has been solved by applying Galerkin's method on linear triangles, cubic triangles, and quadratic isoparametric elements [5, 6]. This application yields

$$\iint_{\mathscr{A}} \left[\frac{\partial^2 \Psi}{\partial X^2} + \frac{\partial^2 \Psi}{\partial Y^2} + A(X, Y)\frac{\partial \Psi}{\partial X} + B(X, Y)\frac{\partial \Psi}{\partial Y} + C(X, Y)\right]\phi_i \, d\mathscr{A} = 0 \qquad (i = 1, \ldots, N) \tag{6.63}$$

where $\phi_i$ is a basis function and $N$ the number of unknowns not necessarily the number of nodes. Green's theorem may be used to remove the second derivatives from the integral expression and obtain

$$\iint_{\mathscr{A}} \left[-\frac{\partial \Psi}{\partial X}\frac{\partial \phi_i}{\partial X} - \frac{\partial \Psi}{\partial Y}\frac{\partial \phi_i}{\partial Y} + A(X, Y)\frac{\partial \Psi}{\partial X}\phi_i + B(X, Y)\frac{\partial \Psi}{\partial Y}\phi_i + C(X, Y)\phi_i\right] d\mathscr{A} + \int_s \nabla\Psi \cdot \mathbf{n}\phi_i \, ds = 0 \qquad (i = 1, \ldots, N) \tag{6.64}$$

Depending on the assumptions made for the representation of $A(X, Y)$, $B(X, Y)$, and $C(X, Y)$ and on the types of basis functions chosen, the integrated approximation to Eq. (6.62) will take on different forms.

*Linear Triangles*

Assume that $\Psi$ is represented by

$$\Psi \simeq \sum_{j=1}^{M} \psi_j \phi_j \tag{6.65}$$

where $M$ is the number of nodes in the grid, $\psi_j$ the finite element solution for $\Psi$ at node $j$, and $\phi_j$ linear basis functions defined over triangles. Various assumptions can be made as to the way in which $A$, $B$, and $C$ vary in the domain. For instance, a constant value for each of these functions can be chosen for each triangular element. Alternatively one could specify some kind of higher-order variation of these functions through the triangle. Perhaps one of the most efficient ways of handling these functions without assuming them to be constant is to assume that they vary in the same manner as the basis functions for the triangles. Thus these coefficients are approximated by

$$A(X, Y) = \sum_{k=1}^{M} A_k \phi_k \tag{6.66}$$

$$B(X, Y) = \sum_{k=1}^{M} B_k \phi_k \tag{6.67}$$

$$C(X, Y) = \sum_{k=1}^{M} C_k \phi_k \tag{6.68}$$

where $A_k$, $B_k$, and $C_k$ are nodal values of $A(X, Y)$, $B(X, Y)$, and $C(X, Y)$ specified a priori.

Because the value of $\Psi$ has been specified as zero on the shore, it will not be necessary to generate Eq. (6.64) when node $i$ is on the boundary. Furthermore, the surface integral will be zero when $i$ is an interior node (because $\phi_i$ evaluated at the boundary of the domain is zero when $i$ is an interior node), and need not be considered in further discussion of linear triangular elements.

Substitution of (6.65)–(6.68) into (6.64) yields

$$\iint_{\mathscr{A}} \left\{ \sum_{j=1}^{M} \left[ -\psi_j \frac{\partial \phi_j}{\partial X} \frac{\partial \phi_i}{\partial X} - \psi_j \frac{\partial \phi_j}{\partial Y} \frac{\partial \phi_i}{\partial Y} + \left( \sum_{k=1}^{M} A_k \phi_k \right) \psi_j \frac{\partial \phi_j}{\partial X} \phi_i + \left( \sum_{k=1}^{M} B_k \phi_k \right) \psi_j \frac{\partial \phi_j}{\partial Y} \phi_i \right] + \left( \sum_{k=1}^{M} C_k \phi_k \right) \phi_i \right\} d\mathscr{A} = 0 \qquad (i = 1, \ldots, N) \tag{6.69}$$

where $N$ is the number of interior nodes on the finite element grid. This equation can be written in matrix form as

$$[K]\{\psi\} = \{E\} \tag{6.70}$$

where $\{E\}$ takes into account the last term in the integral of (6.69) as well as the terms that arise from the partitioning of the coefficient matrix $K$ [see the discussion of Eq. (6.35)]. In this particular problem, since the known values of $\Psi$ have been prescribed as zero on the boundary, the partitioning of $[K]$ contributes nothing to $\{E\}$. Thus $[K]$ is an $N \times N$ matrix and $\{E\}$ is an $N \times 1$ vector with elements

$$k_{i,j} = \iint_{\mathscr{A}} \left[ -\frac{\partial \phi_j}{\partial X} \frac{\partial \phi_i}{\partial X} - \frac{\partial \phi_j}{\partial Y} \frac{\partial \phi_i}{\partial Y} + \left( \sum_{k=1}^{M} A_k \phi_k \right) \frac{\partial \phi_j}{\partial X} \phi_i + \left( \sum_{k=1}^{M} B_k \phi_k \right) \frac{\partial \phi_j}{\partial Y} \phi_i \right] d\mathscr{A} \tag{6.71}$$

$$e_i = - \iint_{\mathscr{A}} \left( \sum_{k=1}^{M} C_k \phi_k \right) \phi_i \, d\mathscr{A} \tag{6.72}$$

Although summations over all nodes appear in these equations, it should be remembered that when doing the numerical calculations, the summation needs to be done only over the region where $\phi_i$ is nonzero.

In contrast to the simplified model considered previously the $K$ matrix is not symmetric. However, the sparse, banded nature of $K$ may be exploited in solving (6.70).

*Cubic Triangles*

The first type of element that comes to mind when considering cubic triangles is a triangular element with ten nodes, one at each vertex, two on each side, and one in the interior (Fig. 6.8a). Formulation of the finite element problem on this type of element would be analogous to the development for linear triangles, the major difference being that the $\phi_j$s would be cubic basis functions.

An alternative approach can be taken which still uses complete cubic polynomials. Instead of defining element nodes at the ten locations described above, only the three vertices and a point at the centroid of the element are considered (Fig. 6.8b). To accomodate the ten terms of a complete cubic, the derivatives of $\Psi$ at the vertices ($\partial\Psi/\partial X$ and $\partial\Psi/\partial Y$ at points 1, 2, and 3) are treated as solution parameters. Thus, at a typical vertex $i$, there are unknown parameters $\psi_i$, $\psi_{Xi}$, and $\psi_{Yi}$ which correspond, respectively, to the finite element solution for $\Psi|_i$, $\partial\Psi/\partial X|_i$, and $\partial\Psi/\partial Y|_i$. The only solution at the centroid node 4 is $\Psi_4$.

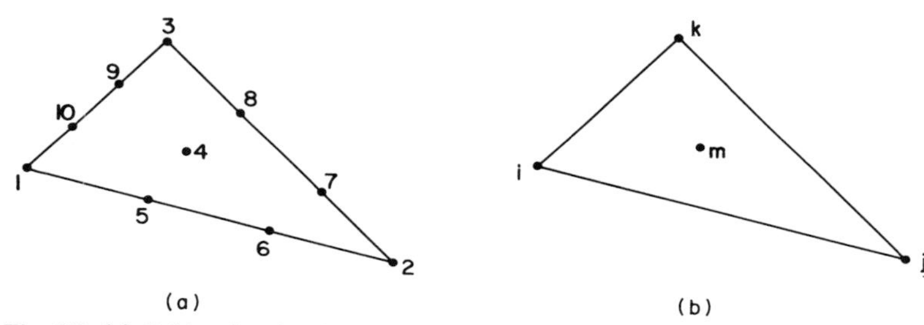


**Fig. 6.8.** (a) Cubic triangle with ten nodes and one unknown per node. (b) Cubic triangle with four nodes and three unknowns at each vertex, one unknown at the centroid.

The introduction of the derivatives of $\Psi$ as solution parameters has two advantageous features. First, the elimination of nodes along the side of the element makes possible a reduction in the band width of the equations. Second, the solution parameters $\psi_{Xi}$ and $\psi_{Yi}$ are directly proportional to the quantities of interest, $\bar{U}$ and $\bar{V}$, so that the latter two functions are obtained without having to differentiate the solution.

The shape function description for $\Psi$ on the triangles in Fig. 6.8b is [7, 9]

$$\Psi = \psi_i\phi_i + \psi_{Xi}\phi_{Xi} + \psi_{Yi}\phi_{Yi} + \psi_j\phi_j + \cdots + \psi_{Yk}\phi_{Yk} + \psi_m\phi_m \quad (6.73)$$

where

$$\begin{aligned}
\phi_i &= L_i^2(L_i + 3L_j + 3L_k) - 7\kappa \\
\phi_{Xi} &= L_i^2(X_{ki}L_k - X_{ij}L_j) + (X_{ij} - X_{ki})\kappa \\
\phi_{Yi} &= L_i^2(Y_{ki}L_k - Y_{ij}L_j) + (Y_{ij} - Y_{ki})\kappa \\
\phi_m &= 27\kappa
\end{aligned} \quad (6.74)$$

and $X_{ij} = X_i - X_j$, $\kappa = L_1 L_2 L_3$, and

$$L_i = [X_j Y_k - X_k Y_j + Y_{jk} X + X_{kj} Y]/2\Delta$$

is the triangular area coordinate with $i$, $j$, and $k$ being the vertices labeled in counterclockwise order, and $\Delta$ is the area of the triangle. Basis functions associated with nodes $j$ and $k$ are obtained by permuting the indices $i$, $j$, and $k$ cyclically in the counterclockwise direction.

One inconvenient aspect of the formulation which results from the use of basis functions (6.74) is the presence of the interior node in the triangle. Because the solution parameter $\psi_4$ appears only when integration is done over that element, it can be eliminated from the element equation before assembly of the element matrix into the global matrix. Thus the element equations, initially of order $10 \times 10$, are reduced to the order $9 \times 9$ immediately before assembly into the global matrix.

In the evaluation of the element coefficients through integration of Eq. (6.64), one could assume the terms $A(X, Y)$, $B(X, Y)$, and $C(X, Y)$ vary linearly within each element. For this case the appropriate expansion would be

$$A = \sum_{k=1}^{M_c} A_k \mu_k \tag{6.75}$$

where $\mu_k$ are the linear triangular basis functions and $M_c$ the number of corner nodes of the finite element grid. Of course it is possible to express $A$, $B$, and $C$ in terms of basis functions (6.74) but this would only complicate the data requirements and the computation.

As a notational convenience, the ten coefficients in terms of basis functions that appear in expansion (6.73) will all be denoted by $\psi_j^*$ and $\phi_j^*$, respectively. Therefore the global representation of $\Psi$ is

$$\Psi \simeq \sum_{j=1}^{N_T} \psi_j^* \phi_j^* \tag{6.76}$$

where $N_T = M_c + M_c + M$, twice the number of corner nodes (to account for $\psi_{xj}$ and $\psi_{yj}$ coefficients) plus the total number of nodes (to account for the $\psi_j$ coefficients). Substitution of (6.75) and (6.76) into (6.64) yields

$$\begin{aligned} &\iint_{\mathscr{A}} \left\{ \sum_{j=1}^{N_T} \left[ -\psi_j^* \frac{\partial \phi_j^*}{\partial X} \frac{\partial \phi_i^*}{\partial X} - \psi_j^* \frac{\partial \phi_j^*}{\partial Y} \frac{\partial \phi_i^*}{\partial Y} + \left( \sum_{j=1}^{M_c} A_k \mu_k \right) \psi_j^* \frac{\partial \phi_j^*}{\partial X} \phi_i^* \right. \right. \\ &\quad \left. \left. + \left( \sum_{k=1}^{M_c} B_k \mu_k \right) \psi_j^* \frac{\partial \phi_j^*}{\partial Y} \phi_i^* \right] + \left( \sum_{k=1}^{M_c} C_k \mu_k \right) \phi_i^* \right\} d\mathscr{A} \\ &\quad + \sum_{j=1}^{N_T} \int_\Gamma \psi_j^* \frac{\partial \phi_j^*}{\partial n} \phi_i^* \, ds = 0 \qquad (i = 1, \ldots, N) \end{aligned} \tag{6.77}$$

where $N = N_T - 2M_E$, $M_E$ is the number of nodes on the boundary of the global flow domain, and $\partial \phi_j^* / \partial n$ the normal gradient of the $j$th basis function.

As mentioned previously the boundary conditions require that the stream function be a constant, selected as zero for convenience, all along the boundary. This is equivalent to requiring that the tangential gradient of the stream function all along the boundary be zero. Along the boundary of the flow domain there are $M_E$ nodes and $3M_E$ parameters. The specification of $\psi_j$ as zero removes $M_E$ of the unknowns, and the specification of a zero tangential gradient defines a relationship between $\psi_{Xj}$ and $\psi_{Yj}$ at a boundary node and removes $M_E$ additional unknowns. These considerations explain why only $N_T - 2M_E$ equations are generated by (6.77).

One further approximation must be made before the system (6.77) can be solved. In the finite element analysis, the boundary is usually approximated by piecewise continuous functions. Whenever the slope of this boundary changes discontinuously (this can occur only at a node) the tangential direction is not uniquely defined. Selection of the tangential direction is presently rather subjective as many possibilities of apparently equal merit exist. For instance, one can argue that on each line approaching the node a different relation exists between $\partial\Psi/\partial X$ and $\partial\Psi/\partial Y$; the only way for both relations to be satisfied at the node is for both $\partial\Psi/\partial X$ and $\partial\Psi/\partial Y$ to be zero at the node. Alternatively one may consider the average direction of the two lines meeting at a node as the tangential direction, or perhaps take the direction indicated by either one of the lines as the required direction, being consistent at each node.

The arguments for setting both $\partial\Psi/\partial X$ and $\partial\Psi/\partial Y$ equal to zero are unsatisfactory. This specification is equivalent to requiring a no-flow boundary condition. Although this condition may at first appear reasonable, the effect of the no-slip condition is physically felt only in a small boundary layer. The scale of this layer is much smaller than that considered numerically and therefore the numerical boundary layer set up by the no-flow specification is inconsistent with the scale of the physical system. Furthermore if another type of transformation function were used such that the boundary geometry was continuous, there would be no need to specify both $\partial\Psi/\partial X$ and $\partial\Psi/\partial Y$ zero at a point. It seems unreasonable that boundary conditions, which should in fact help to define the physical process, depend in turn on the mathematical function used to describe the boundary.

While the other alternatives mentioned for selection of the normal direction may be convenient, they are rather arbitrary. The discussion which follows describes a rational method for selection of the normal direction.

Consider, for example, the calculation of the normal direction at corner node $C$ on the portion of the finite element grid indicated in Fig. 6.9. Because the basis functions associated with node $C$ are nonzero only along the portion of the boundary $DCB$, it is only necessary to consider this segment. If the normal flow to this boundary is zero, analytically the correct boundary condition is $\Psi = \Psi_0$, which is a constant along this boundary. This condition cannot be written exactly along $DCB$ because the finite element expansion for $\Psi$ involves approximation to the functions $\partial\Psi/\partial X$ and $\partial\Psi/\partial Y$ which do not have unique values at node $C$. An alternative to exactly satisfying the boundary condition at every point is to satisfy it in some average sense. To do this one can require that the expansion in $\Psi$ associated with node $C$ integrated over $DCB$ be equal to $\frac{1}{2}$ of $\Psi_0$ integrated along $DCB$ or

$$\int_D^C (\psi_C\phi_C + \psi_{XC}\phi_{XC} + \psi_{YC}\phi_{YC})\, ds + \int_C^B (\psi_C\phi_C + \psi_{XC}\phi_{XC} + \psi_{YC}\phi_{YC})\, ds$$

$$= \tfrac{1}{2}[\Psi_0 \cdot (\overline{DC} + \overline{CB})] \tag{6.78}$$

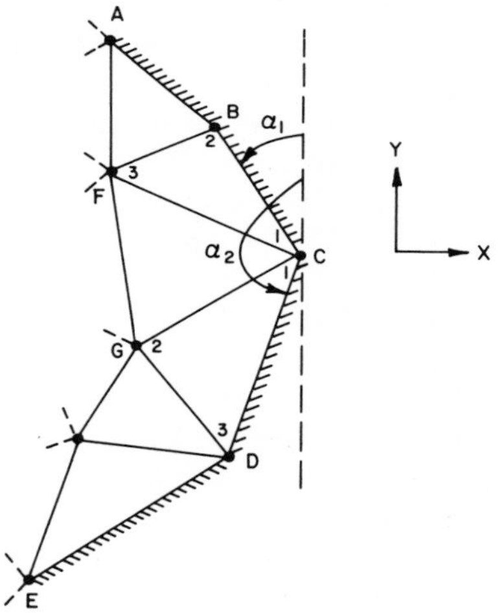

**Fig. 6.9.** Boundary of finite element grid with nonunique normal direction of nodes.

where $\phi_C$, $\phi_{XC}$, and $\phi_{YC}$ are the basis functions associated with node $C$ whose coefficients are $\psi_C$, $\psi_{XC}$, and $\psi_{YC}$, respectively, and $\overline{DC}$ and $\overline{CB}$ are the lengths of the sides $DC$ and $CB$, respectively. The $\frac{1}{2}$ in the right side of (6.78) arises because only one-half of the contribution to the integral of $\Psi$ along $DCB$ is considered to be made by the basis functions and coefficients associated with node $C$, the other half coming from the coefficients and basis functions associated with nodes $B$ and $D$. If triangle $DCG$ is transformed into local coordinates with nodes $C$, $G$, and $D$ corresponding to nodes 1, 2, and 3, respectively, then the first integral in (6.78) takes the form

$$\int_D^C (\psi_C \phi_C + \psi_{XC}\phi_{XC} + \psi_{YC}\phi_{YC})\, ds$$
$$= \overline{DC} \int_0^1 \{\psi_C[L_1^2(3 - 2L_1)] + \psi_{XC}[L_1^2(1 - L_1)(X_D - X_C)]$$
$$+ \psi_{YC}[L_1^2(1 - L_1)(Y_D - Y_C)]\}\, dL_1 \tag{6.79}$$

where use has been made of the fact that along $DC$, $L_2 = 0$ and $L_3 = 1 - L_1$. Evaluation of the right side of (6.79) yields

$$\int_D^C (\psi_C \phi_C + \psi_{XC}\phi_{XC} + \psi_{YC}\phi_{YC})\, ds$$
$$= \overline{DC}[\tfrac{1}{2}\psi_C + \tfrac{1}{12}\psi_{XC}(X_D - X_C) + \tfrac{1}{12}\psi_{YC}(Y_D - Y_C)] \tag{6.80}$$

Similarly the second integral in (6.78) can be evaluated by

$$\int_C^B (\psi_C \phi_C + \psi_{XC}\phi_{XC} + \psi_{YC}\phi_{YC})\, ds = -\overline{CB} \int_1^0 \{\psi_C[L_1^2(3 - 2L_1)]$$
$$- \psi_{XC}[L_1^2(1 - L_1)(X_C - X_B)]$$
$$- \psi_{YC}[L_1^2(1 - L_1)(Y_C - Y_B)]\}\, dL_1$$
$$\tag{6.81}$$

where, in this case, use has been made of the fact that along $CB$, $L_3 = 0$ and $L_2 = 1 - L_1$. Evaluation of the right side of (6.81) yields

$$\int_C^B (\psi_C \phi_C + \psi_{XC}\phi_{XC} + \psi_{YC}\phi_{YC})\, ds$$
$$= \overline{CB}[\tfrac{1}{2}\psi_C - \tfrac{1}{12}\psi_{XC}(X_C - X_B) - \tfrac{1}{12}\psi_{YC}(Y_C - Y_B)] \tag{6.82}$$

Equations (6.80) and (6.82) may now be substituted into (6.78) to obtain

$$\tfrac{1}{2}(\overline{DC} + \overline{CB})\psi_C + \tfrac{1}{12}[\overline{DC}(X_D - X_C) - \overline{CB}(X_C - X_B)]\psi_{XC}$$
$$+ \tfrac{1}{12}[\overline{DC}(Y_D - Y_C) - \overline{CB}(Y_C - Y_B)]\psi_{YC} = \tfrac{1}{2}(\overline{DC} + \overline{CB})\Psi_0 \tag{6.83}$$

Because of the condition that $\Psi = \Psi_0$ on the boundary, it is appropriate to set $\psi_C = \Psi_0$. Thus (6.83) becomes

$$[\overline{DC}(X_D - X_C) - \overline{CB}(X_C - X_B)]\psi_{XC}$$
$$+ [\overline{DC}(Y_D - Y_C) - \overline{CB}(Y_C - Y_B)]\psi_{YC} = 0$$

or

$$[-\overline{DC}^2 \sin \alpha_2 + \overline{CB}^2 \sin \alpha_1]\psi_{XC} + [\overline{DC}^2 \cos \alpha_2 - \overline{CB}^2 \cos \alpha_1]\psi_{YC} = 0 \tag{6.84}$$

Normalization of this equation such that the sum of the squares of the coefficients of $\psi_{XC}$ and $\psi_{YC}$ is unity yields

$$\sin \theta\psi_{XC} - \cos \theta\psi_{YC} = 0 \tag{6.85}$$

where

$$\sin \theta = \frac{\overline{CB}^2 \sin \alpha_1 - \overline{DC}^2 \sin \alpha_2}{[\overline{CB}^4 + \overline{DC}^4 - 2\overline{DC}^2\overline{CB}^2 \cos(\alpha_2 - \alpha_1)]^{1/2}}$$

$$\cos \theta = \frac{\overline{CB}^2 \cos \alpha_1 - \overline{DC}^2 \cos \alpha_2}{[\overline{CB}^4 + \overline{DC}^4 - 2\overline{DC}^2\overline{CB}^2 \cos(\alpha_2 - \alpha_1)]^{1/2}}$$

and $\theta$ is the angle between the $Y$ axis and the effective tangential direction at node $C$. The normal direction is, of course, perpendicular to the tangent line.

Once the tangential direction at each boundary node has been established, a transformation is necessary to introduce into the problem the relationship between $\partial\Psi/\partial X$ and $\partial\Psi/\partial Y$ established by Eq. (6.85). This can be done at either the element level or the global matrix level; one way is outlined below.

Consider the corner node $C$ in Fig. 6.9. At node $C$ there are two basis functions associated with derivatives of $\Psi$; $\phi_{XC}$ associated with $\partial\Psi/\partial X$ and $\phi_{YC}$ associated with $\partial\Psi/\partial Y$. Two equations can be generated by letting $\phi_i^* = \phi_{XC}$ and then $\phi_i^* = \phi_{YC}$ in (6.77) and a third condition is obtained by imposing the boundary condition given in (6.85). Thus with $\Psi$ specified at node $C$ three equations may be formulated at a position where there are only two unknowns. Using abbreviated notation we can denote these equations, respectively, by

$$W(\phi_{XC}) = 0 \tag{6.86a}$$

$$W(\phi_{YC}) = 0 \tag{6.86b}$$

$$\psi_{XC} \sin \theta = \psi_{YC} \cos \theta \tag{6.86c}$$

Because there are more equations than unknowns, steps must be taken to eliminate the extraneous equation. Because the boundary condition is needed, it is not appropriate to discard (6.86c). Now note that this equation is actually developed by considering the normal direction (i.e., this condition requires that the normal flux be zero). It seems reasonable to expect that the other equation to be considered at the node would be developed by investigating the tangential flow direction. A form of (6.77) obtained by letting $\phi_i^*$

be a basis function associated with the tangential direction would seem to fit the requirement. Although no such function is explicitly available, one can be defined as

$$\phi_{\lambda C} = \phi_{XC} \cos \theta + \phi_{YC} \sin \theta \tag{6.87}$$

Thus by multiplying Eq. (6.86a) by cos $\theta$, (6.86b) by sin $\theta$, and adding together the two equations, we effectively weight in Eq. (6.77) with a basis function associated with the tangential direction. The two equations generated at node $C$ for the two unknowns $\psi_{XC}$ and $\psi_{YC}$ are

$$W(\phi_{XC} \cos \theta + \phi_{YC} \sin \theta) = 0 \quad \text{(Galerkin's weighted residual equation)} \tag{6.88a}$$

and

$$\psi_{XC} \sin \theta = \psi_{YC} \cos \theta \quad \text{(boundary condition)} \tag{6.88b}$$

These two equations can either be set up as two separate rows in the matrix equation, or (6.88b) can be used to condense $\psi_{XC}$ or $\psi_{YC}$ out of the global matrix formulation.

Because the coefficients that must be computed from (6.77) have such a variety of meanings—some refer to the stream function at vertices, others to derivatives of the stream function at the vertices, and still others to the stream function at the center of an element—it would be a lengthy procedure to write up the typical terms in the matrix expression

$$[K]\{\psi\} = \{E\} \tag{6.89}$$

that results from (6.77). In lieu of this, a recapitulation will be made which indicates the chief differences that the use of the cubic triangles considered here introduces when compared to linear triangles.

The most obvious change is that the cubics generate three unknowns per node. Thus, if a problem is solved on a given triangular grid, the use of cubic basis functions will result in a $K$ matrix with roughly three times as many rows as obtained with linear basis functions. If the midelement node is not condensed out of the cubic problem at the element level, this ratio will be even higher. Furthermore, the bandwidth of the matrix in the cubic case will be greater. Unless the cubic formulation increases the accuracy of the solution enough to permit a significantly coarser grid than required with the linear formulation, it is computationally uneconomical.

One additional complication of the cubic formulation over the linear formulation is the necessity to evaluate the surface integral in (6.77) when $\phi_i$ is on the boundary. Because the stream function and its tangential gradient (i.e., the normal flux) are set to zero at the boundary, the Galerkin equations weighted with respect to these boundary node basis functions are not generated. However, the Galerkin equation for the tangential flow indicated by (6.88a) is required and it is in this equation that the surface integral arises. Since this integral contains some $\Psi$ coefficients that will be known and some that will be unknown, it must be partitioned, a portion being allocated to the $[K]$ matrix and a portion to the known vector $\{E\}$.

An interesting concept introduced in the cubic formulation is the idea of mixing the types of basis functions used on the same elements. Although $\Psi$ is represented using cubic basis functions, the functional coefficients $A$, $B$, and $C$ have been expanded with linear functions. This was an arbitrary choice based on the hypothesis that the bathymetry and wind stresses of the body of water under consideration could be adequately modeled if $A(X, Y)$, $B(X, Y)$, and $C(X, Y)$ vary linearly. In a problem where this hypothesis is found to be false, it would be a trivial matter to introduce a higher-order polynomial fit of these functions into the problem.

*Quadratic Isoparametrics*

Equation (6.64) has also been solved using quadratic isoparametric elements [6]. This type of element has been previously discussed (Chapter 5) so the details of the coordinate transformations will not be considered here. The quadratic isoparametric element is attractive because it uses the lowest-degree basis function which will describe curved boundaries.

For this type of element, $\Psi$ is represented by

$$\Psi \simeq \sum_{j=1}^{M} \psi_j \phi_j \tag{6.90}$$

where $M$ is the number of nodes and $\phi_j$ the basis functions in global coordinates. Notice that in this case, only the values of $\Psi$ at the nodes and not the derivatives of $\Psi$ are chosen as unknowns. Each element has eight nodes and the approximations to $\Psi$ at each of these nodes are the solution parameters. Since the derivatives of $\Psi$ are not solution parameters, the only required boundary condition is that $\Psi$ be a constant, taken to be zero, along the continuous solid boundary.

Substitution of (6.90) into (6.64) yields

$$\iint_{\mathscr{A}} \left\{ \sum_{j=1}^{M} \left[ -\psi_j \frac{\partial \phi_j}{\partial X} \frac{\partial \phi_i}{\partial X} - \psi_j \frac{\partial \phi_j}{\partial Y} \frac{\partial \phi_i}{\partial Y} + A(X, Y)\psi_j \frac{\partial \phi_j}{\partial X} \phi_i + B(X, Y)\psi_j \frac{\partial \phi_j}{\partial Y} \phi_i \right] + C(X, Y)\phi_i \right\} d\mathscr{A} = 0 \qquad (i = 1, \ldots, N) \tag{6.91}$$

where $N$ is the total number of nodes less the number of nodes on the global boundary of the finite element mesh. Although (6.91) is similar in form to (6.69), the equation is different because quadratic basis functions on isoparametrics are being used rather than linear basis functions on triangles.

Because isoparametric elements are being employed, the integrations specified in (6.91) are over irregularly shaped elements. The integrations are performed by transforming each element in the $XY$ domain into a local $\xi\eta$ coordinate system in which the element becomes a $2 \times 2$ square with origin at the center of the square. The resulting integral is too complicated for explicit evaluation so numerical integration can be applied using Gaussian quadrature. One method of treating the coefficients $A(X, Y)$, $B(X, Y)$, and $C(X, Y)$ is to evaluate them at the origin of the $\xi\eta$ coordinate system in each element and assume that they are constant throughout the element. An alternative to this is to expand $A(X, Y)$, $B(X, Y)$, and $C(X, Y)$ through the use of basis functions and thus obtain a continuous representation of these functions across element boundaries. In the presence of detailed data for a given physical problem it could also be feasible to evaluate $A(X, Y)$, $B(X, Y)$, and $C(X, Y)$ at each of the numerical integration points and thereby obtain yet a better approximation to the equation coefficients. However, the equations which relate the bathymetry to these functions are rather complex and computation time would be expected to increase considerably.

With $A$, $B$, and $C$ assumed constant over each element, Eq. (6.91) reduces to the matrix equation

$$[K]\{\Psi\} = \{E\} \tag{6.92}$$

where

$$k_{i,j} = \iint_{\mathscr{A}} \left[ -\frac{\partial \phi_j}{\partial X}\frac{\partial \phi_i}{\partial X} - \frac{\partial \phi_j}{\partial Y}\frac{\partial \phi_i}{\partial Y} + A_e \frac{\partial \phi_j}{\partial X}\phi_i + B_e \frac{\partial \phi_j}{\partial Y}\phi_i \right] d\mathscr{A} \tag{6.93}$$

and

$$e_i = -\iint_{\mathscr{A}} C_e \phi_i \, d\mathscr{A} \tag{6.94}$$

where $A_e$, $B_e$, and $C_e$ are the values of $A(X, Y)$, $B(X, Y)$, and $C(X, Y)$ for each element, and $[K]$ is an $N \times N$ matrix. The terms involving $\psi_j$ at the boundary may be condensed out of the system since $\psi_j$ is specified at the boundaries. Note that no surface integrals need be evaluated in this formulation.

### *6.4.3 Solutions*

The complex flow model has been used to predict the circulation of Lake Ontario with wind shear prevailing at 7° north of east, the local average direction at Rochester, New York, in February [8, 9]. Values chosen for the parameters in the model are $f = 0.001$ rad/sec, $\varepsilon = 0.02$ m$^2$/sec, $|W|/\rho = 1.0 \times 10^{-4}$ m$^2$/sec$^2$, and $g = 9.8$ m/sec$^2$. The characteristic length $L$ is arbitrarily chosen as $2 \times 10^5$ m, while the characteristic depth $D$ is defined as the maximum nodal depth after the lake is discretized with finite elements. The value of $D$ may vary with the net depending on the locations of the nodes. For the triangular mesh shown in Fig. 6.10a, $D$ is equal to 225 m while with the isoparametric representation, shown in Fig. 6.10b, the value of $D$ is 216.5 m. Note that the triangular mesh is formed from 561 triangular elements with 323 nodes while the isoparametric grid uses only 70 elements and 257 nodes. The geometry and bathymetry of the lake were obtained from the Canada Center for Inland Waters. To avoid computational singularities which arise when the depth of the lake is zero, the flow region studied was bounded by the contour line having a water depth of 12 m rather than the shoreline.

Figures 6.10c and 6.10d show the computed stream functions and the surface velocities at cross section $A$–$A$ of the lake. The agreement between the different sets of results is generally good, although toward the southern shoreline some discrepancies in the stream function are seen between the isoparametric and triangular element solutions. This deviation may be due to the less accurate representation of the bathymetry used with the isoparametric grid. The fact that the stream function inaccuracy does not seriously affect the surface velocity, as demonstrated in Fig. 6.10d, suggests that the bathymetry may influence the surface velocity less than it does the stream function.

### *6.4.4 Summary of Results*

The examples cited in this section demonstrate that both higher-order and isoparametric elements can be useful in the application of the finite element method to lake circulation analysis. The isoparametric element can be particularly useful in studies of lakes with irregular shorelines. Such lakes can of course be described using large numbers of straight-sided elements, but such a formulation utilizes many elements and solution parameters solely for geometric definition. The isoparametric concept allows one to model an irregularly shaped region without introducing these extraneous nodes and solution parameters.

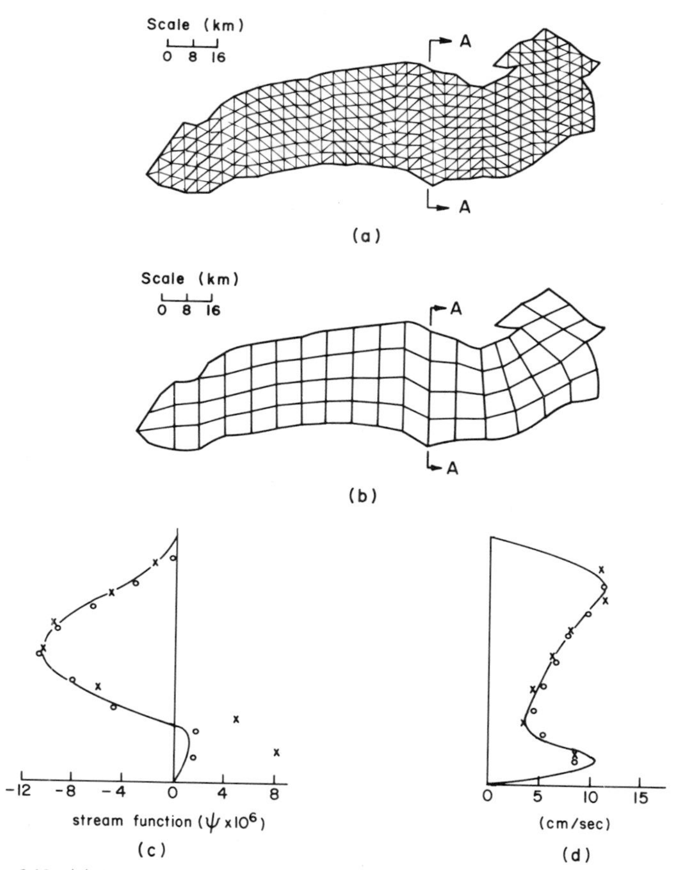


**Fig. 6.10.** (a) Triangular finite element mesh for Lake Ontario (after Gallagher and Chan [9]) with 561 elements and 323 nodes. (b) Isoparametric finite element mesh for Lake Ontario (after Gallagher and Chan [9]) with 70 elements and 257 nodes. (c) Comparison of solutions for the stream function in Lake Ontario at cross section $A$–$A$ (after Gallagher and Chan [9]). (○) Triangular linear elements; (—) triangular cubic elements; (×) isoparametric quadrilaterals. (d)Comparison of solutions for the magnitude of the surface velocity in Lake Ontario at cross section $A$–$A$ (after Gallagher and Chan [9]). (○) Triangular linear elements; (—) triangular cubic elements; (×) quadratic quadrilaterals.

Many fundamental pieces of information about the finite element method are unknown. Questions concerning the efficiency of a method for a specified accuracy are difficult to answer because of the many possible variations on a basic element type. The most efficient and accurate way of handling coefficients of the type $A(X, Y)$, $B(X, Y)$, and $C(X, Y)$ has yet to be established. Although the quadrilateral is popularly used in isoparametric representations, the introduction of curved sides on a triangle is also possible. Perhaps triangles of this form can be used competitively or in conjunction with quadrilaterals. Determination of the optimal order for a basis function is yet another area where definitive information is lacking. (Zienkiewicz [10] suggests that quadratic and cubic basis functions are probably the most efficient.)

It seems reasonable to assume that no unequivocal answers to many of these questions will ever be known. Selection of the "best" representation is clearly problem dependent and will undoubtedly remain an art rather than become a science.

## 6.5 Hydrothermal Model for Lakes

Models have been developed that are useful in simulating the steady state and transient transport of material or energy pollutants in lakes [11–13]. This section deals specifically with the thermal model, although the development for contaminant transport would be parallel.

### 6.5.1 *Restriction and Equations*

The type of lake to be considered in the analysis has the same type of cross section as that used in the more complicated flow model, Fig. 6.7. The surface is level but the lateral topography and bathymetry are arbitrary. Because the model is two-dimensional it is necessary to integrate the three-dimensional thermal energy transport equation through the depth to obtain the appropriate energy balance. By using the techniques outlined in Chapter 5 for two-dimensionalizing a problem, one obtains the following thermal energy equation in dimensional form:

$$\frac{\partial(hT)}{\partial t} + \frac{\partial(huT)}{\partial x} + \frac{\partial(hvT)}{\partial y} - \frac{\partial}{\partial x}\left(h\alpha\frac{\partial T}{\partial x}\right) - \frac{\partial}{\partial y}\left(h\alpha\frac{\partial T}{\partial y}\right) + \frac{Q_0}{\rho C_p} = 0 \qquad (6.95)$$

where $T$ is the vertically averaged temperature, $\rho$ the vertically averaged density assumed constant in the equation, $C_p$ the specific heat of water, assumed constant over the temperature range considered, $\alpha$ the thermometric dispersion coefficient, $Q_0$ the surface heat loss function, and $h$ the total depth of the water body. If the surface heat loss function is assumed proportional to the difference between the lake temperature and the ambient temperature, i.e.,

$$Q_0 = B_0(T - T_0)$$

then (6.95) becomes

$$\frac{\partial(hT)}{\partial t} + \frac{\partial(huT)}{\partial x} + \frac{\partial(hvT)}{\partial y} - \frac{\partial}{\partial x}\left(h\alpha\frac{\partial T}{\partial x}\right) - \frac{\partial}{\partial y}\left(h\alpha\frac{\partial T}{\partial y}\right) + \frac{B_0(T - T_0)}{\rho C_p} = 0 \tag{6.96}$$

The first term in this equation deals with the rate at which heat is added to a differential volume; the second two terms account for the flow of heat into and out of the volume due to the movement of the fluid (convective or advective terms); the fourth and fifth terms are concerned with the transport of heat due to molecular diffusion and turbulent eddies; and finally the last term accounts for heat loss due to heat transfer between the air–water interface at the surface. When a steady state problem is considered, the time derivative of temperature will be zero. Boundary conditions for Eq. (6.96) are the specification of the temperature at points of inflow into the lake and the specification of the heat flux at shoreline boundaries, usually taken to be zero.

To simulate the energy transport problem it is necessary to solve for the velocity and the temperature. In general, the density would be a function of temperature and the temperature field would thus influence the velocity field. This nonlinear interaction requires that the flow and energy transport equations be solved simultaneously. Under the constraints of the problem posed in this section, density is assumed independent of temperature and therefore Eq. (6.62) can be solved for the stream function followed by the solution of (6.96) for the temperature field. Besides adding further complications to the solution procedure, the inclusion of a temperature-dependent density would probably influence the solution only insofar as it gives rise to free convection in the vertical. Thus the assumption of constant density is probably a reasonable approximation in the vertically well-mixed problem being considered.

### 6.5.2 *Determination of the Velocity Field*

The dimensional velocities required as coefficients in Eq. (6.96) are related to the dimensionless stream function by the expressions

$$u = \frac{gD}{fLH}\frac{\partial \Psi}{\partial Y} = \frac{gD^2}{fh}\frac{\partial \Psi}{\partial y} \tag{6.97a}$$

$$v = -\frac{gD}{fLH}\frac{\partial \Psi}{\partial X} = -\frac{gD^2}{fh}\frac{\partial \Psi}{\partial x} \tag{6.97b}$$

The solution procedures for $\Psi$ described in Section 6.4 yield values of $\Phi$ that are continuous across element boundaries. However, the derivatives of $\Psi$, or the velocities, are discontinuous across element boundaries. Even when the cubic triangles discussed in Section 6.4.2 are used the values of $\partial\Psi/\partial X$ and $\partial\Psi/\partial Y$ will be discontinuous along element boundaries except at the nodes. When the velocity field is discontinuous, more heat may be convected out across the side of one element than is convected across the same side into the adjacent element. If convection rather than dispersion is the principal mode of energy transport, serious errors can arise. The usual way to ensure a continuous velocity profile is to obtain approximations to the velocity at the nodes and then expand these estimates in terms of the basis functions. When the cubic triangles are used, $\psi_{Xj}$ and $\psi_{Yj}$ at the corner nodes are solution parameters and are directly related to $u$ and $v$ by Eq. (6.97). Thus expansions for $u$ and $v$ in terms of linear triangular basis functions are readily obtained.

If values of the gradients of $\psi$ are nonunique at nodes throughout the domain (e.g., when linear or quadratic basis functions are used), some weighted average values of $u$ and $v$ obtained by differentiating $\psi$ over different elements must be used at the node. At present there is no definitive averaging rule to use, although active research on this point is being carried out [14, 15]. Once values for $u$ and $v$ at the nodes are obtained, expansion for the velocities in terms of the basis functions may be made and used in thermal energy equations.

### 6.5.3 *Finite Element Formulation of the Energy Transport Equation*

The weighted residual form of Eq. (6.96) is

$$\iint_{\mathscr{A}} \left\{ \frac{\partial(hT)}{\partial t} + \frac{\partial(huT)}{\partial x} + \frac{\partial(hvT)}{\partial y} - \frac{\partial}{\partial x}\left(h\alpha\frac{\partial T}{\partial x}\right) - \frac{\partial}{\partial y}\left(h\alpha\frac{\partial T}{\partial y}\right) + \frac{B_0(T - T_0)}{\rho C_p} \right\} \phi_i \, d\mathscr{A} = 0 \qquad (i = 1, \ldots, M) \tag{6.98}$$

where $M$ is the number of basis functions used in the expansion for $T$. Application of Green's theorem to the convective and dispersive terms in the equation yields

$$\iint_{\mathscr{A}} \left\{ \frac{\partial(hT)}{\partial t}\phi_i - huT\frac{\partial\phi_i}{\partial x} - hvT\frac{\partial\phi_i}{\partial y} + h\alpha\frac{\partial T}{\partial x}\frac{\partial\phi_i}{\partial x} + h\alpha\frac{\partial T}{\partial y}\frac{\partial\phi_i}{\partial y} + \frac{B_0(T - T_0)}{\rho C_p}\phi_i \right\} d\mathscr{A} + \int_s h(\mathbf{v}T + \mathbf{j}_T)\cdot\mathbf{n}\phi_i \, ds = 0 \qquad (i = 1, \ldots, M) \tag{6.99}$$

where $\mathbf{v}$ is the velocity vector, $s$ the boundary of the lake, $\mathbf{n}$ the outwardly directed unit normal vector, and $\mathbf{j}_T$ the dispersive flux. For convenience assume that $T$, $u$, $v$, $\alpha$, $B_0$, and $h$ are expanded in terms of the same set of basis functions. Then

$$T \simeq \sum_{j=1}^{M} T_j(t)\phi_j(x, y), \qquad u \simeq \sum_{k=1}^{M} u_k \phi_k(x, y), \qquad v \simeq \sum_{k=1}^{M} v_k \phi_k(x, y)$$

$$\alpha \simeq \sum_{k=1}^{M} \alpha_k \phi_k(x, y), \qquad B_0 \simeq \sum_{m=1}^{M} B_{0m} \phi_m(x, y), \qquad h \simeq \sum_{m=1}^{M} h_m \phi_m(x, y)$$

where only $T$ is a function of time. Substitution of these expansions into (6.99) yields

$$\sum_{m=1}^{M} \sum_{j=1}^{M} \left\{ \iint_{\mathscr{A}} \frac{dT_j}{dt} h_m \phi_m \phi_j \phi_i - \sum_{k=1}^{M} h_m u_k T_j \phi_m \phi_k \phi_j \frac{\partial \phi_i}{\partial x} \right.$$
$$- \sum_{k=1}^{M} h_m v_k T_j \phi_m \phi_k \phi_j \frac{\partial \phi_i}{\partial y} + \sum_{k=1}^{M} h_m \alpha_k T_j \phi_m \phi_k \frac{\partial \phi_j}{\partial x} \frac{\partial \phi_i}{\partial x}$$
$$\left. + \sum_{k=1}^{M} h_m \alpha_k T_j \phi_m \phi_k \frac{\partial \phi_j}{\partial y} \frac{\partial \phi_i}{\partial y} + \frac{B_{0m}(T_j - T_{0j})}{\rho C_p} \phi_m \phi_j \phi_i \right\} d\mathscr{A}$$
$$+ \int_s h(\mathbf{v}T + \mathbf{j}_T) \cdot \mathbf{n}\phi_i \, ds = 0 \qquad (i = 1, \ldots, M) \tag{6.100}$$

The boundary conditions considered for this problem are $T$ specified along a boundary where the normal flow is nonzero, and the normal heat flux specified as zero at a boundary where the normal flow is zero. When node $i$ is a boundary node where the temperature is specified, instead of using Eq. (6.100), the value of $T$ at node $i$ is set. When node $i$ is situated at a solid boundary, the no-flux boundary condition is invoked and the surface integral is set to zero. Because all nodes not situated on a boundary have basis functions that are zero at the boundary, the surface integral arising in the energy equation need never be evaluated.

The matrix form of (6.100) is

$$[P]\{dT/dt\} + [K]\{T\} = \{E\} \tag{6.101}$$

One could consider $[P]$ and $[K]$ as $N \times N$ matrices where $N$ is the number of unknown $T$ values. Then $\{E\}$ would contain contributions from nonzero prescribed temperature boundaries and from the surface heat loss term. The problem has been formulated this way in the previous sections. Although this procedure saves storage, it has the drawback that some additional book-keeping must be done to keep track of the node number associated with the

$j$th member of the $T$ vector. An alternative to this is to consider $[K]$ and $[P]$ as $M \times M$ matrices where $M$ is the number of nodes and make adjustments to the rows in these matrices to account for set values of $T$. This method has the advantage that it is easy to implement in that the $j$th member of the $T$ vector is associated with the $j$th node. However, it has the drawback that larger $[K]$ and $[P]$ matrices are obtained so more computer storage is used. For illustrative purposes, this second procedure will be used here.

In Eq. (6.101), $[P]$ and $[K]$ are $M \times M$ matrices and $\{E\}$ is an $M \times 1$ vector defined as follows: when node $i$ is not a set temperature node,

$$p_{i,j} = \sum_{m=1}^{M} \iint_{\mathscr{A}} h_m \phi_m \phi_j \phi_i \, d\mathscr{A} \tag{6.102}$$

$$k_{i,j} = \sum_{m=1}^{M} \iint_{\mathscr{A}} \left[ \sum_{k=1}^{M} \left\{ -h_m u_k \phi_m \phi_k \phi_j \frac{\partial \phi_i}{\partial x} - h_m v_k \phi_m \phi_k \phi_j \frac{\partial \phi_i}{\partial y} \right. \right.$$
$$\left. \left. + h_m \alpha_k \phi_m \phi_k \left( \frac{\partial \phi_j}{\partial x} \frac{\partial \phi_i}{\partial x} + \frac{\partial \phi_j}{\partial y} \frac{\partial \phi_i}{\partial y} \right) \right\} + \frac{B_{0m}}{\rho C_p} \phi_m \phi_j \phi_i \right] d\mathscr{A} \tag{6.103}$$

$$e_i = \sum_{m=1}^{M} \sum_{j=1}^{M} \iint_{\mathscr{A}} \frac{B_{0m} T_{0j}}{\rho C_p} \phi_m \phi_j \phi_i \, d\mathscr{A} \tag{6.104}$$

and when node $i$ is a set temperature node,

$$p_{i,j} = \delta_{ij} \tag{6.105}$$

$$k_{i,j} = \delta_{ij} \tag{6.106}$$

$$e_i = T_i(t) + (dT_i(t)/dt) \tag{6.107}$$

where $\delta_{ij}$ is the Kronecker delta.

When considering a steady state problem the $[P]$ matrix is not included in the analysis. Furthermore, the values of $\{E\}$ will be independent of time. The problem to be solved will merely be

$$[K]\{T\} = \{E\}$$

In solving the transient problem many different techniques are available for treating the time derivative. An implicit integration procedure for the time domain suggested by Wilson and Clough [16] has been successfully used in this type of problem [12]. A second-order difference approximation for $dT_j/dt$ is applied when node $j$ is not a specified temperature node:

$$\left. \frac{dT_j}{dt} \right|_{t+\Delta t} = \frac{2(T_j|_{t+\Delta t} - T_j|_t)}{\Delta t} - \left. \frac{dT_j}{dt} \right|_t \tag{6.108}$$

This expression is substituted into the left side of (6.101) to obtain an equation of the form

$$[\hat{K}]\{T\}\Big|_{t+\Delta t} = \{\hat{E}\} \tag{6.109}$$

where

$$[\hat{K}] = [K] + (2/\Delta t)[\hat{P}] \tag{6.110}$$

$$\{\hat{E}\} = \{E\} + (2/\Delta t)[\hat{P}]\{T\}\Big|_{t} + [\hat{P}]\{dT/dt\}\Big|_{t} - [P - \hat{P}]\{dT/dt\}\Big|_{t+\Delta t} \tag{6.111}$$

Matrix $[\hat{P}]$ is simply $[P]$ with the $j$th column zeroed out when $j$ is a node with set temperature, and

$$[P - \hat{P}] = [P] - [\hat{P}] \tag{6.112}$$

Although the known vector $\{E\}$ appears to contain some unknowns due to the last term in (6.111), in fact the only values of $\{dT/dt\}|_{t+\Delta t}$ that have nonzero multipliers in the $[P - \hat{P}]$ matrix are those at set temperature nodes which are known from differentiation of the set temperature data.

To begin the calculations, substitute the known values of $T$ at the initial time $t_0$ into (6.101). Then $\{dT/dt\}$ at time $t_0$ is obtained from

$$\{dT/dt\}\Big|_{t_0} = [P]^{-1}\{E\}\Big|_{t_0} - [P]^{-1}[K]\{T\}\Big|_{t_0} \tag{6.113}$$

Now the right side of (6.109) can be completely determined and a solution for $\{T\}|_{t_0+\Delta t}$ can be obtained. Equation (6.108) is next applied to obtain $\{dT/dt\}|_{t_0+\Delta t}$ and the calculations proceed to the next time step. Ralston [17] has demonstrated that this procedure is convergent and unconditionally stable.

### *6.5.4 Solutions*

The numerical procedure described above has been applied to Horseshoe Lake in Oklahoma. This lake was constructed by damming a dry loop of the North Canadian River to provide a source of cooling water for a power generation station.

Figure 6.11 is an aerial view of the lake and the numbers indicate the locations of temperature probes. Probe 1 is located nearest the discharge and probe 16 is located in the intake. The data from probes 3, 7, 12, and 16 are used for comparison with the numerical model.

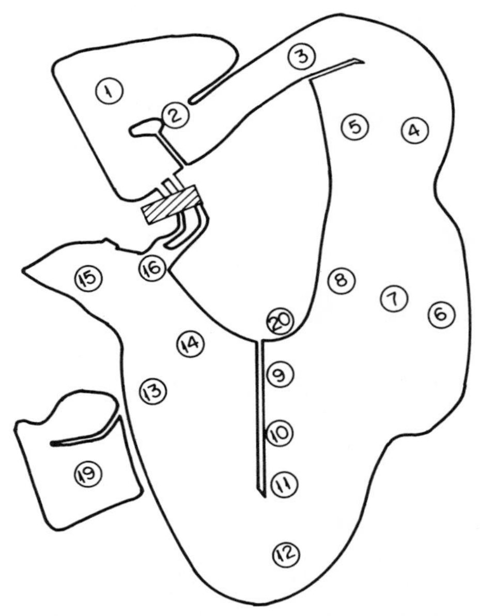


**Fig. 6.11.** Locations of temperature probes in Horseshoe Lake (after Loziuk *et al.* [12]).

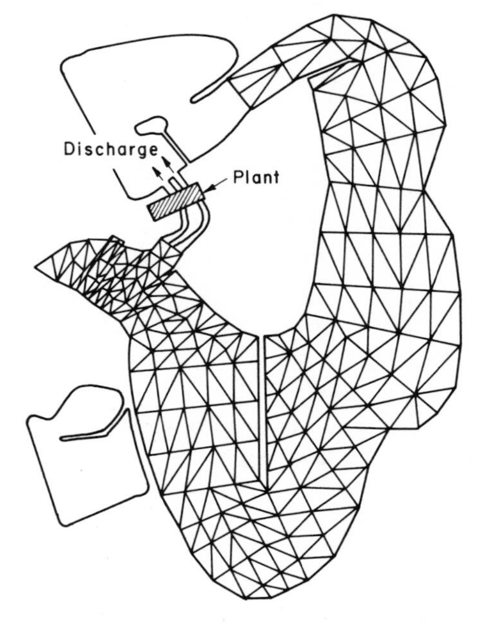


**Fig. 6.12.** Finite element mesh for Horseshoe Lake (after Loziuk *et al.* [12]).

The finite element mesh used in the computation is illustrated in Fig. 6.12. Due to the number of discharge ports from the various generating units, the modeling of the discharge area was not practical, and the open section was considered to be a mixing section.

The velocity field used in the calculation was obtained from a potential flow formulation in which all viscous and turbulent losses were neglected.

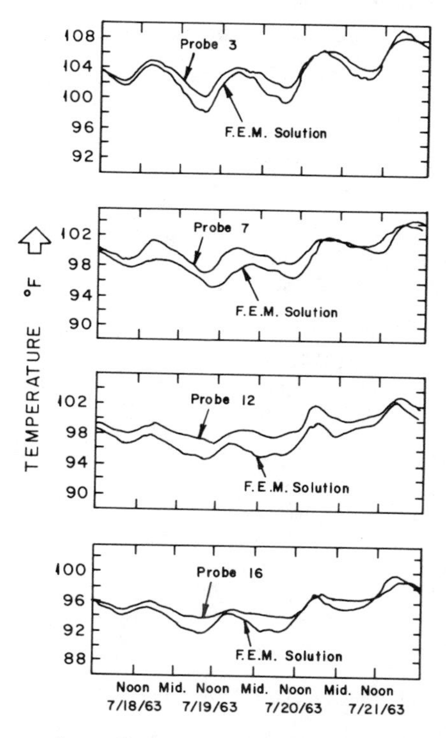


**Fig. 6.13.** Comparison of recorded and computed temperatures at various monitoring stations in Horseshoe Lake (after Loziuk *et al.* [12]).

Furthermore, the depth is considered constant and the coefficient $\alpha$ is chosen as 0.19 $m^2$/sec throughout the lake. The comparison between recorded temperature at the water surface and computed temperature at various temperature probes is shown in Fig. 6.13. Discrepancies between the data and the numerical solution are attributed to errors in estimation of the heat transfer at the water surface and to stratification in the lake which results in the warmest water being at the lake surface. All in all, the solution indicates that the finite element method can be a useful tool in hydrothermal analysis.

## 6.6 Conclusions

Lake circulation and heat transport are processes amenable to the finite element analysis. Various models of differing theoretical and numerical complexities can be considered. Different types of computational grids and basis functions can be useful in the analyses and irregular geometries can be easily incorporated into a finite element solution.

## References

1. J. A. Liggett and C. Hadjitheodorou, Circulation in shallow homogeneous lakes, *J. Hydraul. Div. ASCE* **95**, 609 (1969).
2. J. P. Catchpole and G. Fulford, Dimensionless groups, *Ind. Eng. Chem.* **58**, 46 (1966).
3. G. T. Csanady, Large-scale motion in the Great Lakes, *J. Geophys. Res.* **72**, 4151 (1967).
4. V. W. Ekman, On the influences of the earth's rotation on ocean currents, *Ark. Met. Astr. Fysik* **2** (1905).
5. R. T. Cheng and C. Tung, Wind driven lake circulation by the finite element method, *Proc. Conf. Great Lakes Res., 13th*, 891 (1970).
6. R. T. Cheng, Numerical investigation of lake circulation around islands by the finite element method, *Int. J. Num. Methods Eng.* **5**, 103 (1972).
7. C. A. Felippa, Refined Finite Element Analysis of Linear and Nonlinear Two-dimensional Structures. Rep. SESM 66-22, Structural Eng. Lab., Univ. of California, Berkeley (1966).
8. R. H. Gallagher, J. A. Liggett, and S. T. K. Chan, Finite element shallow lake circulation analysis, *J. Hydraul. Div. ASCE* **99**, 1083 (1973).
9. R. H. Gallagher and S. T. K. Chan, Higher-order finite element analysis of lake circulation, *Comput. Fluids* **1**, 119 (1973).
10. O. C. Zienkiewicz, Isoparametric and Allied Numerically Integrated Elements—A Review, *in* Numerical and Computer Methods in Structural Mechanics" (S. J. Fenves, N. Perrone, A. R. Robinson, and W. C. Schnobrick, eds.), p. 13. Academic Press, New York, 1973.
11. L. A. Loziuk, J. C. Anderson, and T. Belytschko, Hydrothermal analysis by finite element method, *J. Hydraul. Div. ASCE* **98**, 1983 (1972).
12. L. A. Loziuk, J. C. Anderson, and T. Belytschko, Transient hydrothermal analysis of small lakes, *J. Power Div. ASCE* **99**, 349 (1973).
13. P. Tong, Finite element solution of the wind-driven currents and its mean transport in lakes, *in* "Numerical Methods in Fluid Dynamics" (C. A. Brebbia and J. J. Connor, eds.), p. 440. Pentech Press, London, 1974.
14. K. Fischer, On the calculation of higher derivatives in finite elements, *Comput. Methods Appl. Mech. Eng.* **7**, 323 (1976).
15. B. E. Larock and L. R. Herrmann, Improved flux prediction using low order finite elements, *Proc. Int. Conf. Finite Elements Water Resources, Princeton Univ.* Pentech, Press, London, 1976.
16. E. L. Wilson and R. W. Clough, Dynamic response by step-by-step matrix analysis, *Proc. Symp. Use Comput. Civil Eng., Lisbon, Portugal* (1962).
17. A. Ralston, "A First Course in Numerical Analysis." McGraw-Hill, New York, 1965.

# Chapter 7 | Analysis of Model Behavior

## 7.1 Stability and Consistency

Before programming a model, it is important to know whether the computer results will approximate the dynamics of the system under consideration or merely represent a curve fit through some data points. In the present chapter, the method is presented for analyzing the ability of a scheme to predict accurately the amplitude and phase of a wave. Various finite element approximations to the linearized forms of the shallow water equations are examined for stability and consistency. The concept of stability has previously been discussed in Section 2.3.2 and refers to the growth of numerical errors. If the errors introduced by a computational method grow in an unbounded manner, the method is unstable and unsuitable for simulation. As defined here, a method is said to be consistent if, in the limit as the spatial and temporal step sizes go to zero, the solution to the discretized form of the partial differential equation converges to the analytic solution to the equation. An inconsistent scheme may be stable, but the solution it generates will not be a solution to the differential equation.

## 7.2 Equations and Restrictions

The equations for long wave propagation in estuaries can be approximated by either the finite difference or the finite element method. The properties of the numerical technique can then be ascertained by Fourier analysis. The scope of this investigation will be restricted as follows:

(1) Only one-dimensional flow will be considered. More exactly, the analysis will be restricted to the propagation of a wave in only one coordinate direction. Leendertse [1] and Sobey [2] have shown that the celerity or propagation velocity of the wave is affected by the orientation of the direction of propagation relative to the computational grid. However, this effect is not large and does not alter the conclusions that can be drawn from a one-dimensional analysis.

(2) Only linearized forms of the governing partial differential equations will be considered. Once the equations have been written in their linearized form, they may be analyzed by von Neumann's method [3].

(3) The equations considered will be vertically averaged through the depth. Convective acceleration, Coriolis acceleration, and surface stresses will be neglected. When included, the bottom stress will be represented as a linear function of velocity.

Under these conditions, the continuity equation takes the form

$$\frac{\partial \zeta}{\partial t} + h\left(\frac{\partial U}{\partial x} + \frac{\partial V}{\partial y}\right) = 0 \tag{7.1}$$

and the momentum equations are

$$\frac{\partial U}{\partial t} + g\frac{\partial \zeta}{\partial x} + \kappa U = 0 \tag{7.2}$$

and

$$\frac{\partial V}{\partial t} + g\frac{\partial \zeta}{\partial y} + \kappa V = 0 \tag{7.3}$$

where $U$ is the vertically averaged easterly velocity, $V$ the vertically averaged northerly velocity, $h$ the distance from the bottom of the estuary to mean sea level, assumed to be independent of space, $\zeta$ the wave height with respect to mean sea level, $t$ the time coordinate, $x$ the coordinate which is positive eastward, and $y$ the coordinate which is positive northward.

Before examining the behavior of numerical solutions to (7.1)–(7.3), it will be useful to obtain an analytical solution to serve as a basis for comparisons.

## 7.3 Analytical Solution

Assume that the analytical solution to Eqs. (7.1)–(7.3) is represented by the Fourier series

$$U = \sum_{m=-\infty}^{\infty} U_m^0 \exp[\hat{\imath}(\beta_m t + \sigma_{1m} x + \sigma_{2m} y)] \tag{7.4}$$

$$V = \sum_{m=-\infty}^{\infty} V_m^0 \exp[\hat{i}(\beta_m t + \sigma_{1m} x + \sigma_{2m} y)] \tag{7.5}$$

$$\zeta = \sum_{m=-\infty}^{\infty} \zeta_m^0 \exp[\hat{i}(\beta_m t + \sigma_{1m} x + \sigma_{2m} y)] \tag{7.6}$$

where $U_m^0$, $V_m^0$, $\zeta_m^0$ are the amplitudes of the $m$th component, $\beta_m$ the real wave frequency of the $m$th component, $\hat{i}$ is $(-1)^{1/2}$, $\sigma_{1m} = \sigma_m \cos \gamma_m$, $\sigma_{2m} = \sigma_m \sin \gamma_m$, $\sigma_m$ the wave number or spatial frequency of the $m$th component, and $\gamma_m$ the angle between the wave propagation direction and the $x$ axis. Because the equations under consideration are linear, it will only be necessary to examine one term in each of these series. We will consider

$$\zeta \sim \zeta^0 \exp[\hat{i}(\beta t + \sigma_1 x + \sigma_2 y)] \tag{7.7}$$

$$U \sim U^0 \exp[\hat{i}(\beta t + \sigma_1 x + \sigma_2 y)] \tag{7.8}$$

$$V \sim V^0 \exp[\hat{i}(\beta t + \sigma_1 x + \sigma_2 y)] \tag{7.9}$$

where the subscripts $m$ have been dropped for convenience. Substitution of these equations back into (7.1)–(7.3) yields the matrix equation

$$\begin{bmatrix} \hat{i}\beta & \hat{i}h\sigma_1 & \hat{i}h\sigma_2 \\ \hat{i}g\sigma_1 & \hat{i}\beta + \kappa & 0 \\ \hat{i}g\sigma_2 & 0 & \hat{i}\beta + \kappa \end{bmatrix} \begin{Bmatrix} \zeta^0 \\ U^0 \\ V^0 \end{Bmatrix} = \begin{Bmatrix} 0 \\ 0 \\ 0 \end{Bmatrix} \tag{7.10}$$

It can be shown [4] that if $\zeta^0$, $U^0$, and $V^0$ are not all equal to zero, the determinant of the coefficient matrix on the left side of (7.10) must be zero, i.e.,

$$\hat{i}\beta(\hat{i}\beta + \kappa)^2 + (\sigma_1^2 + \sigma_2^2)(gh)(\hat{i}\beta + \kappa) = 0 \tag{7.11}$$

From their definitions it can be ascertained that $\sigma_1^2 + \sigma_2^2 = \sigma^2$ and therefore

$$(\hat{i}\beta + \kappa)[\hat{i}\beta(\hat{i}\beta + \kappa) + \sigma^2 gh] = 0 \tag{7.12}$$

This equation can be solved for the roots of $\beta$:

$$\beta = \hat{i}\kappa \tag{7.13a}$$

$$\beta = \sigma\{(\hat{i}\kappa/2\sigma) + [gh - (\kappa/2\sigma)^2]^{1/2}\} \tag{7.13b}$$

$$\beta = \sigma\{(\hat{i}\kappa/2\sigma) - [gh - (\kappa/2\sigma)^2]^{1/2}\} \tag{7.13c}$$

The first root represents the decay of the flow field due to bottom resistance. When $\kappa = 0$, there is no resistance to the flow field and this root corresponds to steady state flow.

The two roots given by (7.13b) and (7.13c) represent the characteristics of the wave and give relationships between the wave number $\sigma$ and the frequency $\beta$. The progressive and retrogressive gravity waves are indicated by (7.13b) and (7.13c), respectively. The velocity at which the wave propagates, or celerity, is given by the ratio $\mathrm{Re}(\beta/\sigma)$, where Re indicates that only the real part of the $\beta/\sigma$ ratio is considered. When $\kappa = 0$, corresponding to no frictional losses, the celerity of the wave is

$$\mathrm{Re}(\beta/\sigma) = \pm(gh)^{1/2} \tag{7.14}$$

The wave number $\sigma$ is defined for a particular wave as $2\pi/L$, where $L$ is the length of the wave. Therefore

$$\beta = (\hat{i}\kappa/2) \pm (2\pi/L)[gh - (\kappa L/4\pi)^2]^{1/2} \tag{7.15}$$

As previously indicated this analysis will be restricted to wave propagation in only one coordinate direction. When $\gamma = 0°$, the direction of wave propagation is parallel with the $x$ axis and Eqs. (7.7)–(7.9) take the form

$$\begin{Bmatrix} \zeta \\ U \\ V \end{Bmatrix} = \begin{Bmatrix} \zeta^0 \\ U^0 \\ V^0 \end{Bmatrix} \exp\left(-\frac{\kappa t}{2}\right) \exp\left\{\pm\hat{i}\frac{2\pi t}{L}\left[gh - \left(\frac{\kappa L}{4\pi}\right)^2\right]^{1/2}\right\} \exp\left(\hat{i}\frac{2\pi x}{L}\right) \tag{7.16}$$

If this result is substituted into Eq. (7.3), we see that

$$V^0\left\{-\frac{\kappa}{2} \pm \hat{i}\frac{2\pi}{L}\left[gh - \left(\frac{\kappa L}{4\pi}\right)^2\right]^{1/2}\right\} + \kappa V^0 = 0 \tag{7.17}$$

The only way for this to be true is if $V^0$ is zero. Therefore, if wave propagation parallel to the $x$ axis in a two-dimensional system is considered, the behavior of the wave can be determined by considering only Eqs. (7.1) and (7.2) with $V$ set to zero. Conversely, if propagation in a direction parallel to the $y$ axis is considered, the behavior of the wave is determined by (7.1) and (7.3) with $U$ set to zero.

Equation (7.16) implies that if a spatial harmonic wave is taken as an initial condition, it will decay with time. Alternatively if a harmonic motion is introduced at a particular point in space, it will decay as it propagates away from the point. After a time $\Delta t$, the amplitude of the wave will be reduced to $\exp(-\kappa\,\Delta t/2)$ of its initial size. Furthermore, it will have advanced a distance $\Delta t[gh - (\kappa/2\sigma)^2]^{1/2}$.

## 7.4 Numerical Solution

By proposing a numerical scheme and comparing the calculated wave amplitude with the analytical amplitude and the calculated wave velocity with the analytical velocity, one can obtain an idea of the dissipative and dispersive properties of the discretized equations.

As in the discussion of numerical methods and the transport equation in Chapter 5, a scheme will be said to be dissipative if the amplitude of the numerical wave is smaller than that of the physical wave. A conservative scheme is one in which the numerical and physical waves have the same amplitude, and a nonconservative scheme is one which is either dissipative or computes a wave amplitude greater than that of the physical wave. In the absence of physical dissipation (i.e., when $\kappa = 0$ in our system of equations) a nonconservative, nondissipative scheme can be shown to be unstable.

If the present examination is confined to the behavior of the three dependent variables at a fixed point in space, we can see that

$$\begin{Bmatrix} \zeta_{t+\Delta t} \\ U_{t+\Delta t} \\ V_{t+\Delta t} \end{Bmatrix} = [G] \begin{Bmatrix} \zeta_t \\ U_t \\ V_t \end{Bmatrix}$$

where $G$ is called the amplification matrix [1]. The eigenvalues of the amplification matrix will be denoted by $\lambda$ and it can be easily shown that $\lambda = \exp(\hat{i}\beta\ \Delta t)$. The von Neumann necessary condition for stability states that the time step $\Delta t$ must be finite and that for all wave numbers the eigenvalues of the amplification matrix must satisfy

$$|\lambda| \leq 1 + O(\Delta t) \tag{7.18}$$

The special case when $|\lambda|$ is exactly equal to one is referred to as neutral stability.

In general components of all wavelengths may be present. Even when a problem in which some wavelengths are not initially present or introduced by the boundary conditions is considered, computer round-off error will probably introduce them. Then as computation proceeds through time, if $|\lambda| > 1$ for these wavelengths, the error will grow without bounds and overwhelm the true solution to the problem.

The concept of a complex propagation factor has previously been used to study the behavior of a concentration profile. Here the complex propagation factor $T(\sigma\ \Delta x)$, where $\sigma$ is the wave number and $\Delta x$ the grid step size in the spatial domain, will be used to study the propagation of a gravity wave. This factor is the ratio of the numerically computed wave to the physical wave after the time interval in which the physical wave has propagated through one of its wavelengths or

$$T(\sigma\ \Delta x) = \exp[\hat{i}(\beta' t + \sigma x)]/\exp[\hat{i}(\beta t + \sigma x)] \tag{7.19}$$

where $x = 2\pi/\sigma = L$, $L$ is the wavelength, $t = 2\pi/\mathrm{Re}(\beta)$, and $\beta'$ is the approximation to $\beta$ given by the numerical solution. Equation (7.19) can be rearranged as

$$T(\sigma\ \Delta x) = \exp[\hat{i}(\beta' t - \beta t)] \tag{7.20}$$

and then substitution of the value of $t$ as given above yields

$$T(\sigma\,\Delta x) = \exp[2\pi\hat{\imath}(\beta' - \beta)/\mathrm{Re}(\beta)] \tag{7.21}$$

The argument of $T(\sigma\,\Delta x)$ represents the phase lag of the computed wave relative to the physical wave and the modulus of $T(\sigma\,\Delta x)$ is the relative magnitude of the numerical wave to the physical wave.

If a scheme is considered in which the time step is denoted by $\Delta t$, the number of operations $N$ that must be performed for the physical wave to propagate through one wavelength is the ratio of the wavelength to the absolute value of the celerity multiplied by the time step, or

$$N = \frac{L}{|\mathrm{Re}(\beta/\sigma)\,\Delta t|} = \frac{(2\pi/\sigma)}{|\mathrm{Re}(\beta/\sigma)\,\Delta t|} = \frac{2\pi}{|\mathrm{Re}(\beta)\,\Delta t|} \tag{7.22}$$

From Eq. (7.13b) or (7.13c), $\mathrm{Re}(\beta)$ can be obtained and therefore

$$N = \frac{2\pi}{\sigma\,\Delta t[gh - (\kappa/2\sigma)^2]^{1/2}} \tag{7.23}$$

Because the time for the wave to propagate through its wavelength is $t = N\,\Delta t$, (7.20) may be rewritten as

$$T(\sigma\,\Delta x) = \exp[\hat{\imath}N\,\Delta t(\beta' - \beta)] \tag{7.24}$$

For convenience in the following studies, the progressive wave will be considered although all conclusions reached will apply as well to the retrogressive wave. Therefore $\beta$ is given by (7.13b) and Eq. (7.24) becomes

$$T(\sigma\,\Delta x) = [\exp(\hat{\imath}\beta'\,\Delta t)/\exp(-\tfrac{1}{2}\kappa\,\Delta t)]^N \tag{7.25}$$

This definition will be applied to the finite element scheme to obtain an estimate of the behavior of the scheme under certain idealized conditions. The effects of boundary conditions and of the irregularities in nodal locations will be beyond the scope of the analyses.

## 7.5 Central Implicit Scheme on Linear Elements

The first scheme to be examined makes use of Eqs. (7.1) and (7.2) on a regular one-dimensional finite element grid in space, oriented parallel to the $x$ axis. Nodes are located at the endpoints of the one-dimensional elements, and basis functions that vary linearly across each element are selected. From the discussion on the influence of the $y$ velocity on wave propagation in the $x$ direction, one would expect that the results of this analysis would also apply to wave propagation in the $y$ direction on a two-dimensional square grid oriented with element sides parallel to the coordinate axes.

For one-dimensional wave propagation, the governing linearized partial differential equations become

$$\frac{\partial \zeta}{\partial t} + h\frac{\partial U}{\partial x} = 0 \tag{7.26}$$

$$\frac{\partial U}{\partial t} + g\frac{\partial \zeta}{\partial x} + \kappa U = 0 \tag{7.27}$$

Expansions for $\zeta$ and $U$ in terms of the basis functions can be made such that

$$\zeta \simeq \sum_{j=1}^{M} \zeta_j(t)\phi_j(x) \tag{7.28}$$

$$U \simeq \sum_{j=1}^{M} U_j(t)\phi_j(x) \tag{7.29}$$

where $\zeta_j(t)$ and $U_j(t)$ are, respectively, the approximations to $\zeta$ and $U$ at node $j$ and time $t$, and $\phi_j(x)$ is the basis function associated with node $j$.

For convenience, the nodes are assumed to be numbered consecutively with node $j + 1$ adjacent to node $j$ in the positive $x$ direction. Application of Galerkin's method to (7.26) and (7.27) by weighting with respect to the $i$th basis function yields

$$\sum_{j=1}^{M} \int_x \left[\frac{d\zeta_j}{dt}\phi_j\phi_i + hU_j\frac{d\phi_j}{dx}\phi_i\right] dx = 0 \tag{7.30}$$

$$\sum_{j=1}^{M} \int_x \left[\frac{dU_j}{dt}\phi_j\phi_i + g\zeta_j\frac{d\phi_j}{dx}\phi_i + \kappa U_j\phi_j\phi_i\right] dx = 0 \tag{7.31}$$

If the length of each element is denoted by $\Delta x$ and it is recognized that $\phi_j\phi_i$ and $\phi_i(d\phi_j/dx)$ are nonzero only for $j = i - 1$, $i$, or $i + 1$, the integrals in these equations can be evaluated to obtain

$$\frac{1}{6}\left(\frac{d\zeta_{i-1}}{dt} + 4\frac{d\zeta_i}{dt} + \frac{d\zeta_{i+1}}{dt}\right) + h\frac{U_{i+1} - U_{i-1}}{2\,\Delta x} = 0 \tag{7.32}$$

$$\frac{1}{6}\left(\frac{dU_{i-1}}{dt} + 4\frac{dU_i}{dt} + \frac{dU_{i+1}}{dt}\right) + g\frac{\zeta_{i+1} - \zeta_{i-1}}{2\,\Delta x} + \frac{\kappa}{6}(U_{i-1} + 4U_i + U_{i+1}) = 0 \tag{7.33}$$

The time derivatives can be finite differenced over the time step $\Delta t$ and the other terms in these equations can be averaged over the two time levels to yield

$$\frac{1}{6}\left[\frac{\zeta_{i-1,t+\Delta t}-\zeta_{i-1,t}}{\Delta t}+4\frac{\zeta_{i,t+\Delta t}-\zeta_{i,t}}{\Delta t}+\frac{\zeta_{i+1,t+\Delta t}-\zeta_{i+1,t}}{\Delta t}\right]$$
$$+\frac{h}{2}\left[\frac{U_{i+1,t+\Delta t}-U_{i-1,t+\Delta t}}{2\,\Delta x}+\frac{U_{i+1,t}-U_{i-1,t}}{2\,\Delta x}\right]=0 \qquad (7.34)$$

$$\frac{1}{6}\left[\frac{U_{i-1,t+\Delta t}-U_{i-1,t}}{\Delta t}+4\frac{U_{i,t+\Delta t}-U_{i,t}}{\Delta t}+\frac{U_{i+1,t+\Delta t}-U_{i+1,t}}{\Delta t}\right]$$
$$+\frac{g}{2}\left[\frac{\zeta_{i+1,t+\Delta t}-\zeta_{i-1,t+\Delta t}}{2\,\Delta x}+\frac{\zeta_{i+1,t}-\zeta_{i-1,t}}{2\,\Delta x}\right]$$
$$+\frac{\kappa}{12}[U_{i-1,t+\Delta t}+4U_{i,t+\Delta t}+U_{i+1,t+\Delta t}$$
$$+U_{i-1,t}+4U_{i,t}+U_{i+1,t}]=0 \qquad (7.35)$$

Now let

$$U=U^0\exp(\hat{i}\beta' t+\hat{i}\sigma x) \qquad (7.36)$$

and

$$\zeta=\zeta^0\exp(\hat{i}\beta' t+\hat{i}\sigma x) \qquad (7.37)$$

where $\hat{i}=(-1)^{1/2}$, and $\beta'$ is the numerical approximation to $\beta$. Equations (7.34) and (7.35) reduce to

$$\frac{1}{6\,\Delta t}[\exp(\hat{i}\beta'\,\Delta t)-1][\exp(-\hat{i}\sigma\,\Delta x)+4+\exp(\hat{i}\sigma\,\Delta x)]\zeta^0$$
$$+\frac{h}{4\,\Delta x}[\exp(\hat{i}\beta'\,\Delta t)+1][\exp(\hat{i}\sigma\,\Delta x)-\exp(-\hat{i}\sigma\,\Delta x)]U^0=0 \qquad (7.38)$$

$$\frac{1}{6\,\Delta t}[\exp(\hat{i}\beta'\,\Delta t)-1][\exp(-\hat{i}\sigma\,\Delta x)+4+\exp(\hat{i}\sigma\,\Delta x)]U^0$$
$$+\frac{g}{4\,\Delta x}[\exp(\hat{i}\beta'\,\Delta t)+1][\exp(\hat{i}\sigma\,\Delta x)-\exp(-\hat{i}\sigma\,\Delta x)]\zeta^0$$
$$+\frac{\kappa}{12}[\exp(\hat{i}\beta'\,\Delta t)+1][\exp(-\hat{i}\sigma\,\Delta x)+4+\exp(\hat{i}\sigma\,\Delta x)]U^0=0 \qquad (7.39)$$

Since there are two equations in the two coefficients $\zeta^0$ and $U^0$, they may be eliminated to obtain an expression for $\exp(-\hat{i}\beta' \Delta t)$ of the form

$$\exp(\hat{i}\beta' \Delta t) = \frac{1 - A_{FE} \pm \hat{i}[4A_{FE} - (\frac{1}{2}\kappa \Delta t)^2]^{1/2}}{1 + A_{FE} + \frac{1}{2}\kappa \Delta t} \tag{7.40}$$

where

$$A_{FE} = \frac{1}{4}\left\{\frac{3}{2 + \cos(\sigma \Delta x)}\right\}^2 \frac{\Delta t^2}{\Delta x^2} gh \sin^2(\sigma \Delta x) \tag{7.41}$$

Recall that $\exp(\hat{i}\beta' \Delta t)$ is also $\lambda$, the eigenvalue of the amplification matrix. For convenience in determining the stability of the scheme, the amplification factor, and the phase shift, it will be convenient to rearrange (7.40) into the form

$$\exp(\hat{i}\beta' \Delta t) = |\lambda| \exp(\hat{i}\theta) \tag{7.42}$$

The magnitude of $\lambda$ can be obtained from Eq. (7.40) as

$$|\lambda| = \frac{\{(1 - A_{FE})^2 + [4A_{FE} - (\frac{1}{4} \Delta t^2 \kappa^2)]\}^{1/2}}{1 + A_{FE} + \frac{1}{2} \Delta t \kappa} \tag{7.43}$$

or

$$|\lambda| = \left(\frac{1 + A_{FE} - \frac{1}{2} \Delta t \kappa}{1 + A_{FE} + \frac{1}{2} \Delta t \kappa}\right)^{1/2} \tag{7.44}$$

The value for $\theta$ in (7.42) can be derived by making use of the identity

$$\exp(\hat{i}\theta) = \cos \theta + \hat{i} \sin \theta \tag{7.45}$$

Therefore

$$|\lambda| \exp(\hat{i}\theta) = \exp(\hat{i}\beta' \Delta t) = |\lambda| \cos \theta + \hat{i} |\lambda| \sin \theta \tag{7.46}$$

By equating the real part of Eq. (7.40) to $|\lambda| \cos \theta$ we obtain the expression for $\theta$:

$$\left[\frac{1 - A_{FE}}{1 + A_{FE} + \frac{1}{2} \Delta t \kappa}\right] = \left[\frac{1 + A_{FE} - \frac{1}{2} \Delta t \kappa}{1 + A_{FE} + \frac{1}{2} \Delta t \kappa}\right]^{1/2} \cos \theta \tag{7.47}$$

or

$$\theta = \cos^{-1}\left[\frac{1 - A_{FE}}{[(1 + A_{FE} + \frac{1}{2} \Delta t \kappa)(1 + A_{FE} - \frac{1}{2} \Delta t \kappa)]^{1/2}}\right] \tag{7.48}$$

Substitution of the expression for $\exp(\hat{i}\beta' \Delta t)$ back into (7.25) yields

$$T(\sigma \Delta x) = \left\{ \left[ \frac{1 + A_{FE} - \frac{1}{2} \Delta t \kappa}{1 + A_{FE} + \frac{1}{2} \Delta t \kappa} \right]^{1/2} \exp(\tfrac{1}{2}\kappa \Delta t) \right\}^N \times \exp\left\{ \hat{i} N \cos^{-1} \left[ \frac{1 - A_{FE}}{[(1 + A_{FE} + \frac{1}{2} \Delta t \kappa)(1 + A_{FE} - \frac{1}{2} \Delta t \kappa)]^{1/2}} \right] \right\} \tag{7.49}$$

From this expression the modulus of $T(\sigma \Delta x)$ may be obtained as

$$|T(\sigma \Delta x)| = \left\{ \left[ \frac{1 + A_{FE} - \frac{1}{2} \Delta t \kappa}{1 + A_{FE} + \frac{1}{2} \Delta t \kappa} \right] \exp(\kappa \Delta t) \right\}^{N/2} \tag{7.50}$$

Recall that this ratio is the relative magnitude of the computed wave to the physical wave after the amount of time required by the physical wave to propagate one wavelength. The argument of $T(\sigma \Delta x)$ is

$$\arg[T(\sigma \Delta x)] = N \cos^{-1} \left[ \frac{1 - A_{FE}}{[(1 + A_{FE} + \frac{1}{2} \Delta t \kappa)(1 + A_{FE} - \frac{1}{2} \Delta t \kappa)]^{1/2}} \right] \tag{7.51}$$

This term is the ratio of the celerity of the numerical wave to the celerity of the physical wave. The phase lag in degrees is obtained from this expression as

$$\text{phase lag} = 360^\circ \{\arg[T(\sigma \Delta x)] - 2\pi\}/2\pi \tag{7.52}$$

If this number is negative, the computed wave propagates slower than the real wave; if the number is positive, the numerical wave propagates faster than the physical wave.

### *7.5.1 Central Implicit Scheme without Friction*

As a first example, consider the case when there is no frictional dissipation, i.e., $\kappa = 0$. From Eq. (7.44) the magnitude of the eigenvalue of the amplification matrix is 1, and the scheme is neutrally stable. Equation (7.49) indicates that the modulus of $T(\sigma \Delta x)$ is 1 so the magnitude of the computed wave will be equal to the magnitude of the physical wave regardless of the spatial and temporal grid step sizes. The solid lines in Fig. 7.1 indicate the phase angle of the propagation factor for various ratios of time step to spatial step as a function of spatial grid discretization. The abscissa $L/\Delta x$ is the number of grid steps per wavelength and is a measure of the coarseness of the grid. The numerically computed wave celerity is seen to be always less than the physical wave celerity, although as the coarseness of the grid decreases, the phase lag of the numerical wave becomes very small. It can also be seen that as the ratio of the time step to the spatial step increases, the ability of the method to propagate the smaller wavelengths deteriorates.

For comparison purposes, the central implicit scheme was also examined when used with the finite difference method. The basic difference between the equations generated and the finite element equations, (7.34) and (7.35), is that the time derivatives at $i + 1$ and $i - 1$ do not appear but are replaced by time derivatives at $i$, and the velocities in the dissipation term are not evaluated at $i + 1$ and $i - 1$ but only at $i$. In other words, the averaging in the spatial domain introduced by the finite element method does not arise in the finite difference method.

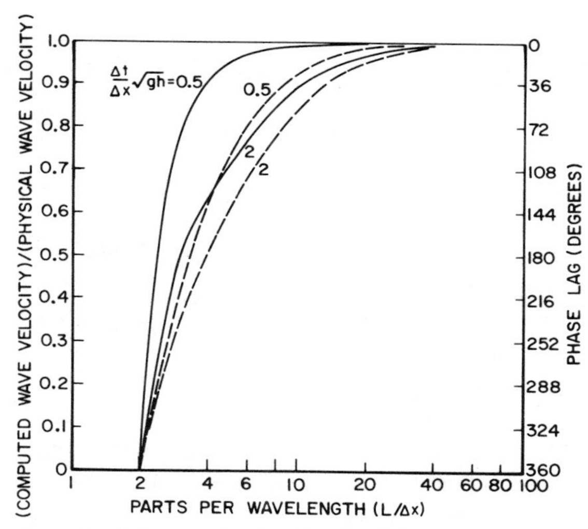


**Fig. 7.1.** Phase angle of the propagation factor of the central implicit method with no frictional dissipation for finite element (——) and finite difference (- - -) methods.

Leendertse [1] has examined the propagation factor for this finite difference scheme and the expression obtained for $T(\sigma\,\Delta x)$ will be identical to Eq. (7.49) if we replace $A_{\mathrm{FE}}$ by $A_{\mathrm{FD}}$ where

$$A_{\mathrm{FD}} = \tfrac{1}{4}(\Delta t^2/\Delta x^2)gh\,\sin^2(\sigma\,\Delta x) \tag{7.53}$$

Comparison with Eq. (7.41) indicates that

$$A_{\mathrm{FE}} = A_{\mathrm{FD}}\left(\frac{3}{2 + \cos(\sigma\,\Delta x)}\right)^2 \tag{7.54}$$

Therefore, in the limit where $\cos(\sigma\,\Delta x) \simeq 1$ (this is the case within a percent for $L/\Delta x > 25$), the finite element and finite difference methods propagate the waves identically.

For this finite difference procedure, $|\lambda|$ and $|T(\sigma \Delta x)|$ are both equal to 1 as with the finite element method. The plots of the phase angle of the propagation factor, indicated by the dashed lines in Fig. 7.1 show that when the number of mesh spaces per wavelength is large, the finite difference method propagates the wave as well as the finite element method. However, the smaller wavelengths are propagated much better by the finite element method than by the finite difference method. This indicates that it may be possible to use a coarser grid with the finite elements than with the finite differences to obtain wave celerity of equal accuracy.

In summary, the central implicit method with no dissipation has been shown to be neutrally stable and consistent in that, as the mesh is refined (or the ratio of $L/\Delta x$ becomes large), the numerical solution approaches the analytical solution.

### *7.5.2 Central Implicit Scheme with Friction*

From Eq. (7.44) it can be seen that when friction is included in the differential equations, $\kappa > 0$, the magnitude of the eigenvalue of the amplification matrix will be less than 1. Therefore, the central implicit scheme is unconditionally stable. This is true both for the finite element and the finite difference methods.

The modulus of the propagation factor, given in Eq. (7.50), is not equal to 1 when $\kappa > 0$. Figure 7.2 is a plot of $|T(\sigma \Delta x)|$ versus the number of grid steps per wavelength for a dimensionless dissipation constant of 0.4. Values for the finite difference plots were obtained from (7.50) with $A_{FD}$ defined in (7.53) substituted in for $A_{FE}$. From this figure it can be seen that in both the finite element and finite difference methods the magnitude of the computed wave is always greater than the magnitude of the physical wave. In other words, as computations proceed through time, the numerical wave decreases in amplitude more slowly than the physical wave. This discrepancy is aggravated as the $\Delta t/\Delta x$ ratio increases. Furthermore, the finite difference method is seen to represent the wave amplitude more accurately than the finite element method. As the $L/\Delta x$ ratio increases, both methods model the physical system quite well.

The ratio of the celerity of the numerical wave to the celerity of the physical wave obtained from Eq. (7.51) is plotted in Fig. 7.3. As was the case when dissipation was neglected, the finite element method propagates the wave more accurately than the finite difference method. A comparison of Figs. 7.1 and 7.3 indicates that the frictional dissipation has only a very

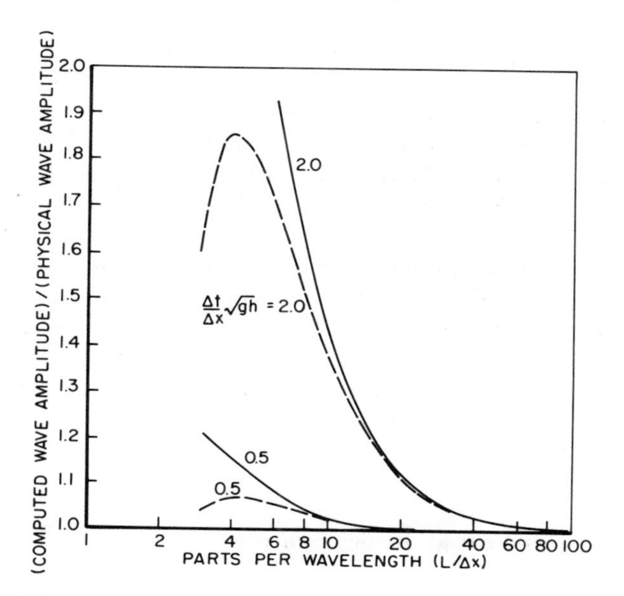


**Fig. 7.2.** Modulus of the propagation factor of the central implicit method with frictional dissipation for finite element (——) and finite difference (- - -) methods. The constant $\kappa/\sigma(gh)^{1/2} = 0.4$.

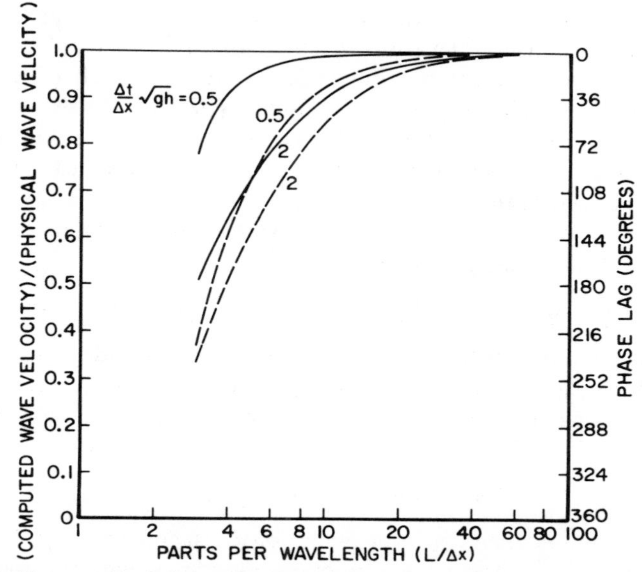


**Fig. 7.3.** Phase angle of the propagation factor of the central implicit method with frictional dissipation for finite element (——) and finite difference (- - -) methods. The constant $\kappa/\sigma(gh)^{1/2} = 0.4$.

small effect on the phase shift of the wave. In fact with $\Delta t(gh)^{1/2}/\Delta x = 0.5$, the difference in phase shift between the two plots is less than 0.5° for $L/\Delta x = 5$; and with $\Delta t(gh)^{1/2}/\Delta x = 2.0$, the difference is on the order of 5° with $L/\Delta x = 5$. As $L/\Delta x$ becomes large, the importance of frictional dissipation to phase shift is negligible. Interestingly, the phase lag is less with dissipation present than when there is no dissipation.

The central implicit scheme has been shown to be unconditionally stable. Although the finite element method gives a more accurate value of the wave celerity at small wavelengths, the finite difference method gives a more accurate value of the wave amplitude. Waves of large wavelength with respect to the grid spacing are modeled equally well by the two numerical methods. The central implicit scheme is consistent with the partial differential equation in that as $L/\Delta x$ becomes large and $\Delta t(gh)^{1/2}/\Delta x$ becomes small, the numerical solution approaches the analytical solution.

## 7.6 Higher-Order Central Implicit Scheme

When applying the finite element method along with a particular scheme for marching through time, it is possible to specify a variation other than linear between spatial nodes. In the example to be considered here, the velocity will be assumed to vary quadratically between the end nodes of the element, but the surface elevation will be allowed to vary only linearly.

Consider two adjacent one-dimensional elements each of length 2 $\Delta x$ with midelement nodes as well as end nodes. The nodes at the endpoints of the elements are important for the description of both the velocity function and the surface elevation function. The two midelement nodes are important only for the description of the velocity. If the five nodes are numbered consecutively from $j - 2$ to $j + 2$, in the two-element region under consideration, the velocity field will be described by

$$U = U_{j-2}\mu_{j-2} + U_{j-1}\mu_{j-1} + U_j\mu_j + U_{j+1}\mu_{j+1} + U_{j+2}\mu_{j+2} \quad (7.55)$$

where $U_j$ is the time-dependent approximation to $U$ at node $j$, and the $\mu_j$s are the one-dimensional quadratic Lagrangian basis functions. On the other hand, the linear surface elevation field will require only a three-term expansion

$$\zeta = \zeta_{j-2}\phi_{j-2} + \zeta_j\phi_j + \zeta_{j+2}\phi_{j+2} \quad (7.56)$$

where $\zeta_j$ is the time-dependent approximation to $\zeta$ at end node $j$, and the $\phi_j$s are the one-dimensional linear basis functions associated with the end nodes of the elements. Note that since the expansion for $\zeta$ is linear, there are no basis functions associated with the midelement nodes $j - 1$ and $j + 1$.

In the two-element region under consideration, there are four coefficients associated with basis functions which are zero in all other elements: $\zeta_j$, $U_{j-1}$, $U_j$, and $U_{j+1}$. To obtain a sufficient number of equations on which to perform the analysis of the behavior of the numerical method, four weighted residual equations must be generated. These are obtained by requiring continuity equation (7.26) to be orthogonal to $\phi_j$ and momentum equation (7.27) to be orthogonal to $\mu_{j-1}$, $\mu_j$, and $\mu_{j+1}$. An obvious alternative to this would be to require the momentum equation to be orthogonal to $\phi_j$. When this latter method is used, however, the time derivative integrates out of the system and the equations that result admit only a trivial solution. The former method can be rationalized by thinking of the continuity equation as the equation for $\zeta$ and the momentum equation as the equation for $U$. Then Galerkin's method requires that the equation for $\zeta$ be orthogonal to the basis functions used in the expansion for $\zeta$ and the equation for $U$ be orthogonal to the basis functions used in the expansion for $U$.

Application of the orthogonality requirements and evaluation of the integral equation yields, from the continuity equation,

$$\frac{1}{6}\left(\frac{d\zeta_{j-2}}{dt} + 4\frac{d\zeta_j}{dt} + \frac{d\zeta_{j+2}}{dt}\right) + \frac{h}{3}\left(\frac{U_{j+2} - U_{j-2}}{4\,\Delta x} + 2\frac{U_{j+1} - U_{j-1}}{2\,\Delta x}\right) = 0 \tag{7.57}$$

from the momentum equation weighted with respect to $\mu_{j-1}$,

$$\frac{1}{10}\left(\frac{dU_{j-2}}{dt} + 8\frac{dU_{j-1}}{dt} + \frac{dU_j}{dt}\right) + g\left(\frac{\zeta_j - \zeta_{j-2}}{2\,\Delta x}\right) + \frac{\kappa}{10}(U_{j-2} + 8U_{j-1} + U_j) = 0 \tag{7.58}$$

from the momentum equation weighted with respect to $\mu_j$,

$$\frac{1}{10}\left(-\frac{dU_{j-2}}{dt} + 2\frac{dU_{j-1}}{dt} + 8\frac{dU_j}{dt} + 2\frac{dU_{j+1}}{dt} - \frac{dU_{j+2}}{dt}\right) + g\left(\frac{\zeta_{j+2} - \zeta_{j-2}}{4\,\Delta x}\right) + \frac{\kappa}{10}(-U_{j-2} + 2U_{j-1} + 8U_j + 2U_{j+1} - U_{j+2}) = 0 \tag{7.59}$$

and from the momentum equation weighted with respect to $\mu_{j+1}$,

$$\frac{1}{10}\left(\frac{dU_j}{dt} + 8\frac{dU_{j+1}}{dt} + \frac{dU_{j+2}}{dt}\right) + g\left(\frac{\zeta_{j+2} - \zeta_j}{2\,\Delta x}\right) + \frac{\kappa}{10}(U_j + 8U_{j+1} + U_{j+2}) \tag{7.60}$$

In performing a Fourier analysis here, one assumes that the solution in each region is obtained in an identical manner to the solution in every other region. For the two-element region under consideration, the assumption is valid. Therefore, Eqs. (7.57)–(7.60) can be used to study the behavior of the numerical scheme. Note, however, that within this region $U_j$s at midelement nodes are in different surroundings from values of $U_j$s at end nodes. Consequently these two types of $U_j$ are calculated from different numerical approximations to the differential equations. In other words, even though each two-element region is the same as every other two-element region, within each region different equations are applied to obtain solutions for the same function. This local heterogeneity can be exploited to simplify the algebraic manipulation necessary to examine the properties of the numerical schemes, although it is not necessary to do so.

In Eqs. (7.57)–(7.60), $U_{j+1}$ and $U_{j-1}$ are in different environments from $U_{j-2}$, $U_j$, and $U_{j+2}$, but in similar environments to each other. For convenience designate $U_{j-1}$ and $U_{j+1}$ by $W_{j-1}$ and $W_{j+1}$, respectively, to emphasize this difference. Replacement of the continuous time derivatives by finite difference approximations, averaging the spatial derivatives and friction terms over the two time levels, and substitution of

$$U = U^0 \exp(\hat{i}\beta' t + \hat{i}\sigma x) \tag{7.61a}$$

$$\zeta = \zeta^0 \exp(\hat{i}\beta' t + \hat{i}\sigma x) \tag{7.61b}$$

$$W = W^0 \exp(\hat{i}\beta' t + \hat{i}\sigma x) \tag{7.61c}$$

into (7.57), (7.58) or (7.60), and (7.59), yields, respectively,

$$\frac{1}{6\,\Delta t}[\exp(\hat{i}\beta'\,\Delta t) - 1][2\cos(2\sigma\,\Delta x) + 4]\zeta^0 + \frac{h}{6\,\Delta x}[\exp(\hat{i}\beta'\,\Delta t) + 1]\left[\frac{U^0\hat{i}\sin(2\sigma\,\Delta x)}{2} + 2W^0\hat{i}\sin(\sigma\,\Delta x)\right] = 0 \tag{7.62}$$

$$\frac{1}{10\,\Delta t}[\exp(\hat{i}\beta'\,\Delta t) - 1][2U^0\cos(\sigma\,\Delta x) + 8W^0] + \frac{g}{2\,\Delta x}[\exp(\hat{i}\beta'\,\Delta t) + 1][\hat{i}\sin(\sigma\,\Delta x)]\zeta^0 + \frac{\kappa}{20}[\exp(\hat{i}\beta'\,\Delta t) + 1][2U^0\cos(\sigma\,\Delta x) + 8W^0] = 0 \tag{7.63}$$

$$\frac{1}{10\,\Delta t}[\exp(\hat{i}\beta'\,\Delta t) - 1][-2\cos(2\sigma\,\Delta x) + 8]U^0 + [4\cos(\sigma\,\Delta x)]W^0$$

$$+ \frac{g}{4\,\Delta x}[\exp(\hat{i}\beta'\,\Delta t) + 1][\hat{i}\sin(2\sigma\,\Delta x)]\zeta^0$$

$$+ \frac{\kappa}{20}[\exp(\hat{i}\beta'\,\Delta t) + 1]\{[-2\cos(2\sigma\,\Delta x) + 8]U^0$$

$$+ [4\cos(\sigma\,\Delta x)]W^0\} = 0 \qquad (7.64)$$

These three equations can be solved simultaneously to eliminate the three parameters $\zeta^0$, $U^0$, and $W^0$ and obtain an expression for $\exp(i\beta'\,\Delta t)$. Because the algebra involved is quite tedious, only the result will be presented here:

$$\exp(i\beta'\,\Delta t) = \frac{1 - A_{HO} \pm \hat{i}(4A_{HO} - \frac{1}{4}\,\Delta t^2\kappa^2)^{1/2}}{1 + A_{HO} + \frac{1}{2}\,\Delta t\kappa} \qquad (7.65)$$

where

$$A_{HO} = \frac{1}{4}\left\{\frac{8 - 2\cos(2\sigma\,\Delta x)}{[3 - \cos(2\sigma\,\Delta x)][\cos(2\sigma\,\Delta x) + 2]}\right\}\frac{\Delta t^2 gh}{\Delta x^2}\sin^2(\sigma\,\Delta x) \qquad (7.66)$$

Note that these equations are quite similar to (7.40) and (7.41) obtained for a linear grid. Moreover the equation for $|T(\sigma\,\Delta x)|$ and for the phase lag are the same as (7.50)–(7.52) with $A_{HO}$ replacing $A_{FE}$. The magnitude of the eigenvalues of the amplification matrix are given in the same way by (7.44). Since $|\lambda|$ will always be less than or equal to 1, the higher-order central implicit scheme will always be stable.

### *7.6.1 Higher-Order Scheme without Friction*

In the absence of frictional losses, $|T(\sigma\,\Delta x)|$ will be equal to 1, so the computed wave and the physical wave will have the same magnitude. The plot of the phase shift for this case (Fig. 7.4) reveals a very interesting phenomenon. If the time derivative is treated as a continuous function, the computed wave velocity is greater than the physical wave velocity for $L/\Delta x > 3$. Instead of a phase lag, the numerical solution yields a phase lead. With the continuous time derivative, the celerity ratio approaches 1 from above as $L/\Delta x$ increases. When $\Delta t(gh)^{1/2}/\Delta x = 0.5$, the celerity ratio becomes greater than 1 as $L/\Delta x$ increases from 2, and then less than 1. In the limit of large $L/\Delta x$, the celerity ratio approaches 1 from below. For a value of $\Delta t(gh)^{1/2}/\Delta x = 2.0$, the celerity ratio is always less than 1. In this case a numerical wave never propagates faster than the physical wave.

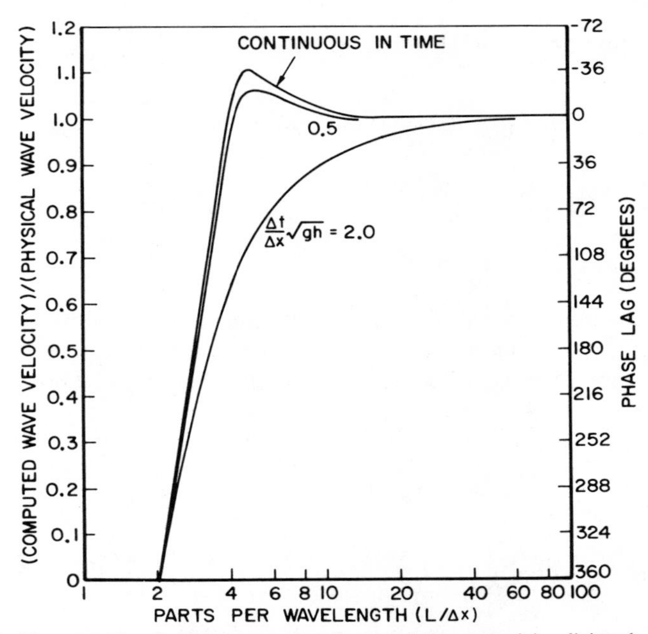


**Fig. 7.4.** Phase angle of the propagation factor of the central implicit scheme using a higher-order finite element method with no frictional dissipation.

### *7.6.2 Higher-Order Scheme with Friction*

An analysis was also performed with friction included. The modulus of the propagation factors is plotted in Fig. 7.5 for $\kappa/\sigma(gh)^{1/2} = 0.4$. Comparison with Fig. 7.2 indicates that $|T(\sigma \Delta x)|$ is larger for the higher-order elements than for the linear elements. In both instances, the numerical wave has a larger amplitude than the physical wave. As with the linear case, the phase shift (Fig. 7.6) is only slightly affected by the inclusion of dissipation in the analysis.

### *7.6.3 Conclusions*

The behavior of the higher-order scheme has been shown to be generally similar to the linear scheme. Considering only the evidence presented here, one might opt for the linear scheme because the matrices that will be generated will have a smaller bandwidth and can be solved on the computer with less work. However, the accuracy of the higher-order scheme is expected to deteriorate less rapidly than that of the linear scheme as the grid becomes deformed.

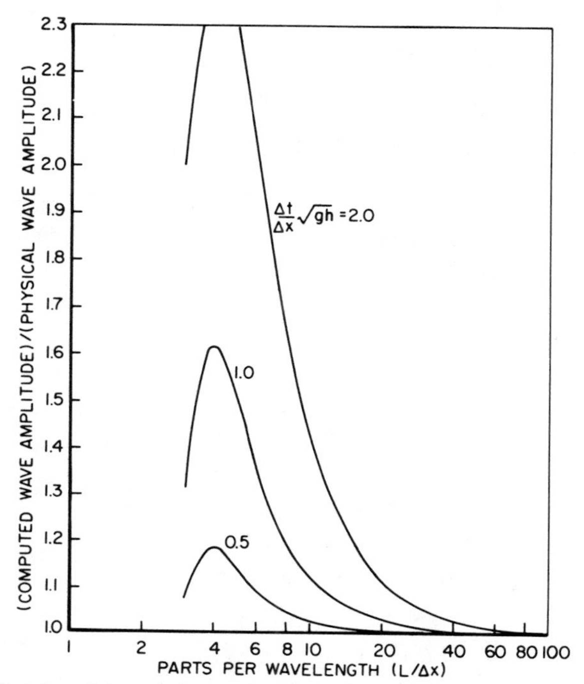


**Fig. 7.5.** Modulus of the propagation factor of the central implicit scheme using a higher-order finite element method with frictional dissipation included. The constant $\kappa/\sigma(gh)^{1/2} = 0.4$.

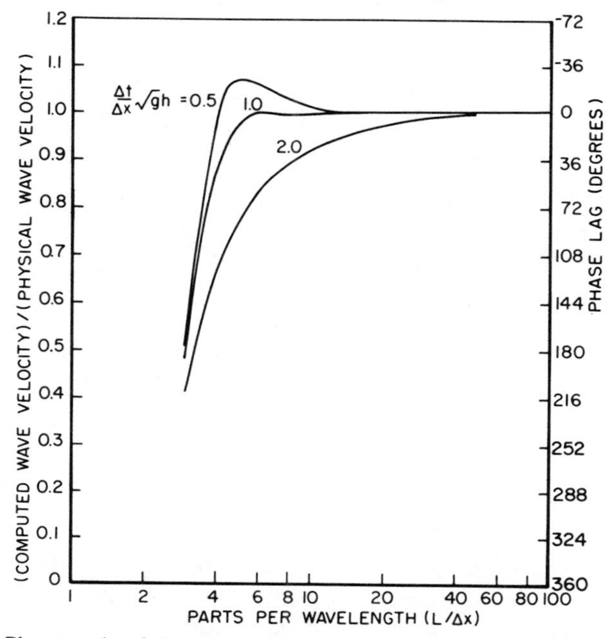


**Fig. 7.6.** Phase angle of the propagation factor of the central implicit scheme using a higher-order finite element method with frictional dissipation included. The constant $\kappa/\sigma(gh)^{1/2} = 0.4$.

## 7.7 Alternative Time Stepping Scheme

In the discussion of lake models in Section 6.5 a time stepping scheme was introduced whereby, instead of finite differencing the time derivative, the following expansion was used:

$$\frac{dU_{j,t+\Delta t}}{dt} + \frac{dU_{j,t}}{dt} = \frac{2}{\Delta t}(U_{j,t+\Delta t} - U_{j,t}) \tag{7.67}$$

with

$$U = U^0 \exp(\hat{i}\beta' t + \hat{i}\sigma x) \tag{7.68a}$$

and

$$dU/dt = (dU^0/dt)\exp(\hat{i}\beta' t + \hat{i}\sigma x) \tag{7.68b}$$

where $(dU^0/dt)$ is simply a constant. Substitution of (7.68b) into the left side of (7.67) and (7.68a) into the right side yields

$$\frac{dU^0}{dt} = \frac{2}{\Delta t}\frac{\exp(\hat{i}\beta' \Delta t) - 1}{\exp(\hat{i}\beta' \Delta t) + 1} U^0 \tag{7.69}$$

so that

$$\frac{dU}{dt} = \frac{2}{\Delta t}\frac{\exp(\hat{i}\beta' \Delta t) - 1}{\exp(\hat{i}\beta' \Delta t) + 1} U^0 \exp(\hat{i}\beta' \Delta t + \hat{i}\sigma x) \tag{7.70}$$

In like manner it can be shown that

$$\frac{d\zeta}{dt} = \frac{2}{\Delta t}\frac{\exp(\hat{i}\beta' \Delta t) - 1}{\exp(\hat{i}\beta' \Delta t) + 1} \zeta^0 \exp(\hat{i}\beta' \Delta t + \hat{i}\sigma x) \tag{7.71}$$

If we consider the case where all basis functions are linear, the finite element integrations of the continuity and momentum equations yield Eqs. (7.32) and (7.33). Substitution of (7.70) and (7.71) into these equations along with expansions for $U$ and $\zeta$ yields Eqs. (7.38) and (7.39). Thus the time stepping scheme of Eqs. (7.67) using the first derivative in time is identical to the central implicit scheme. Although developed in the context of linear basis functions in space, this conclusion is independent of the type of basis functions chosen in the spatial domain.

## 7.8 Uncentered Implicit Scheme

The amplitude ratios in Figs. 7.2 and 7.5 indicate that the inclusion of friction in the numerical scheme does not decrease the amplitude of the computed wave as quickly as it does the amplitude of the physical wave. One might reason that if a scheme can be found which yields values of $|T(\sigma\,\Delta x)|$ which are too small in the absence of friction, $|T(\sigma\,\Delta x)|$ might be closer to 1 when friction is included in the problem. It is often the case that the more implicit a numerical scheme is, the smaller the ratio of a computed wave to a physical wave. Keeping these arguments in mind and considering a linear finite element grid, consider an uncentered averaging of the approximation of $g\,\partial\zeta/\partial x$ in (7.35) and a centered averaging of the continuity equation (7.34). Therefore, with friction neglected Eq. (7.35) takes the form

$$\frac{1}{6}\left[\frac{U_{i-1,t+\Delta t}-U_{i-1,t}}{\Delta t}+4\frac{U_{i,t+\Delta t}-U_{i,t}}{\Delta t}+\frac{U_{i+1,t+\Delta t}-U_{i+1,t}}{\Delta t}\right]$$

$$+g\left[\alpha\frac{\zeta_{i+1,t+\Delta t}-\zeta_{i-1,t+\Delta t}}{2\,\Delta x}+(1-\alpha)\frac{\zeta_{i+1,t}-\zeta_{i-1,t}}{2\,\Delta x}\right]=0 \qquad (7.72)$$

where $\alpha$ is a measure of the implicit nature of the equation with $\alpha = 0.5$ corresponding to the centered implicit case and the equation becoming more implicit as $\alpha$ increases. Substitution of (7.36) and (7.37) into this equation yields

$$\frac{1}{6\,\Delta t}[4+2\cos(\sigma\,\Delta x)][\exp(\hat{i}\beta'\,\Delta t)-1]U^0$$

$$+\frac{g}{2\,\Delta x}[\alpha\exp(\hat{i}\beta'\,\Delta t)+1-\alpha][2\hat{i}\sin(\sigma\,\Delta x)]\zeta^0=0 \qquad (7.73)$$

This equation can be solved simultaneously with (7.38), the centered continuity equation representation, to obtain

$$\exp(\hat{i}\beta'\,\Delta t)=\frac{1-A_{\mathrm{FE}}\pm\hat{i}[4A_{\mathrm{FE}}-A_{\mathrm{FE}}^2(2\alpha-1)]^{1/2}}{1+2\alpha A_{\mathrm{FE}}} \qquad (7.74)$$

where $A_{\mathrm{FE}}$ is given by

$$A_{\mathrm{FE}}=\frac{1}{4}\left[\frac{3}{2+\cos(\sigma\,\Delta x)}\right]^2\frac{\Delta t^2}{\Delta x^2}gh\sin^2(\sigma\,\Delta x) \qquad (7.75)$$

The magnitude of the eigenvalue of the amplification matrix is the magnitude of the right side of (7.74) or

$$|\lambda|=\left[\frac{1+2A_{\mathrm{FE}}+2(1-\alpha)A_{\mathrm{FE}}^2}{1+4\alpha A_{\mathrm{FE}}+4\alpha^2A_{\mathrm{FE}}^2}\right]^{1/2} \qquad (7.76)$$

For the solution scheme to be stable $|\lambda|\leq 1$ or

$$1+4\alpha A_{\mathrm{FE}}+4\alpha^2A_{\mathrm{FE}}^2\geq 1+2A_{\mathrm{FE}}+2(1-\alpha)A^2{}_{\mathrm{FE}} \qquad (7.77)$$

which can be rearranged to show that

$$0 \leq A_{\mathrm{FE}}(2\alpha - 1)[1 + A_{\mathrm{FE}}(\alpha + 1)] \tag{7.78}$$

Because $A_{\mathrm{FE}}$ is always greater than or equal to 0, the scheme will be stable if $\alpha \geq 0.5$ and unstable for $\alpha < 0.5$.

The modulus of the propagation factor is plotted in Fig. 7.7 for various values of $\alpha$ with $\Delta t(gh)^{1/2}/\Delta x = 0.5$. The behavior hoped for in this scheme is

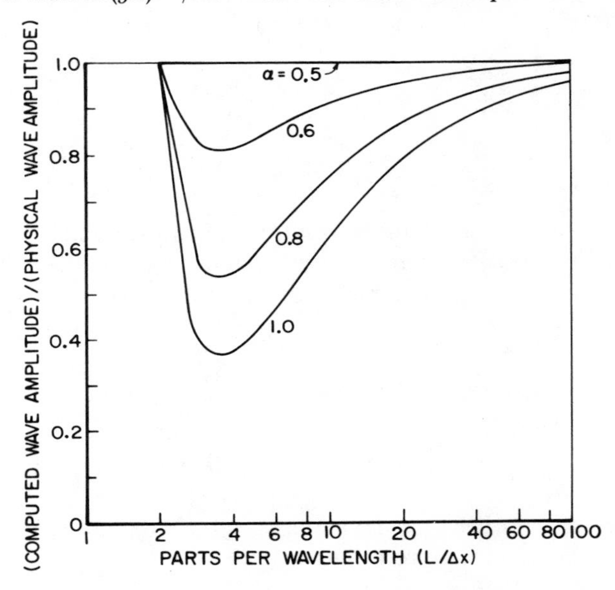


**Fig. 7.7.** Modulus of the propagation factor of the uncentered implicit scheme with no frictional dissipation and $\Delta t(gh)^{1/2}/\Delta x = 0.5$.

evident. The small wavelengths are damped and in the limit as $L/\Delta x$ becomes large, the long waves are not damped. Therefore the increase in $|T(\sigma\,\Delta x)|$ when friction is present would be expected to improve the solution.

In Fig. 7.8, the celerity ratio is plotted for various values of $\alpha$ with $\Delta t(gh)^{1/2}/\Delta x = 0.5$. As can be seen, in the limit of large $L/\Delta x$, when $\alpha > 0.5$, the numerical solution converges to a celerity greater than that of the physical wave and produces a phase lead. From additional numerical experiments it was found that the celerity ratio converged to is independent of $\Delta t(gh)^{1/2}/\Delta x$ and is a function only of $\alpha$. Thus we are forced to conclude that the uncentered implicit scheme is inconsistent with the partial differential equations and not suitable for wave propagation problems.

The conclusion reached in this exercise can not be extrapolated to the statement that all uncentered schemes are unsuitable for surface water analyses. The chief point to be made here is that a seemingly minor modification of a numerical scheme can give results inconsistent with the physical problem although the scheme may remain stable. A preliminary analysis of a numerical scheme before application to a real-world problem can shed a great deal of light on the chances of success for the scheme.

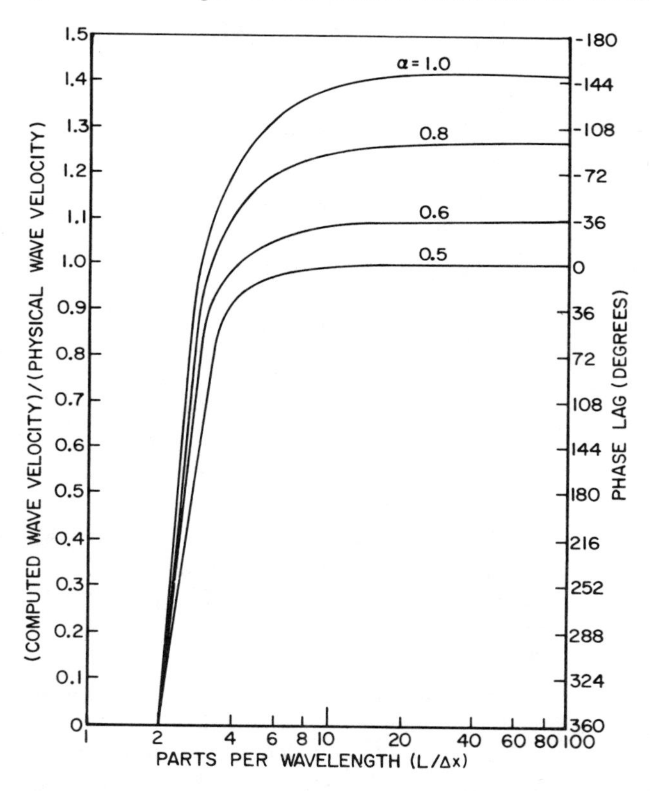


**Fig. 7.8.** Phase angle of the propagation factor of the uncentered implicit scheme with no frictional dissipation and $\Delta t(gh)^{1/2}/\Delta x = 0.5$.

## References


1. J. J. Leendertse, Aspects of a Computational Model for Long-Period Water-Wave Propagation. Rand Memorandum RM-5294-PR, Santa Monica, California (1967).
2. A. J. Sobey, Finite-Difference Schemes Compared for Wave-Deformation Characteristics in Mathematical Modeling of two-Dimensional Long-Wave Propagation. U.S. Army Corps of Eng., Coastal Eng. Res. Center Tech. Memo #32 (1970).
3. B. Carnahan, H. A. Luther, and J. O. Wilkes, "Applied Numerical Methods." Wiley, New York, 1969.
4. C. R. Wylie, Jr., "Advanced Engineering Mathematics." McGraw-Hill, New York, 1966.

# Chapter 8 | Estuaries and Coastal Regions

## 8.1 Scope of the Study

Although finite difference tidal models have been used for some time, tidal models that make use of the finite element method have been a relatively recent development. In this chapter, the equations required for modeling two-dimensional hydrodynamics and transport in estuaries will be developed. This foundation allows examination of the various ways in which these equations have been solved by the finite element method.

The type of estuary that will be considered is drawn in cross section in Fig. 8.1. Because the estuary is allowed to have arbitrary bathymetry, $h$, the distance from mean sea level to the bottom of the estuary, is a function of $x$ and $y$. Moreover, due to the tidal motion, the surface elevation from mean sea level $\zeta$ is a function of time as well as $x$ and $y$. The total distance from the bottom of the estuary to the surface is denoted by $H$ and is equal to $\zeta + h$. The equation for the surface of the estuary is denoted by $F_S(x, y, z, t) = 0$.

## 8.2 Basic Equations

The two-dimensional equations needed for estuary modeling have been carefully derived by Pritchard [1]. The development presented here follows along the same lines as his work. A set of equations appropriate for areal modeling of vertically well-mixed water bodies is obtained by averaging the three-dimensional equations over the depth. It is further assumed that the fluid density is a constant. The reader interested in the form of the equations which take into account density variation due to temperature or concentration gradients is referred to Pritchard's work. In the subsequent development of the equations, the following notation will be used: $\rho$ is the density, $u$, $v$, and $w$ are the velocity components in the $x$, $y$, and $z$ directions, respectively, $x$ is positive eastward, $y$ positive northward, $z$ positive upward, and $t$ is time.

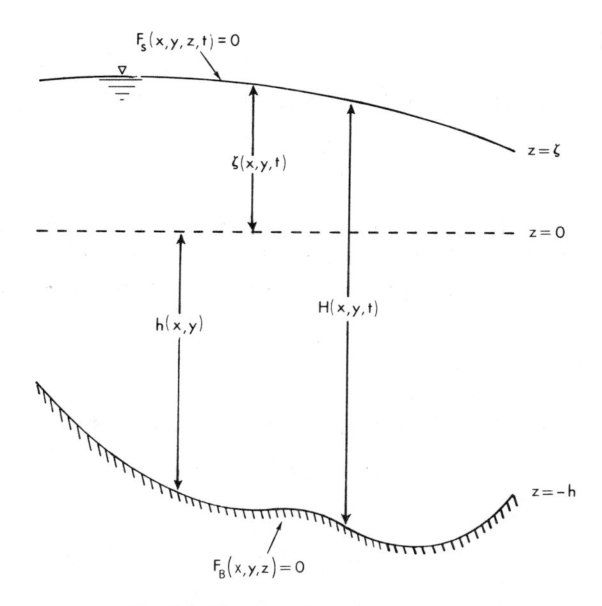


**Fig. 8.1.** Cross section of a typical estuary.

### *8.2.1 Boundary Conditions on Velocity*

When the three-dimensional flow equations are integrated over the depth, the need arises for conditions on the flow at the surface and bottom of the estuary. If the equation of a boundary is $F(x, y, z, t) = 0$, then at every point on the boundary we must have [2]

$$\frac{DF}{Dt} = \frac{\partial F}{\partial t} + u\frac{\partial F}{\partial x} + v\frac{\partial F}{\partial y} + w\frac{\partial F}{\partial z} = 0 \tag{8.1}$$

which indicates that $F$ remains the equation for the boundary through time. At the surface of the estuary,

$$F_S(x, y, z, t) = z - \zeta(x, y, t) \tag{8.2}$$

so Eq. (8.1) becomes

$$w_S - \frac{\partial \zeta}{\partial t} - u_S \frac{\partial \zeta}{\partial x} - v_S \frac{\partial \zeta}{\partial y} = 0 \tag{8.3}$$

where the subscript S refers to a velocity at the surface. At the bottom of the estuary

$$F_B(x, y, z) = z + h(x, y) \tag{8.4}$$

and is independent of time. Substitution into (8.1) yields

$$w_B + u_B \frac{\partial h}{\partial x} + v_B \frac{\partial h}{\partial y} = 0 \tag{8.5}$$

where the subscript B refers to a velocity at the bottom of the estuary.

Equations (8.3) and (8.5) are the conditions at the boundary needed for the integration of the flow equations to obtain their two-dimensional form.

### 8.2.2 *Continuity Equation*

The general equation for the conservation of mass is

$$\frac{\partial \rho}{\partial t} + \frac{\partial(\rho u)}{\partial x} + \frac{\partial(\rho v)}{\partial y} + \frac{\partial(\rho w)}{\partial z} = 0 \tag{8.6}$$

For an incompressible fluid, $D\rho/Dt = 0$ and (8.6) reduces to

$$\frac{\partial u}{\partial x} + \frac{\partial v}{\partial y} + \frac{\partial w}{\partial z} = 0 \tag{8.7}$$

This equation may be integrated over the depth to obtain

$$\int_{-h}^{\zeta} \frac{\partial u}{\partial x} dz + \int_{-h}^{\zeta} \frac{\partial v}{\partial y} dz + \int_{-h}^{\zeta} \frac{\partial w}{\partial z} dz = 0 \tag{8.8}$$

Application of Leibnitz' rule for interchanging the order of differentiation and integration in the first two integrals and direct evaluation of the third integral yield

$$\frac{\partial}{\partial x}\int_{-h}^{\zeta} u\, dz + \frac{\partial}{\partial y}\int_{-h}^{\zeta} v\, dz + \left(w_S - u_S \frac{\partial \zeta}{\partial x} - v_S \frac{\partial \zeta}{\partial y}\right)$$

$$- \left(w_B + u_B \frac{\partial h}{\partial x} + v_B \frac{\partial h}{\partial y}\right) = 0 \tag{8.9}$$

Boundary conditions (8.3) and (8.5) can be applied to this equation to reduce it to the form

$$\frac{\partial \zeta}{\partial t} + \frac{\partial}{\partial x}\int_{-h}^{\zeta} u\, dz + \frac{\partial}{\partial y}\int_{-h}^{\zeta} v\, dz = 0 \tag{8.10}$$

If average values of $u$ and $v$ are defined by

$$U = (\zeta + h)^{-1}\int_{-h}^{\zeta} u\, dz \tag{8.11a}$$

$$V = (\zeta + h)^{-1}\int_{-h}^{\zeta} v\, dz \tag{8.11b}$$

Eq. (8.10) becomes

$$\frac{\partial \zeta}{\partial t} + \frac{\partial (HU)}{\partial x} + \frac{\partial (HV)}{\partial y} = 0 \tag{8.12}$$

where use has been made of the fact that $H = \zeta + h$. Because $h$ is independent of time, this equation can alternatively be expressed as

$$\frac{\partial H}{\partial t} + \frac{\partial (HU)}{\partial x} + \frac{\partial (HV)}{\partial y} = 0 \tag{8.13}$$

Either one of these last two forms of the continuity equation is appropriate for use in an areal estuary model.

### *8.2.3 Momentum Equations*

For a constant density fluid, the equations of motion in the $x$, $y$, and $z$ directions are

$$\frac{\partial u}{\partial t} + \frac{\partial (uu)}{\partial x} + \frac{\partial (uv)}{\partial y} + \frac{\partial (uw)}{\partial z} - fv + \frac{1}{\rho}\frac{\partial p}{\partial x} - \frac{1}{\rho}\left(\frac{\partial \tau_{xx}}{\partial x} + \frac{\partial \tau_{xy}}{\partial y} + \frac{\partial \tau_{xz}}{\partial z}\right) = 0 \tag{8.14}$$

$$\frac{\partial v}{\partial t} + \frac{\partial (uv)}{\partial x} + \frac{\partial (vv)}{\partial y} + \frac{\partial (vw)}{\partial z} + fu + \frac{1}{\rho}\frac{\partial p}{\partial y} - \frac{1}{\rho}\left(\frac{\partial \tau_{yx}}{\partial x} + \frac{\partial \tau_{yy}}{\partial y} + \frac{\partial \tau_{yz}}{\partial z}\right) = 0 \tag{8.15}$$

$$\frac{\partial w}{\partial t} + \frac{\partial (uw)}{\partial x} + \frac{\partial (vw)}{\partial y} + \frac{\partial (ww)}{\partial z} + \frac{1}{\rho}\frac{\partial p}{\partial z} + g - \frac{1}{\rho}\left(\frac{\partial \tau_{zx}}{\partial x} + \frac{\partial \tau_{zy}}{\partial y} + \frac{\partial \tau_{zz}}{\partial z}\right) = 0 \tag{8.16}$$

where $p$ is pressure, $f$ the Coriolis parameter, $g$ gravity, and $\tau_{xx}$, $\tau_{xy}$, etc., are shear stresses.

These equations are unnecessarily general for use in modeling a vertically well-mixed estuary. Equation (8.16) can be simplified by assuming that the vertical accelerations are negligible and that the shear stresses are negligible compared to gravity and the vertical pressure gradient. These assumptions are equivalent to stating that the pressure in the $z$ direction is hydrostatic so (8.16) becomes

$$\frac{1}{\rho}\frac{\partial p}{\partial z} + g = 0 \tag{8.17}$$

If the pressure is assumed to be atmospheric at the surface, the solution to this equation is

$$p - p_{\mathrm{A}} = \rho g(\zeta - z) \tag{8.18}$$

where $p_{\mathrm{A}}$ is the atmospheric pressure. Substitution for $p$ into Eqs. (8.14) and (8.15) with $p_{\mathrm{A}}$ assumed constant yields

$$\frac{\partial u}{\partial t} + \frac{\partial(uu)}{\partial x} + \frac{\partial(uv)}{\partial y} + \frac{\partial(uw)}{\partial z} - fv + g\frac{\partial \zeta}{\partial x} - \frac{1}{\rho}\left(\frac{\partial \tau_{xx}}{\partial x} + \frac{\partial \tau_{xy}}{\partial y} + \frac{\partial \tau_{xz}}{\partial z}\right) = 0 \tag{8.19}$$

and

$$\frac{\partial v}{\partial t} + \frac{\partial(uv)}{\partial x} + \frac{\partial(vv)}{\partial y} + \frac{\partial(vw)}{\partial z} + fu + g\frac{\partial \zeta}{\partial y} - \frac{1}{\rho}\left(\frac{\partial \tau_{yx}}{\partial x} + \frac{\partial \tau_{yy}}{\partial y} + \frac{\partial \tau_{yz}}{\partial z}\right) = 0 \tag{8.20}$$

The vertically averaged forms of these equations are obtained by a series of virtually identical steps. For the sake of brevity, the specific arguments will be applied only to Eq. (8.19) with the integrated form of (8.20) deduced by analogy. Integration of (8.19) over $z$ yields

$$\int_{-h}^{\zeta} \frac{\partial u}{\partial t}\,dz + \int_{-h}^{\zeta} \frac{\partial(uu)}{\partial x}\,dz + \int_{-h}^{\zeta} \frac{\partial(uv)}{\partial y}\,dz + \int_{-h}^{\zeta} \frac{\partial(uw)}{\partial z}\,dz - \int_{-h}^{\zeta} fv\,dz + g\int_{-h}^{\zeta} \frac{\partial \zeta}{\partial x}\,dz - \int_{-h}^{\zeta} \frac{1}{\rho}\frac{\partial \tau_{xx}}{\partial x}\,dz - \int_{-h}^{\zeta} \frac{1}{\rho}\frac{\partial \tau_{xy}}{\partial y}\,dz - \frac{1}{\rho}\int_{-h}^{\zeta} \frac{\partial \tau_{xz}}{\partial z}\,dz = 0 \tag{8.21}$$

With $\rho$ assumed constant, application of Leibnitz' rule to the first three and the seventh and eighth integrals, and integration of the derivatives with respect to $z$ directly, yields

$$\frac{\partial}{\partial t}\int_{-h}^{\zeta} u\,dz + \frac{\partial}{\partial x}\int_{-h}^{\zeta} uu\,dz + \frac{\partial}{\partial y}\int_{-h}^{\zeta} uv\,dz - \int_{-h}^{\zeta} fv\,dz$$
$$+ gH\frac{\partial \zeta}{\partial x} - \frac{1}{\rho}\tau_{xz}\Big|_{\zeta} + \frac{1}{\rho}\tau_{xz}\Big|_{-h} - \frac{1}{\rho}\frac{\partial}{\partial x}\int_{-h}^{\zeta}\tau_{xx}\,dz$$
$$-\frac{1}{\rho}\frac{\partial}{\partial y}\int_{-h}^{\zeta}\tau_{xy}\,dz + u_S\left(-\frac{\partial \zeta}{\partial t} - u_S\frac{\partial \zeta}{\partial x} - v_S\frac{\partial \zeta}{\partial y} + w_S\right)$$
$$- u_B\left(u_B\frac{\partial h}{\partial x} + v_B\frac{\partial h}{\partial y} + w_B\right) + \frac{1}{\rho}\tau_{xx}\Big|_{\zeta}\frac{\partial \zeta}{\partial x} + \frac{1}{\rho}\tau_{xx}\Big|_{-h}\frac{\partial h}{\partial x}$$
$$+ \frac{1}{\rho}\tau_{xy}\Big|_{\zeta}\frac{\partial \zeta}{\partial y} + \frac{1}{\rho}\tau_{xy}\Big|_{-h}\frac{\partial h}{\partial y} = 0 \qquad (8.22)$$

The term in this equation premultiplied by $u_S$ is zero by condition (8.3) and the term premultiplied by $u_B$ is zero from Eq. (8.5). The last four terms in this equation deal with horizontal transport of momentum across the sloping surface or bottom of the estuary and are assumed neglible.

We can relate $u$ and $v$ to their averages defined in (8.11) by

$$u = U[1 + f_u(z, t)], \qquad v = V[1 + f_v(z, t)] \qquad (8.23)$$

where $f_u$ and $f_v$ are vertical distribution functions which have the property that

$$\int_{-h}^{\zeta} f_u(z, t)\,dz = \int_{-h}^{\zeta} f_v(z, t)\,dz = 0 \qquad (8.24)$$

However, the integrals of $f_u^2$, $f_v^2$, and $f_u f_v$ are not necessarily zero, and the following definitions can be made:

$$\alpha_{uu}(t) = (1/H)\int_{-h}^{\zeta} [1 + f_u(z, t)][1 + f_u(z, t)]\,dz \qquad (8.25a)$$

$$\alpha_{uv}(t) = (1/H)\int_{-h}^{\zeta} [1 + f_u(z, t)][1 + f_v(z, t)]\,dz \qquad (8.25b)$$

$$\alpha_{vv}(t) = (1/H)\int_{-h}^{\zeta} [1 + f_v(z, t)][1 + f_v(z, t)]\,dz \qquad (8.25c)$$

Substitution of these relations into (8.22) yields

$$\frac{\partial(HU)}{\partial t} + \frac{\partial}{\partial x}(\alpha_{uu}HUU) + \frac{\partial}{\partial y}(\alpha_{uv}HUV) - fHV + gH\frac{\partial\zeta}{\partial x} - \frac{1}{\rho}\tau_{xz}\bigg|_{\zeta}$$

$$+ \frac{1}{\rho}\tau_{xz}\bigg|_{-h} - \frac{1}{\rho}\frac{\partial}{\partial x}\int_{-h}^{\zeta}\tau_{xx}\,dz - \frac{1}{\rho}\frac{\partial}{\partial y}\int_{-h}^{\zeta}\tau_{xy}\,dz = 0 \tag{8.26}$$

The last four terms in this equation are related to the frictional losses and are modeled by empirical correlations. The stress at the surface due to wind friction is often given the form [3]

$$\frac{\tau_{xz}}{\rho}\bigg|_{\zeta} = KW^2 \cos\psi$$

where $K$ is a dimensionless coefficient that is a function of wind speed, $W$ the wind velocity, and $\psi$ the angle between the wind velocity vector and the $x$ axis. The bottom stress is generally expressed as

$$\frac{\tau_{xz}}{\rho}\bigg|_{-h} = gU(U^2 + V^2)^{1/2}/C^2$$

where $C$ is the Chezy coefficient with units $(\text{length})^{1/2}$ per time. The last two terms in the equation deal with horizontal transport of momentum and are correlated by

$$(1/H)\int_{-h}^{\zeta}\tau_{xx}\,dz = \varepsilon\,\partial U/\partial x \tag{8.27a}$$

$$(1/H)\int_{-h}^{\zeta}\tau_{xy}\,dz = \varepsilon\,\partial U/\partial y \tag{8.27b}$$

where $\varepsilon$ is a coefficient of eddy viscosity. Substitution of these last four relations into (8.23) yields

$$\frac{\partial(HU)}{\partial t} + \frac{\partial}{\partial x}(\alpha_{uu}HUU) + \frac{\partial}{\partial y}(\alpha_{uv}HUV) - fHV + gH\frac{\partial\zeta}{\partial x} - KW^2\cos\psi$$

$$+ \frac{gU(U^2+V^2)^{1/2}}{C^2} - \frac{1}{\rho}\frac{\partial}{\partial x}\left(\varepsilon H\frac{\partial U}{\partial x}\right) - \frac{1}{\rho}\frac{\partial}{\partial y}\left(\varepsilon H\frac{\partial U}{\partial y}\right) = 0 \tag{8.28}$$

On the basis of available evidence, the correlation functions $\alpha_{uu}$, $\alpha_{uv}$, and $\alpha_{vv}$ will, for most estuaries and for most times during the tidal cycle, range between 0.5 and 1.5. The more nearly the estuary is vertically homogeneous in velocity, the closer these functions will be to unity. In this discussion, the value of these functions will be taken to be unity.

The last two terms in Eq. (8.28) are generally small but do have the desirable effect of increasing the computational stability. In the analysis considered here, these terms will be neglected. The two-dimensional form of the momentum equation is therefore

$$\frac{\partial(HU)}{\partial t} + \frac{\partial}{\partial x}(HUU) + \frac{\partial}{\partial y}(HUV) - fHV + gH\frac{\partial \zeta}{\partial x} - KW^2 \cos\psi + \frac{gU(U^2 + V^2)^{1/2}}{C^2} = 0 \quad (8.29)$$

In some models [3, 4] the products $HU$ and $HV$ are considered as independent variables $Q_x$ and $Q_y$, respectively, and the equations of momentum and continuity are solved for $H$, $Q_x$, and $Q_y$. In the finite element models to be considered in detail here, $H$, $U$, and $V$ will be treated as dependent variables $Q_x$ and $Q_y$, respectively, and the equations of momentum equation and could be used in a finite element model. However, if the time derivative and convective terms are expanded by the chain rule, the continuity equation (8.13) is substituted into (8.29), and the resulting equation is divided by $H$, we are left with the somewhat simpler expression

$$\frac{\partial U}{\partial t} + U\frac{\partial U}{\partial x} + V\frac{\partial U}{\partial y} - fV + g\frac{\partial \zeta}{\partial x} - \frac{KW^2}{H}\cos\psi + \frac{gU(U^2 + V^2)^{1/2}}{HC^2} = 0 \quad (8.30)$$

From an analogous development, the $y$ component of the momentum equation is

$$\frac{\partial V}{\partial t} + U\frac{\partial V}{\partial x} + V\frac{\partial V}{\partial y} + fU + g\frac{\partial \zeta}{\partial y} - \frac{KW^2}{H}\sin\psi + \frac{gV(U^2 + V^2)^{1/2}}{HC^2} = 0 \quad (8.31)$$

These last two momentum balance equations form the governing hydrodynamic equations that will be considered in the following finite element models.

### *8.2.4 Species Transport*

Modeling of contaminant movement can be accomplished by coupling the solution of the hydrodynamic equation to the equations that describe convection and dispersion of a chemical species. The appropriate equation for two-dimensional transport is obtained by averaging the three-dimensional equation through the depth. The conservation equation for a chemical species a is

$$\frac{\partial \rho_a}{\partial t} + \frac{\partial}{\partial x}(\rho_a u) + \frac{\partial}{\partial y}(\rho_a v) + \frac{\partial}{\partial z}(\rho_a w) + \frac{\partial j_x}{\partial x} + \frac{\partial j_y}{\partial y} + \frac{\partial j_z}{\partial z} - r_a M_a = 0 \tag{8.32}$$

where $\rho_a$ is the mass concentration of species a, $M_a$ the molecular weight of species a, $j_x$, $j_y$, and $j_z$ the dispersive mass fluxes of a in the $x$, $y$, and $z$ directions, respectively, and $r_a$ the rate of production of species a by chemical reaction, moles per volume-time.

Following the method used for the continuity and momentum equations, (8.32) is integrated over $z$ and Leibnitz' rule is applied to yield

$$\begin{aligned}
&\frac{\partial}{\partial t}\int_{-h}^{\zeta} \rho_a\,dz + \frac{\partial}{\partial x}\int_{-h}^{\zeta} \rho_a u\,dz + \frac{\partial}{\partial y}\int_{-h}^{\zeta} \rho_a v\,dz + \frac{\partial}{\partial x}\int_{-h}^{\zeta} j_x\,dz \\
&+ \frac{\partial}{\partial y}\int_{-h}^{\zeta} j_y\,dz - \int_{-h}^{\zeta} r_a M_a\,dz \\
&- \rho_a\Big|_{\zeta}\left(\frac{\partial \zeta}{\partial t} + u_S\frac{\partial \zeta}{\partial x} + v_S\frac{\partial \zeta}{\partial y} - w_S\right) - \rho_a\Big|_{-h}\left(u_B\frac{\partial h}{\partial x} + v_B\frac{\partial h}{\partial y} + w_B\right) \\
&+ \left(-j_x\Big|_{\zeta}\frac{\partial \zeta}{\partial x} - j_y\Big|_{\zeta}\frac{\partial \zeta}{\partial y} - j_x\Big|_{-h}\frac{\partial h}{\partial x} - j_y\Big|_{-h}\frac{\partial h}{\partial y} + j_z\Big|_{\zeta} - j_z\Big|_{-h}\right) = 0
\end{aligned} \tag{8.33}$$

The last group of terms in this series deals with diffusive transport across an interface. These terms could be important in a re-aeration problem or in deposition of the species a at the bottom of the estuary. For the present discussion, however, they will be neglected. The terms in parentheses premultiplied by $\rho_a|_{\zeta}$ and $\rho_a|_{-h}$ are zero by boundary conditions (8.3) and (8.5). The average mass concentration $\mathscr{P}_a$ is defined by

$$\mathscr{P}_a = (1/H)\int_{-h}^{\zeta} \rho_a\,dz \tag{8.34}$$

Also, a vertical distribution function for concentration $f_\rho$, similar to that defined for velocity in (8.23), is given by

$$\rho_a = \mathscr{P}_a[1 + f_\rho(z, t)] \tag{8.35}$$

where

$$\int_{-h}^{\zeta} f_\rho(z, t)\,dz = 0 \tag{8.36}$$

Analogously to (8.25), correlation functions $\alpha_{u\rho}$ and $\alpha_{v\rho}$ are defined by

$$\alpha_{u\rho}(t) = (1/H) \int_{-h}^{\zeta} [1 + f_u(z, t)][1 + f_\rho(z, t)]\, dz \tag{8.37a}$$

$$\alpha_{v\rho}(t) = (1/H) \int_{-h}^{\zeta} [1 + f_v(z, t)][1 + f_\rho(z, t)]\, dz \tag{8.37b}$$

where values of $\alpha_{u\rho}$ and $\alpha_{v\rho}$ will be close to unity for a vertically well-mixed estuary. Deviation of $\alpha_{u\rho}$ and $\alpha_{v\rho}$ from unity indicate that a chemical species is being convected laterally at a velocity which differs from the vertically averaged velocity. Substitution of (8.34)–(8.37) into (8.33) yields

$$\frac{\partial(H\mathscr{P}_a)}{\partial t} + \frac{\partial(\alpha_{u\rho} HU\mathscr{P}_a)}{\partial x} + \frac{\partial(\alpha_{v\rho} HV\mathscr{P}_a)}{\partial y} + \frac{\partial}{\partial x}(HJ_x) + \frac{\partial}{\partial y}(HJ_y) - HR_a M_a = 0 \tag{8.38}$$

where $J_x$, $J_y$, and $R_a$ are the values of $j_x$, $j_y$, and $r_a$ averaged over the depth $H$. If $\alpha_{u\rho}$ and $\alpha_{v\rho}$ are assumed to be unity and $J_x$ and $J_y$ are approximated by Ficks' law as

$$J_x = -D\, \partial\mathscr{P}_a/\partial x \tag{8.39a}$$

$$J_y = -D\, \partial\mathscr{P}_a/\partial y \tag{8.39b}$$

where $D$ is the eddy dispersion coefficient, then the species balance equation assumes the form

$$\frac{\partial(H\mathscr{P}_a)}{\partial t} + \frac{\partial(HU\mathscr{P}_a)}{\partial x} + \frac{\partial(HV\mathscr{P}_a)}{\partial y} - \frac{\partial}{\partial x}\left(HD\frac{\partial\mathscr{P}_a}{\partial x}\right) - \frac{\partial}{\partial y}\left(HD\frac{\partial\mathscr{P}_a}{\partial y}\right) - HR_a M_a = 0 \tag{8.40}$$

Often this balance equation is expressed in terms of an average concentration $C_a = \mathscr{P}_a/M_a$. This change of variables causes no complication for the incompressible form of the equation derived here.

## 8.3 Linearized Hydrodynamic Model

After developing the governing differential equations, the next step in building a finite element model is to develop a computational algorithm for the solution of the equations. The algorithm discussed in this section is based on linearizing the continuity equation (8.12) and the momentum equations, (8.30) and (8.31).

### 8.3.1 *Linearized Equations for Finite Element Computation*

In this model the continuity equation and the two components of the momentum equation are solved simultaneously. In order to avoid iteration in solving these equations and to reduce the computer time requirements, the equations are linearized by a series of approximations [5].

Continuity equation (8.12) is expanded by noting that $H = h + \zeta$ to obtain

$$\frac{\partial \zeta}{\partial t} + \frac{\partial (hU)}{\partial x} + \frac{\partial (\zeta U)}{\partial x} + \frac{\partial (hV)}{\partial y} + \frac{\partial (\zeta V)}{\partial y} = 0 \tag{8.41}$$

The third and fifth terms in this equation contain products of functions that vary through time. If small perturbations in time are assumed, these terms can be linearized for computation and (8.41) can be written as

$$\frac{\partial \zeta}{\partial t} + \frac{\partial (hU)}{\partial x} + \frac{\partial (\zeta U^*)}{\partial x} + \frac{\partial (hV)}{\partial y} + \frac{\partial (\zeta V^*)}{\partial y} = 0 \tag{8.42}$$

where $U^*$ and $V^*$ are values of $U$ and $V$ known from an already computed time level.

The influence of the nonlinear convective terms in the momentum equations (8.30) and (8.31) is considered to be small so that elementwise averaged values of known flow variables $U^*$ and $V^*$ provide an adequate approximation to $U$ and $V$. Elementwise average values of $h$, $\zeta^*$, $W^*$, $U^*$, and $V^*$ are also introduced into the friction terms so the linearized forms of (8.30) and (8.31) are

$$\frac{\partial U}{\partial t} + \bar{U}\frac{\partial U}{\partial x} + \bar{V}\frac{\partial U}{\partial y} - fV + g\frac{\partial \zeta}{\partial x} - \frac{K\bar{W}W}{\bar{h} + \bar{\zeta}}\cos\psi + \frac{gU(\bar{U}^2 + \bar{V}^2)^{1/2}}{(\bar{h} + \bar{\zeta})C^2} = 0 \tag{8.43}$$

$$\frac{\partial V}{\partial t} + \bar{U}\frac{\partial V}{\partial x} + \bar{V}\frac{\partial V}{\partial y} + fU + g\frac{\partial \zeta}{\partial y} - \frac{K\bar{W}W}{\bar{h} + \bar{\zeta}}\sin\psi + \frac{gV(\bar{U}^2 + \bar{V}^2)^{1/2}}{(\bar{h} + \bar{\zeta})C^2} = 0 \tag{8.44}$$

where overbars are used to indicate the elementwise averaged quantities at an already computed time level.

Many different types of elements and basis functions could be used to obtain a solution for $\zeta$, $U$, and $V$ in Eqs. (8.42)–(8.44). The discussion here will be restricted to space–time elements which have been considered by Grotkop [5].

### 8.3.2 Finite Element Solution

The flow region is subdivided into triangles because they combine economy in computation with adequate flexibility to represent geometrically complex areas. Nodes are located only at the corners of the triangles and linear basis functions are chosen for the spatial interpolation. The time axis is orthogonal to the triangle and is divided into time steps $\Delta t$, thereby generating a triangular-based prism element of height $\Delta t$ (see Section 4.8.2 and Fig. 4.12 for a discussion and illustration of space–time elements). Linear basis functions are selected for the time domain such that each prism element contains six nodes. Variables $U$, $V$, and $\zeta$ must be calculated simultaneously in a computational step.

Approximate expansions for the variables in the problem can be made in terms of the time- and space-dependent basis functions:

$$\zeta(x, y, t) \simeq \sum_{k=1}^{2} \hat{\zeta}_k(x, y)\mu_k(t) = \sum_{j=1}^{M} \sum_{k=1}^{2} \zeta_{j,k}\phi_j(x, y)\mu_k(t) \tag{8.45a}$$

$$U(x, y, t) \simeq \sum_{k=1}^{2} \hat{U}_k(x, y)\mu_k(t) = \sum_{j=1}^{M} \sum_{k=1}^{2} U_{j,k}\phi_j(x, y)\mu_k(t) \tag{8.45b}$$

$$V(x, y, t) \simeq \sum_{k=1}^{2} \hat{V}_k(x, y)\mu_k(t) = \sum_{j=1}^{M} \sum_{k=1}^{2} V_{j,k}\phi_j(x, y)\mu_k(t) \tag{8.45c}$$

$$h(x, y) \simeq \sum_{j=1}^{M} h_j\phi_j(x, y) \tag{8.45d}$$

$$U^*(x, y) \simeq \sum_{j=1}^{M} U_{j,1}\phi_j(x, y) \tag{8.45e}$$

$$V^*(x, y) \simeq \sum_{j=1}^{M} V_{j,1}\phi_j(x, y) \tag{8.45f}$$

$$W(x, y, t) \simeq \sum_{k=1}^{2} \hat{W}_k(x, y)\mu_k(t) = \sum_{j=1}^{M} \sum_{k=1}^{2} W_{j,k}\phi_j(x, y)\mu_k(t) \tag{8.45g}$$

where $\phi_j$ are the spatial basis functions, $\mu_k$ the time basis functions with $k = 1$ being the known time level and $k = 2$ the unknown time level, and $M$ the number of spatial nodes.

Note that since there are $2M$ basis functions of the form $\phi_j\mu_k$, $2M$ weighted residual approximations for the continuity and momentum equations may be formed. However, because the solution for $\zeta$, $U$, and $V$ is known at the time level corresponding to $k = 1$, we only need $M$ weighted residual forms of the continuity and momentum equations to obtain the $M$

unknown parameters at the $k = 2$ level. The $2M$ equations normally generated can be condensed to $M$ equations in a general but systematic fashion by weighting with respect to $\phi_i[\gamma\mu_2 + (1-\gamma)\mu_1]$, where $\gamma$ is an arbitrary parameter.

Substitution of (8.45) into (8.42)–(8.44) and application of Galerkin's procedure yields

$$\sum_{j=1}^{M}\sum_{k=1}^{2}\int_t\int_y\int_x\Bigg\{\zeta_{j,k}\phi_j\frac{d\mu_k}{dt}+\sum_{l=1}^{M}\bigg[h_l U_{j,k}\mu_k\frac{\partial}{\partial x}(\phi_j\phi_l)$$
$$+\zeta_{j,k}U_{l,1}\mu_k\frac{\partial}{\partial x}(\phi_j\phi_l)+h_l V_{j,k}\mu_k\frac{\partial}{\partial y}(\phi_j\phi_l)+\zeta_{j,k}V_{l,1}\mu_k\frac{\partial}{\partial y}(\phi_j\phi_l)\bigg]\Bigg\}$$
$$\times\ \phi_i[\gamma\mu_2+(1-\gamma)\mu_1]\,dx\,dy\,dt=0\qquad(i=1,\ldots,M)\tag{8.46}$$

$$\sum_{j=1}^{M}\sum_{k=1}^{2}\int_t\int_y\int_x\Bigg\{U_{j,k}\phi_j\frac{d\mu_k}{dt}+\bar U_e U_{j,k}\mu_k\frac{\partial\phi_j}{\partial x}+\bar V_e U_{j,k}\mu_k\frac{\partial\phi_j}{\partial y}-fV_{j,k}\phi_j\mu_k$$
$$+g\zeta_{j,k}\mu_k\frac{\partial\phi_j}{\partial x}-\frac{K\bar W_e}{\bar h_e+\bar\zeta_e}W_{j,k}\mu_k\phi_j\cos\psi+\frac{gU_{j,k}\phi_j\mu_k(\bar U_e^2+\bar V_e^2)^{1/2}}{(\bar h_e+\bar\zeta_e)C^2}\Bigg\}$$
$$\times\ \phi_i[\gamma\mu_2+(1-\gamma)\mu_1]\,dx\,dy\,dt=0\qquad(i=1,\ldots,M)\tag{8.47}$$

$$\sum_{j=1}^{M}\sum_{k=1}^{2}\int_t\int_y\int_x\Bigg\{V_{j,k}\phi_j\frac{d\mu_k}{dt}+\bar U_e V_{j,k}\mu_k\frac{\partial\phi_j}{\partial x}+\bar V_e V_{j,k}\mu_k\frac{\partial\phi_j}{\partial y}+fU_{j,k}\phi_j\mu_k$$
$$+g\zeta_{j,k}\mu_k\frac{\partial\phi_j}{\partial y}-\frac{K\bar W_e}{\bar h_e+\bar\zeta_e}W_{j,k}\mu_k\phi_j\sin\psi+\frac{gV_{j,k}\phi_j\mu_k(\bar U_e^2+\bar V_e^2)^{1/2}}{(\bar h_e+\bar\zeta_e)C^2}\Bigg\}$$
$$\times\ \phi_i[\gamma\mu_2+(1-\gamma)\mu_1]\,dx\,dy\,dt=0\qquad(i=1,\ldots,M)\tag{8.48}$$

where $\bar U_e$, $\bar V_e$, $\bar\zeta_e$, etc., are elementwise average values of $U$, $V$, $\zeta$, etc., at time position 1. Because the elements being used here are linear triangles in the spatial domain, an elementwise average of a function is simply the sum of the values of the function at the vertices divided by 3.

A selection of the value for $\gamma$ determines where in time the spatial derivatives and friction and body force terms are located according to the formula

$$\varepsilon = \tfrac{1}{3}(\gamma + 1)\tag{8.49}$$

where $\varepsilon$ is the fractional location between the known values and the unknown values. For example, $\gamma = 1$ corresponds to $\varepsilon$ of $\frac{2}{3}$. Therefore, the spatial derivatives are centered $\frac{2}{3}\,\Delta t$ beyond the old time level. This is automatically accomplished in Eqs. (8.46)–(8.48) wherein, with $\gamma = 1$, the approximation to the spatial derivatives is calculated as $\frac{2}{3}$ of the term evaluated at the new time level plus $\frac{1}{3}$ of the term evaluated at the old time level. Because many numerical schemes become conditionally stable when spatial derivatives are located less than halfway beyond the old time level toward the new level, most numerical models consider only the case of $\varepsilon \geq 0.5$. By Eq. (8.49) this would restrict the cases considered to $\gamma \geq 0.5$. However, as indicated by the Fourier analyses presented earlier, when $\varepsilon$ is much greater than 0.5 the numerical damping introduced makes accurate simulation of the physical system difficult.

### *8.3.3 Boundary Conditions*

In seas and estuaries there are two different types of boundaries: the fixed boundary given by coast lines, and the open boundary which normally consists of a set of straight lines that act as somewhat artificial limits to either other parts of the open sea or to the mouth of a river. On fixed boundaries, the normal velocity is zero; on open boundaries, either the normal velocity or the value of $\zeta$ must be specified. To incorporate a set depth or stage boundary condition at node $i$ into the formulation one merely sets the value of $\zeta_i$ instead of generating Eq. (8.46).

The difficulties of determining the normal direction at a fixed boundary node $i$ where two lines with different slopes meet has already been discussed for the case when the solution is to be obtained in terms of the stream function [see the discussion of Fig. 6.9 in Section 6.4.2]. In the present problem, even though the solution is in terms of velocity, the normal direction is obtained by the same procedure. If two straight line segments $AB$ and $BC$ meet at point $B$, and triangular elements with linear basis functions are used, Wang and Connor [6] have shown that the direction of the normal which conserves mass is the direction perpendicular to straight line segment $AC$.

Once the normal direction at node $i$ has been determined, the equation for tangential momentum transfer can be formed. If $\theta$ is the angle between the negative $y$ axis and the tangential direction, the tangential velocity $V_\lambda$ is related to $U$ and $V$ by

$$V_{\lambda_i} = U_i \sin\theta - V_i \cos\theta \tag{8.50}$$

and the normal velocity $V_{n_i}$ is related to $U_i$ and $V_i$ by

$$V_{n_i} = U_i \cos\theta + V_i \sin\theta \tag{8.51}$$

When node $i$ is a solid boundary node, instead of assembling (8.47) and (8.48) as two rows in the global matrix, (8.47) is multiplied by $\sin\theta$, (8.48) is multiplied by $-\cos\theta$, and the two equations are then added and assembled as one row in the matrix. This row of the matrix is the tangential component of the momentum equation for node $i$. The boundary condition of zero normal velocity is then prescribed into the other row of the matrix by setting

$$U_i \cos \theta = -V_i \sin \theta \tag{8.52}$$

The set of matrix equations that arise from (8.46)–(8.48) and the boundary conditions can now be solved for $\zeta$, $U$, and $V$ at the new time level. This new set of values serves as the initial conditions for computations in the next layer of elements in the time domain.

### *8.3.4 Applications*

To demonstrate the use of this model, the oscillations of the North Sea due to the semidiurnal lunar $M_2$ tide based on prescribed sea levels at the open boundaries have been calculated [5]. The area of the North Sea is subdivided rather roughly into 97 triangles (Fig. 8.2) with 69 nodes. The

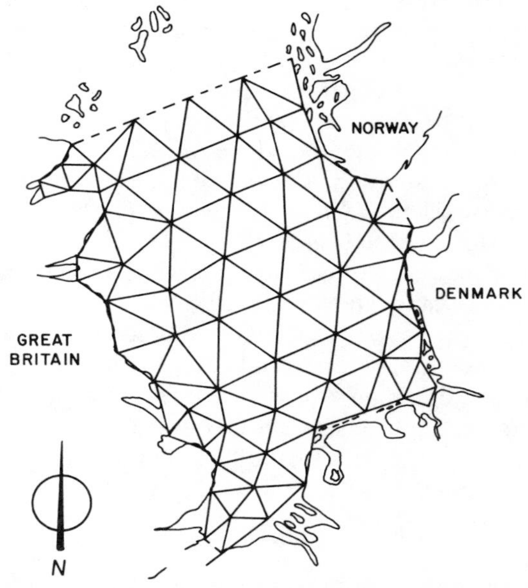


**Fig. 8.2.** Finite element grid of the North Sea (after Grotkop [5]).

maximum distance between nodes is 150 km in areas of constant depth and about 30–50 km in shallow coastal areas. The tidal heights are specified at the three different open boundaries.

Because the water motion is independent of the initial conditions after a certain time and becomes influenced only by the specified oscillations of the boundary values, computations can be begun from an initial condition of

$\zeta = 0$, $U = 0$, $V = 0$ at all points. After several periods of the tide, the effect of the initial conditions should have vanished and the flow will show a steady periodic behavior.

The tide was calculated through seven periods [5], the first four with a time step of 30 min and the last three with a time step of 15 min. After six periods, or 200 time steps, the flow had nearly reached its dynamic steady state. Differences between calculated values of the water levels and the velocities at the sixth and seventh period were less than 0.05 m and 0.02 m/sec, respectively.

The calculated tidal ranges (differences between high and low water), and cotidal lines (lines with the same high-water time) are shown in Fig. 8.3. The results are in accord with observed values, with a few minor discrepancies which can be attributed to the coarseness of the finite element grid.

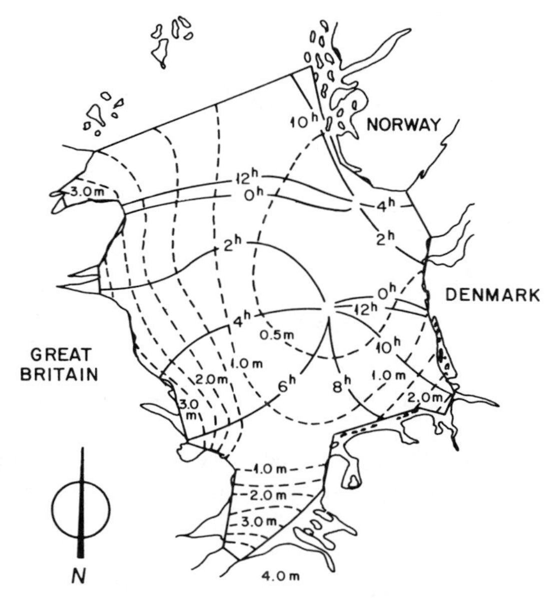


Fig. 8.3. Calculated tidal ranges and cotidal lines for the North Sea (after Grotkop [5]).

## 8.4 Implicit Nonlinear Model

If the hydrodynamic equations are not linearized and an implicit formulation in the time domain is used, it is necessary to iterate on the solution to obtain convergence. Generally, the stability of the implicit procedure is not limited by time step size. Because the nonlinearities in the governing equations are preserved, the physics of the flow should be represented by the simulation. Implicit finite element schemes have been successfully applied by Wang and Connor [6] and by Taylor and Davis [7, 8] among others.

### 8.4.1 Finite Element Formulation

The equations used to model the flow in an estuary are the continuity equation (8.12) and momentum equations (8.30) and (8.31). The finite element solution to (8.12), (8.30), and (8.31) is obtained by first making the expansions

$$U \simeq \sum_{j=1}^{M} U_j(t)\phi_j(x, y) \tag{8.53a}$$

$$V \simeq \sum_{j=1}^{M} V_j(t)\phi_j(x, y) \tag{8.53b}$$

$$\zeta \simeq \sum_{j=1}^{M} \zeta_j(t)\phi_j(x, y) \tag{8.53c}$$

$$h \simeq \sum_{j=1}^{M} h_j\phi_j(x, y) \tag{8.53d}$$

$$f \simeq \sum_{j=1}^{M} f_j\phi_j(x, y) \tag{8.53e}$$

$$KW^2 \cos \psi \simeq \sum_{j=1}^{M} K_j W_j^2[\cos \psi_j]\phi_j(x, y) \tag{8.53f}$$

$$KW^2 \sin \psi \simeq \sum_{j=1}^{M} K_j W_j^2[\sin \psi_j]\phi_j(x, y) \tag{8.53g}$$

Note that in contrast to the model discussed in Section 8.3, the basis functions $\phi_j(x, y)$ are functions only of space and the time dependence is incorporated into the coefficients. $M$ is the number of nodes in the finite element domain.

Equations (8.53) are substituted into the continuity and momentum equations and the resulting equations are made orthogonal to $\phi_i$ according to Galerkin's procedure. The equations that result are

$$\sum_{j=1}^{M} \iint_{\mathscr{A}} \left\{ \frac{d\zeta_j}{dt} \phi_j\phi_i + \frac{\partial}{\partial x}\left[(h_j + \zeta_j)\phi_j\left(\sum_{k=1}^{M} U_k\phi_k\right)\right]\phi_i + \frac{\partial}{\partial y}\left[(h_j + \zeta_j)\phi_j\left(\sum_{k=1}^{M} V_k\phi_k\right)\right]\phi_i \right\} dx\, dy = 0 \qquad (i = 1, 2, \ldots, M) \tag{8.54}$$

$$\sum_{j=1}^{M} \iint_{\mathscr{A}} \left\{ \frac{dU_j}{dt} \phi_j \phi_i + \left( \sum_{k=1}^{M} U_k \phi_k \right) U_j \frac{\partial \phi_j}{\partial x} \phi_i + \left( \sum_{k=1}^{M} V_k \phi_k \right) U_j \frac{\partial \phi_j}{\partial y} \phi_i \right.$$

$$- \left( \sum_{k=1}^{M} f_k \phi_k \right) V_j \phi_j \phi_i + g \zeta_j \frac{\partial \phi_j}{\partial x} \phi_i$$

$$+ g \frac{[(\sum_{k=1}^{M} U_k \phi_k)^2 + (\sum_{k=1}^{M} V_k \phi_k)^2]^{1/2}}{C^2[\sum_{k=1}^{M} (\zeta_k + h_k)\phi_k]} U_j \phi_j \phi_i$$

$$\left. - \frac{K_j W_j^2 (\cos \psi_j) \phi_j}{[\sum_{k=1}^{M} (\zeta_k + h_k)\phi_k]} \phi_i \right\} dx\, dy = 0 \qquad (i = 1, 2, \ldots, M) \quad (8.55)$$

$$\sum_{j=1}^{M} \iint_{\mathscr{A}} \left\{ \frac{dV_j}{dt} \phi_j \phi_i + \left( \sum_{k=1}^{M} U_k \phi_k \right) V_j \frac{\partial \phi_j}{\partial x} \phi_i + \left( \sum_{k=1}^{M} V_k \phi_k \right) V_j \frac{\partial \phi_j}{\partial y} \phi_i \right.$$

$$+ \left( \sum_{k=1}^{M} f_k \phi_k \right) U_j \phi_j \phi_i + g \zeta_j \frac{\partial \phi_j}{\partial y} \phi_i$$

$$+ g \frac{[(\sum_{k=1}^{M} U_k \phi_k)^2 + (\sum_{k=1}^{M} V_k \phi_k)^2]^{1/2}}{C^2[\sum_{k=1}^{M} (\zeta_k + h_k)\phi_k]} V_j \phi_j \phi_i$$

$$\left. - \frac{K_j W_j^2 (\sin \psi_j) \phi_j}{[\sum_{k=1}^{M} (\zeta_k + h_k)\phi_k]} \phi_i \right\} dx\, dy = 0 \qquad (i = 1, 2, \ldots, M) \quad (8.56)$$

The boundary conditions applied to the flow equations are the specification of $\zeta$ or the normal flux at an open sea boundary. Along fixed boundaries, a zero normal velocity boundary condition should be applied.

The integrated Galerkin equations may be assembled in matrix form as

$$[KI]\{d\zeta/dt\} = \{E_\zeta\} \qquad (8.57a)$$

$$[KI]\{dU/dt\} = \{E_U\} \qquad (8.57b)$$

$$[KI]\{dV/dt\} = \{E_V\} \qquad (8.57c)$$

where the elements of $[KI]$ are the multipliers of the time derivatives in each balance equation given by

$$ki_{i,j} = \iint_{\mathscr{A}} \phi_j \phi_i \, d\mathscr{A} \qquad (8.58)$$

and the $\{E\}$ vectors account for all other terms in the balances. The resulting nonlinear algebraic equations are solved iteratively. The time derivatives may be finite differenced over a time step $\Delta t$ and the $\{E\}$ vectors averaged over $\Delta t$ to obtain

$$[KI]\{\zeta\}_{t+\Delta t} = [KI]\{\zeta\}_t + \tfrac{1}{2}\,\Delta t\{E_\zeta\}_{t+\Delta t} + \tfrac{1}{2}\,\Delta t\{E_\zeta\}_t = \{R_\zeta\} \tag{8.59a}$$

$$[KI]\{U\}_{t+\Delta t} = [KI]\{U\}_t + \tfrac{1}{2}\,\Delta t\{E_U\}_{t+\Delta t} + \tfrac{1}{2}\,\Delta t\{E_U\}_t = \{R_U\} \tag{8.59b}$$

$$[KI]\{V\}_{t+\Delta t} = [KI]\{V\}_t + \tfrac{1}{2}\,\Delta t\{E_V\}_{t+\Delta t} + \tfrac{1}{2}\,\Delta t\{E_V\}_t = \{R_V\} \tag{8.59c}$$

From the known solution at $t$, estimates are made for $\zeta$, $U$, and $V$ at $t + \Delta t$. These estimates are used to approximate $\{E_\zeta\}_{t+\Delta t}$, $\{E_U\}_{t+\Delta t}$, and $\{E_V\}_{t+\Delta t}$ and thus completely determine the right-hand sides of (8.59). Through iteration, better and better estimates of the right-hand sides of (8.59) are obtained until convergence is achieved. After a converged solution is obtained, calculation may proceed to the next time step.

The boundary condition must be applied to (8.59) before these equations can be solved. When modifying $[KI]$ to take into account the boundary conditions, its symmetry should be preserved in order to keep computer storage down and to improve computational efficiency.

Application of a specified depth boundary condition is straightforward. If $\zeta_{i_{t+\Delta t}}$ is specified, the following steps are required:

(a) $ki_{i,j}$ is set to zero for $j = 1, \ldots, M$,
(b) $ki_{m,i}\zeta_{i_{t+\Delta t}}$ is subtracted from $r_{\zeta_m}$ for $m = 1, \ldots, M$,
(c) $ki_{m,i}$ is set to zero for $m = 1, \ldots, M$,
(d) $ki_{i,i}$ is set to unity,
(e) $r_{\zeta_i}$ is set to $\zeta_{i,t+\Delta t}$

If $U_i$, $V_i$ or both are specified at time $t + \Delta t$, an exactly analogous procedure to the one described for $h_i$ can be performed to preserve the symmetry of the coefficient matrices in Eqs. (8.59b) and (8.59c). However, in a problem when the normal velocity is specified at a node where the normal direction is not coincident with either the $x$ or $y$ axes, a different procedure must be adopted to maintain a symmetric coefficient matrix.

The first step in this procedure is to combine (8.59b) and (8.59c) into one matrix equation of the form

$$[KJ]\{UV\}_{t+\Delta t} = \{R\} \tag{8.60}$$

where $[KJ]$ is an $M \times M$ matrix with $2 \times 2$ elements such that

$$KJ_{i,j} = \begin{bmatrix} ki_{i,j} & 0 \\ 0 & ki_{i,j} \end{bmatrix}$$

and $\{UV\}$ and $\{R\}$ are $M \times 1$ vectors with $2 \times 1$ elements

$$UV_i = \begin{Bmatrix} U_i \\ V_i \end{Bmatrix}, \qquad R_i = \begin{Bmatrix} r_{U_i} \\ r_{V_i} \end{Bmatrix}$$

If velocity is to be specified at a node $k$, a procedure is applied to Eq. (8.60) similar to that applied to (8.57a) when depth is specified. However, where the velocity is specified, two rows and two columns must now be zeroed out for each node to account for both the $x$ and $y$ components. When the normal velocity is specified at a boundary node, the velocity components at that node are computed in terms of their tangential and normal components rather than in terms of their $x$ and $y$ components. Recall from Eqs. (8.50) and (8.51) that the tangential and normal velocity components at a node can be related to the $x$ and $y$ components by

$$\begin{Bmatrix} V_{\lambda i} \\ V_{n_i} \end{Bmatrix} = \begin{bmatrix} \sin\theta_i & -\cos\theta_i \\ \cos\theta_i & \sin\theta_i \end{bmatrix} \begin{Bmatrix} U_i \\ V_i \end{Bmatrix} \tag{8.61}$$

where $\theta_i$ is the angle between the $x$ axis and the normal direction. An $M \times M$ diagonal matrix $[A]$ may now be formed which is composed of $2 \times 2$ elements such that

$$A_{i,i} = \begin{bmatrix} 1 & 0 \\ 0 & 1 \end{bmatrix}$$

if no normal velocity boundary is specified at node $i$; if a normal velocity boundary condition is specified at node $i$, the elements of $[A]$ are given by

$$A_{i,i} = \begin{bmatrix} \sin\theta_i & -\cos\theta_i \\ \cos\theta_i & \sin\theta_i \end{bmatrix}$$

Because matrix $[A]$ has the property that $[A]^{\mathrm{T}} = [A]^{-1}$, (8.60) may be rewritten as

$$[A][KJ][A]^{\mathrm{T}}[A]\{UV\}_{t+\Delta t} = [A]\{R\} \tag{8.62}$$

Next, define

$$[KJ^*] = [A][KJ][A]^{\mathrm{T}}, \qquad \{UV^*\}_{t+\Delta t} = [A]\{UV\}_{t+\Delta t}, \qquad \{R^*\} = [A]\{R\}$$

so that (8.62) becomes

$$[KJ^*]\{UV^*\}_{t+\Delta t} = \{R^*\} \tag{8.63}$$

Matrix $[KJ^*]$ is symmetric and $V_{\lambda_k}$ and $V_{n_k}$ rather than $U_k$ and $V_k$ compose those elements of the unknown vector $\{UV^*\}_{t+\Delta t}$ associated with node $k$ where the normal velocity has been specified. Because the unknown vector contains the normal velocity, it is now possible to impose the constant normal velocity condition in a straightforward manner. If the normal velocity is specified at node $k$, row $2k$ and column $2k$ of $[KJ^*]$ can be zeroed out in the appropriate manner, $KJ^*_{2k,2k}$ is set to unity and $R^*_{2k}$ is set to the value of $V_{n_k}$. After (8.63) is solved, the tangential and normal velocity components obtained can be readily converted back to $x$ and $y$ velocities.

The chief difficulty with this formulation lies in obtaining the appropriate values for sin $\theta$ and cos $\theta$ at nodes along curved sides or at nodes where the slope of the boundary is discontinuous. While this problem is considered in Chapter 6, and in the previous section describing the linearized flow model when linear triangular elements are used, the case of an element of arbitrary shape has yet to be investigated.

Consider the boundary of the estuary which is indicated by the hatched line in Fig. 8.4. The elements I, II, III, and IV are located along this boundary. For simplicity, let the normal velocity along this boundary be zero (use of a nonzero value for the normal velocity will not alter the results). We will require that the normal velocity condition be satisfied in an average sense

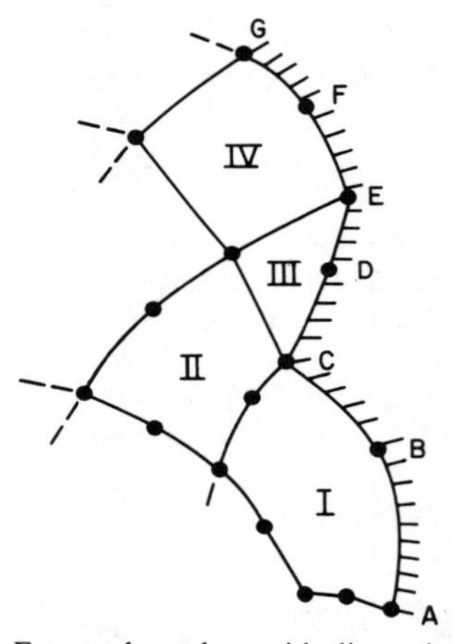


**Fig. 8.4.** Estuary boundary with discontinuous slope.

but not necessarily hold exactly at every point. For node $F$, this condition can be stated as

$$\int_s [U_F \phi_F \cos\theta + V_F \phi_F \sin\theta]\, ds = 0 \tag{8.64}$$

where $\phi_F$ is the basis function in the velocity expansion associated with node $F$, and $U_F \cos\theta + V_F \sin\theta$ is the velocity normal to $s$. Application of the divergence theorem to (8.64) yields

$$\iint_{\mathscr{A}} [U_F(\partial\phi_F/\partial x) + V_F(\partial\phi_F/\partial y)]\, d\mathscr{A} = 0$$

The only region where $\partial\phi_F/\partial x$, and $\partial\phi_F/\partial y$ are nonzero is in element IV and therefore element IV is the area over which the integration must be performed. After the integral is evaluated, the coefficient of $U_F$ must be proportional to $\cos\theta$ at node $F$ and the coefficient of $V_F$ must be proportional to $\sin\theta$ at node $F$. Normalization of these coefficients will yield $\cos\theta$ and $\sin\theta$ such that for node $F$

$$\cos\theta_F = \frac{\int_{\text{IV}} (\partial\phi_F/\partial x)\, d\mathscr{A}}{\{[\int_{\text{IV}}(\partial\phi_F/\partial x)\, d\mathscr{A}]^2 + [\int_{\text{IV}}(\partial\phi_F/\partial y)\, d\mathscr{A}]^2\}^{1/2}} \tag{8.65a}$$

$$\sin\theta_F = \frac{\int_{\text{IV}} (\partial\phi_F/\partial y)\, d\mathscr{A}}{\{[\int_{\text{IV}}(\partial\phi_F/\partial x)\, d\mathscr{A}]^2 + [\int_{\text{IV}}(\partial\phi_F/\partial y)\, d\mathscr{A}]^2\}^{1/2}} \tag{8.65b}$$

Thus the sine and cosine functions needed for matrix $[A]$ can be easily evaluated even at points of discontinuous slope. For example, to obtain values of $\sin\theta$ and $\cos\theta$ at node $C$, $\partial\phi_C/\partial x$ and $\partial\phi_C/\partial y$ should be used in Eqs. (8.65) rather than $\partial\phi_F/\partial x$ and $\partial\phi_F/\partial y$, and the integrations should be performed over elements I, II, and III.

### *8.4.2 Application of an Implicit Model*

Taylor and Davis [8] have applied an implicit hydrodynamic model to the southern North Sea. The region considered was coarsely divided into nine cubic isoparametric elements (Fig. 8.5a). Some typical velocity vectors and the cotidal and corange lines are given in Figs. 8.5b and 8.5c. Boundary conditions employed were: tidal elevations at sea boundaries, zero velocity along land boundaries, and specified discharge at river boundaries. The results obtained are essentially in agreement with measured data.

## 8.5 Contaminant Transport in Estuaries

### *8.5.1 Finite Element Formulation*

When the presence of a pollutant does not affect the hydrodynamics of an estuary or coastal region, the conservation equation for a chemical species can be solved using the computed flow field. In the model to be discussed here, the convective dispersive equation is based on (8.40); however, mass density $\mathscr{P}_a$ is set equal to the molar concentration $C_a$ times the molecular weight of species $M_a$. Furthermore, the continuity equation is

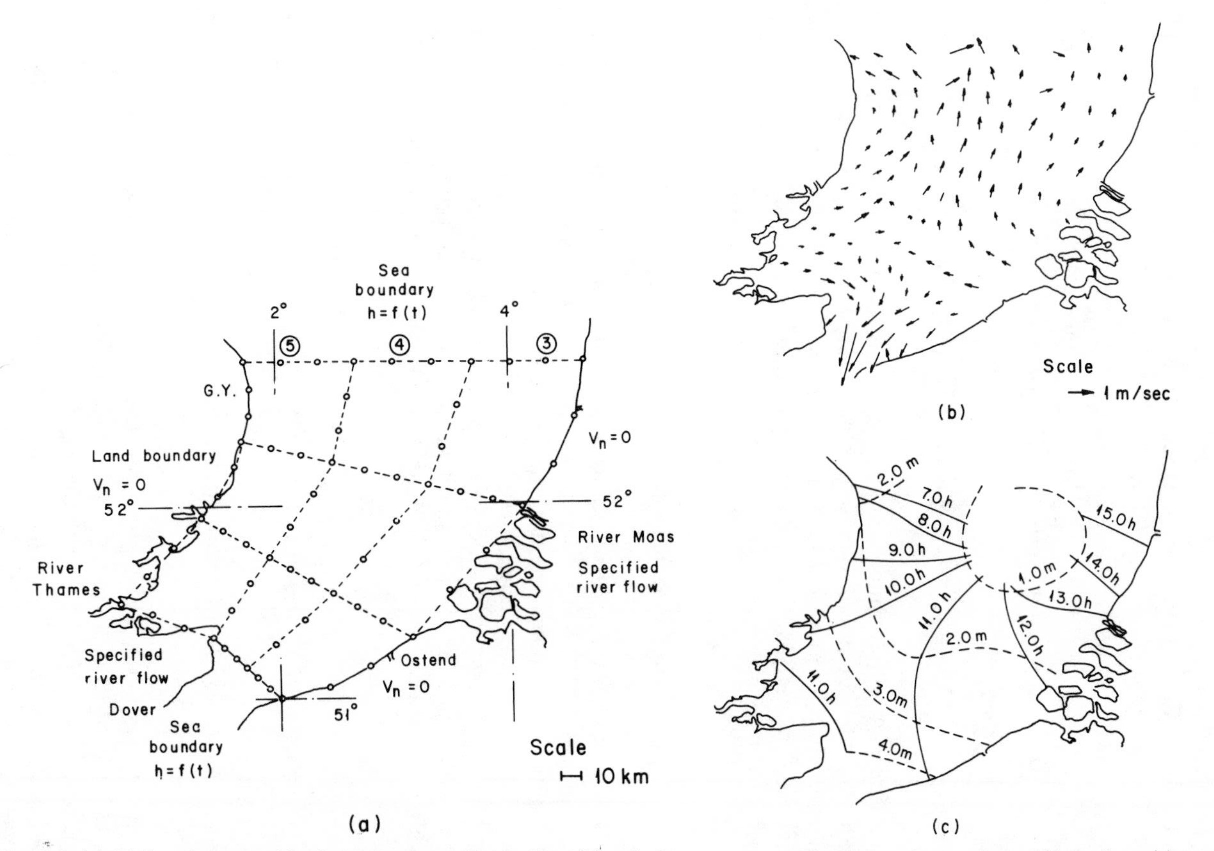


**Fig. 8.5.** (a) Finite element grid of the southern North Sea; (b) velocity vectors computed for the southern North Sea; (c) cotidal (——) and corange (- - -) lines computed for the southern North Sea. (After Taylor and Davis [8].)

substituted into (8.40) to obtain the simplified form†

$$\frac{\partial C_a}{\partial t} + U\frac{\partial C_a}{\partial x} + V\frac{\partial C_a}{\partial y} - \frac{1}{H}\left[\frac{\partial}{\partial x}\left(HD_x\frac{\partial C_a}{\partial x}\right) + \frac{\partial}{\partial y}\left(HD_y\frac{\partial C_a}{\partial y}\right)\right] - F_a - S_a C_a = 0 \tag{8.66}$$

where the dispersivity $D$ has been replaced by $D_x$ and $D_y$ to indicate that dispersion is not necessarily isotropic and $R_a$, the rate of production of species a has been replaced by $F_a$, a source term, and $S_a C_a$, a chemical rate expression. For simplicity, the principal axes of the dispersion tensor have been assumed coincident with the $x$ and $y$ axes.

If the velocities $U$ and $V$ and the depth $H$ (which is equal to $h + \zeta$) are expanded as in Eqs. (8.53) and if $C_a$, $D_x$, $D_y$, $F_a$, and $S_a$ are also expanded in terms of basis functions $\phi_j$ which only maintain continuity of the function and not any derivatives across element boundaries, the following Galerkin expression is obtained:

$$\begin{aligned}\sum_{j=1}^{M} \iint_{\mathscr{A}} \Bigg\{ & \frac{dC_j}{dt}\phi_j\phi_i + \left(\sum_{k=1}^{M} U_k\phi_k\right)C_j\frac{\partial\phi_j}{\partial x}\phi_i + \left(\sum_{k=1}^{M} V_k\phi_k\right)C_j\frac{\partial\phi_j}{\partial y}\phi_i \\ & + \frac{\partial}{\partial x}\left[\frac{\phi_i}{\sum_{k=1}^{M}(\zeta_k + h_k)\phi_k}\right]\left[\sum_{k=1}^{M}(\zeta_k + h_k)\phi_k\right]\left[\sum_{k=1}^{M} D_{xk}\phi_k\right]C_j\frac{\partial\phi_j}{\partial x} \\ & + \frac{\partial}{\partial y}\left[\frac{\phi_i}{\sum_{k=1}^{M}(\zeta_k + h_k)\phi_k}\right]\left[\sum_{k=1}^{M}(\zeta_k + h_k)\phi_k\right]\left[\sum_{k=1}^{M} D_{yk}\phi_k\right]C_j\frac{\partial\phi_j}{\partial y} \\ & - F_j\phi_j\phi_i - \left(\sum_{k=1}^{M} S_k\phi_k\right)C_j\phi_j\phi_i\Bigg\}\,dx\,dy - \int_{\Gamma}\mathbf{n}\cdot[\mathbf{D}\cdot\nabla C_a]\phi_i\,d\Gamma = 0\end{aligned}$$

$$(i = 1, \ldots, M) \tag{8.67}$$

where the surface integral involves the dispersive flux in the direction normal to the boundary and arises from application of Green's theorem to the dispersive terms to remove the second derivatives. If a dispersive flux boundary condition is specified, it is applied to the finite element formulation through this surface integral. If the concentration is specified at boundary

† Equation (8.66) is called the nonconservative form of the chemical species balance. The conservative form would be (8.40) with $M_a C_a$ substituted in for $\mathscr{P}_a$. Since $H$, $U$, and $V$ in the model are determined independently of the concentration of species a from the hydrodynamic model, solution of (8.40) by the finite element method would amount to solving the linear equation in $C_a$ with known, nonconstant coefficients $H$, $U$, and $V$, a task no more difficult than solving (8.66). An analogous situation occurs in solving the heat transfer equation and the conservative form is considered in Section 6.5.3.

node $i$, (8.67) generated using $\phi_i$ as a weighting function is replaced by that boundary condition.

Usually this equation is solved using a procedure analogous to that introduced for the energy equation in Section 6.5.3. While this approach requires no iteration, the coefficient matrix is not symmetric and therefore requires more storage. One way to reduce the storage requirements at the expense of execution time is to solve the equations iteratively. By analogy with Eq. (8.57a), (8.67) can be written as

$$[KI]\{dC_a/dt\} = \{E_C\} \tag{8.68}$$

where again the elements of the symmetric matrix $[KI]$ are given by (8.58) and $\{E_C\}$ accounts for the terms in (8.67) which do not multiply the time derivative. Finite differencing in the time domain yields

$$[KI]\{C_a\}_{t+\Delta t} = [KI]\{C_a\}_t + \tfrac{1}{2}\,\Delta t\{E_C\}_{t+\Delta t} + \tfrac{1}{2}\,\Delta t\{E_C\}_t \tag{8.69}$$

This equation is solved using the known values of $\{C_a\}_t$ and estimates of $\{C_a\}_{t+\Delta t}$ to evaluate the terms on the right side of (8.69). Then, after applying the boundary conditions, an updated estimate of $\{C_a\}_{t+\Delta t}$ may be obtained. The vector $\{C_a\}_{t+\Delta t}$ is solved for iteratively until successive iterations agree within a specified tolerance. This solution method minimizes computer storage requirements because only a half bandwidth of the symmetric coefficient matrix $[KI]$ must be stored. However, the number of iterations required for convergence may increase the computer time needed to such an extent that the scheme becomes uneconomical.

### 8.5.2 *Example of Application*

Taylor and Davis [8] have used the method outlined above to model the dispersion of a conservative pollutant in a vertically well-mixed hypothetical tidal estuary as shown in Fig. 8.6. The 15,000 m-long estuary varies in width from 1000 m upstream to 4000 m at the mouth and was represented by six

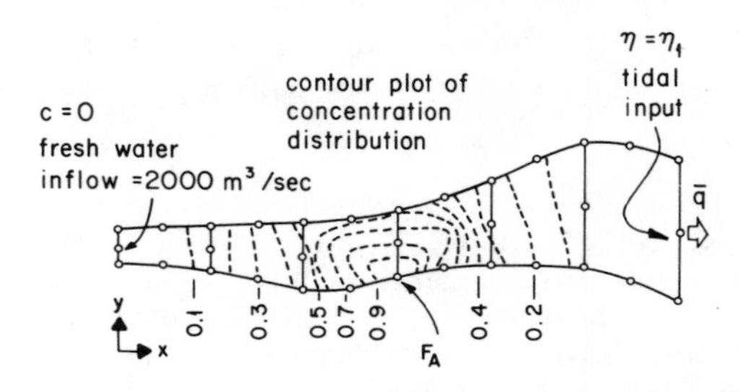


**Fig. 8.6.** Concentration contours in a hypothetical estuary (after Taylor and Davis [8]).

quadratic isoparametric elements. At the upstream boundary, the inflow is assumed to be fresh water and at the downstream boundary, the tidal flow is set. The dispersion coefficients were assumed to be $D_x = 1000$ and $D_y = 800$ m$^2$/sec. The point designated as $F_A$ represents the continuous discharge of a conservative waste with a concentration of 1.0. The concentration contours in Fig. 8.6 are typical of those predicted by the model.

## 8.6 Summary

This chapter summarizes the present state of modeling tidal regions with finite elements. The types of models presented are meant to be indicative of the techniques being used rather than an exhaustive survey.

In modeling hydrodynamics by finite elements a banded matrix problem must be solved. In contrast, finite difference modeling either does not involve a matrix equation at all, as in the case of an explicit formulation [3], or requires the solution of a tridiagonal matrix when an implicit procedure is used. Weare [9] argues that because band algorithms are currently employed for finite element models, this method is not economically competitive with the finite difference method. The chief advantage of the finite element method is the ability to describe lateral topography and bathymetry accurately. This ability arises because of the freedom to place nodes at arbitrary locations and to use elements with curved sides. At present, techniques are under development (e.g. Gray [10]) that will reduce the cost of the finite element method for tidal simulation and perhaps make this method economically competitive with finite differences in some instances.

## References


1. D. W. Pritchard, *in* "Estuarine Modeling: An Assessment" (G. H. Ward, Jr. and W. H. Epsey, Jr., eds.). Nat. Tech. Inform. Serv. Publ. #PB-206-807, 1971.
2. J. J. Dronkers, "Tidal Computations in Rivers and Coastal Water." North-Holland Publ., Amsterdam, distributed by Wiley (Interscience), New York, 1964.
3. R. O. Reid and B. R. Bodine, Numerical model for storm surges in Galveston Bay, *J. Waterways Harbors Div ASCE* **94**, 33 (1968).
4. F. D. Masch and R. J. Brandes, Tidal Hydrodynamic Simulation in Shallow Estuaries. Tech. Rep. #HYD 12-7102, Hydraulic Eng. Lab., Univ. Texas, Austin, 1971.
5. G. Grotkop, Finite element analysis of long-period water waves, *Comput. Methods Appl. Mech. Eng.* **2**, 147 (1973).
6. J. D. Wang and J. J. Connor, Mathematical Modeling of Near Coastal Circulation. Ralph M. Parsons Lab. for Water Resources and Hydrodynamics, Rep #200 (1975).

7. C. Taylor and J. M. Davis, Tidal and long wave propagation—A finite element approach, *Comput. Fluids* **3**, 125 (1975).
8. C. Taylor and J. M. Davis, Tidal propagation and dispersion in estuaries, *in* "Finite Elements in Fluids" (R. H. Gallagher, J. T. Oden, C. Taylor, and O. C. Zienkiewicz, eds.), Vol. 1, p. 290. Wiley, New York, 1975.
9. T. J. Weare, Finite element or finite difference methods for the two-dimensional shallow water equations? *Comput. Methods Appl. Mech Eng.* **7**, 351 (1976).
10. W. G. Gray, An efficient finite element scheme for two-dimensional surface water computation, *Proc. Int. Conf Finite Elements Water Resources, Princeton Univ., July 1976* Pentech Press, London (in press).

# Index

## R



## S



## T



## U

UNI